Advanced Topics in Human Genetics

Advanced Topics in Human Genetics

Editor: Lesley Easton

R CALLISTO REFERENCE

www.callistoreference.com

Callisto Reference,
118-35 Queens Blvd., Suite 400,
Forest Hills, NY 11375, USA

Visit us on the World Wide Web at:
www.callistoreference.com

ISBN: 978-1-64116-245-6 (Hardback)

Cataloging-in-Publication Data

Advanced topics in human genetics / edited by Lesley Easton.
    p. cm.
Includes bibliographical references and index.
ISBN 978-1-64116-245-6
1. Human genetics. 2. Human biology. 3. Genetic engineering. I. Easton, Lesley.
QH431 .A38 2020
599.935--dc23

Table of Contents

Preface

The human genome refers to the complete set of nucleic acid sequences in humans. These are encoded as DNA within the chromosomes in cell nuclei and the DNA molecule in each mitochondrion. The genome is organized into 22 pairs of chromosomes, plus a combination of X and X chromosomes for females, and a combination of X and Y chromosomes for males. The human genome comprises of both protein-coding DNA genes and noncoding DNA. There are approximately between 19,000-20,000 protein-coding genes. These account for only 1.5% of the genome, while the rest is associated with non-coding RNA molecules, introns, SINEs, LINEs, regulatory DNA sequences and those sequences for which no function has been determined as yet. Studies have shown that most noncoding DNA within the human genome have associated biochemical activities such as regulation of gene expression, control of epigenetic inheritance and organization of chromosome architecture. The human genome is the first vertebrate to be completely sequenced. The resulting data has applications in anthropology, biomedical science, forensics, etc. This book presents researches and studies performed by experts across the globe on the human gene. It elucidates the concepts and innovative models around prospective developments with respect to this area of study. It attempts to assist those with a goal of delving into genetics.

The researches compiled throughout the book are authentic and of high quality, combining several disciplines and from very diverse regions from around the world. Drawing on the contributions of many researchers from diverse countries, the book's objective is to provide the readers with the latest achievements in the area of research. This book will surely be a source of knowledge to all interested and researching the field.

In the end, I would like to express my deep sense of gratitude to all the authors for meeting the set deadlines in completing and submitting their research chapters. I would also like to thank the publisher for the support offered to us throughout the course of the book. Finally, I extend my sincere thanks to my family for being a constant source of inspiration and encouragement.

Editor

miR-203 and miR-320 Regulate Bone Morphogenetic Protein-2-Induced Osteoblast Differentiation by Targeting Distal-Less Homeobox 5 (*Dlx5*)

Navya Laxman [1,2,3,*], **Hans Mallmin** [4], **Olle Nilsson** [4] **and Andreas Kindmark** [1,2]

1   Department of Medical Sciences, Uppsala University, Uppsala 75185, Sweden; andreas.kindmark@medsci.uu.se
2   Science for Life Laboratory, Department of Medical Sciences, Uppsala University Hospital, Uppsala 75185, Sweden
3   Science for Life Laboratory, Department of Biochemistry and Biophysics, Stockholm University, Stockholm 17121, Sweden
4   Department of Surgical Sciences, Uppsala University, Uppsala 75185, Sweden; Hans.Mallmin@akademiska.se (H.M.); Olof.Nilsson@akademiska.se (O.N.)
*   Correspondence: navya.laxman@medsci.uu.se

Academic Editor: George A. Calin

Abstract: MicroRNAs (miRNAs) are a family of small, non-coding RNAs (17–24 nucleotides), which regulate gene expression either by the degradation of the target mRNAs or inhibiting the translation of genes. Recent studies have indicated that miRNA plays an important role in regulating osteoblast differentiation. In this study, we identified miR-203 and miR-320b as important miRNAs modulating osteoblast differentiation. We identified *Dlx5* as potential common target by prediction algorithms and confirmed this by knock-down and over expression of the miRNAs and assessing *Dlx5* at mRNA and protein levels and specificity was verified by luciferase reporter assays. We examined the effect of miR-203 and miR-320b on osteoblast differentiation by transfecting with pre- and anti-miRs. Over-expression of miR-203 and miR-320b inhibited osteoblast differentiation, whereas inhibition of miR-203 and miR-320b stimulated alkaline phosphatase activity and matrix mineralization. We show that miR-203 and miR-320b negatively regulate BMP-2-induced osteoblast differentiation by suppressing *Dlx5*, which in turn suppresses the downstream osteogenic master transcription factor *Runx2* and *Osx* and together they suppress osteoblast differentiation. Taken together, we propose a role for miR-203 and miR-320b in modulating bone metabolism.

Keywords: miRNA; *Dlx5*; BMP-2; osteoblast differentiation

1. Introduction

Bone Morphogenetic Proteins (BMPs) are powerful cytokines that are capable of inducing ectopic bone formation [1]. BMPs are a part of the transforming growth factor-β (TGFβ) superfamily [2] and they play a role in bone and cartilage formation by inducing the differentiation of mesenchymal progenitors into the osteoblast lineage. BMPs stimulate osteoblast differentiation through the modulation of BMP receptors and signal transducers Smad1 and Smad5 [3–5] and osteoblast-specific transcription factor Distal-less Homeobox 5 (*Dlx5*) [6,7], Runt-related transcription factor 2 (*Runx2*) [8,9], and Osterix (*Osx*) [10].

The expression of the bone-inducing transcription factor, *Dlx5*, is associated with osteoblast differentiation [11] and exhibits its highest expression when extracellular matrix mineralizes at the final stage of osteoblast differentiation. Studies show that in BMP-2-induced osteoblast differentiation, *Dlx5* plays an essential role in upregulation of the downstream osteogenic master transcription factor *Runx2* and *Osx*, whose expression in turn sequentially regulates the expression of osteoblast specific genes to induce osteoblast differentiation [8,12]. Concomitantly, *Dlx5* inhibits adipogenic differentiation by inhibiting *PPARγ* (Peroxisome proliferator-activated receptor gamma) expression in bone marrow mesenchymal stem cells (MSCs) [13].

Runx2 is the master transcription factor for osteoblast differentiation. *Runx2* expression is substantially upregulated by BMP2 through the activation of Smad signaling [9,14]. The physical interaction of *Runx2* with *Smad1* and *Smad5* is necessary to enhance its transcriptional and osteogenic activity. Studies have shown that the BMP2-Smad-Runx2 axis is important for osteoblast differentiation. This complex also induces alkaline phosphatase (ALP) activity and the expression of other osteoblast-specific genes.

Osterix is a novel zinc finger-containing transcription factor, which acts downstream of Runx2, and its expression is essential for osteoblast differentiation. During the differentiation of mesenchymal cells into osteoblasts, the expression of Osterix is considerably upregulated by BMP-2, suggesting the BMP2 acts upstream of Osterix during osteoblast differentiation [10].

MicroRNAs (miRNAs) are an abundant class of small, single-stranded non-coding RNAs (~22 nucleotides long). They have emerged as important post-transcriptional regulators in diverse processes, e.g., cell proliferation and differentiation [15,16]. MiRNAs anneal to the $3'$ UTR of their target genes and regulate protein translation and/or mRNA stability [17]. There is an increasing number of miRNAs identified recently that contribute to the regulation of osteoblast differentiation and bone formation. MiRNAs may target negative regulators of osteogenesis and as a result operate as positive regulators, or they may act as negative regulators by targeting important osteogenic factors. As a result, miRNAs exert control over skeletal gene expression [18–21]. Studies show that several bone-inducing transcription factors and signaling molecules that are involved in the function and differentiation of MSCs to osteoblasts are targets of miRNAs. Recent studies reported many bone-regulating miRNAs ("osteomiRs") that orchestrate BMP2-induced osteogenesis, such as miR-141 and miR-200a by downregulating *Dlx5* represses BMP2-induced pre-osteoblast differentiation [22]. miR-133 directly targets *Runx2*, and miR-135 attenuates Smad5 pathway and inhibits osteoblast differentiation [23]. Studies also show that mir-93 attenuates osteoblast mineralization by directly targeting *Osx* [24], whereas mir-206 targets connexin 43 and inhibits osteogenesis [25].

In the present study, we aimed to identify miRNAs involved in BMP2-induced osteogenesis using primary human osteoblasts (HOBs). Our results show that miR-203 and miR-320b target *Dlx5*, a bone-inducing transcription factor, and together they suppress osteoblast differentiation. These results were confirmed by Western blot analysis, real-time PCR, and luciferase reporter assays. The activity of miR-203 and miR-320b were modulated using an antimiR oligonucleotide, which markedly increased osteogenic differentiation in vitro, whereas miR-203 and miR-320b overexpression reversed these effects. This study confirmed earlier reports that *Dlx5* is a common upstream regulator of *Runx2* and *Osx*, and both genes are regulated independently by *Dlx5*. These results strongly suggest that miR-203 and miR-320b suppresses BMP-induced osteogenic differentiation by suppressing *Dlx5* and its downstream signaling.

2. Materials and Methods

2.1. Bone Cell Culture

Primary human osteoblast (HOB) cells were isolated from human trabecular bone collected from 3 donors undergoing total hip replacement as published previously [26–28]. The bone chips were washed thoroughly and minced with PBS. The minced bone chips were cultured in medium containing α-MEM (Sigma-Aldrich, Haverhill, UK) supplemented with 2 mmol/L L-glutamine, 100 U/mL penicillin, 100 mg/mL streptomycin and 10% fetal bovine serum (Sigma-Aldrich) at 37 °C with 5% CO_2 until confluence was reached. The culture medium was changed twice weekly. The study was approved by the local ethics committee (Ethical approval # Ups 03-561).

Cells from second passage were trypsinized and seeded at a density of 35,000 cells/well in 24 well plates and grown to 60%–80% confluency for transfection. miR-203 and miR-320 were over expressed and inhibited by transfecting the cells with mirVana hsa-miR-203 mimic (Ambion, Catalog No. 4464066), mirVana hsa-miR-203 inhibitor (Ambion, Catalog No. 4464084), mirVana hsa-miR-320 mimic (Ambion, Catalog No. 4464066), mirVana hsa-miR-320 inhibitor (Ambion, Catalog No. 4464084) using mirVana miRNA inhibitor negative control #1 (Ambion, Catalog No. 4464077) and mirVana miRNA mimic negative control #1 (Ambion, Catalog No. 4464061) at 40 nM concentrations with Magnet Assisted Transfection (MATra-si) reagent (IBA GmbH, Göttingen, Germany) according to the manufacturers' protocols. Each transfection was performed in triplicate. The cells were incubated for 24 h and stimulated with 300ng/mL of recombinant human BMP-2 (InductOs, Pfizer, New York, NY, US) and control cells were left untreated. Cells were harvested at the over a period of 7 days at intervals of 2 h, 12 h, 1 day, 2 days, 3 days, 5 days and 7 days.

2.2. Total RNA Extraction

The HOB cells were harvested at the seven different time points. The cell lysates were homogenized using QIAshredder (Qiagen, Hilden, Germany). RNA was extracted from the cell lysates using the RNeasy Mini Kit (Qiagen). Agilent 2100 BioAnalyzer (Agilent Technologies, Palo Alto, CA, USA) was used to confirm high RNA quality for all samples, RIN values in our study were between 8.2 and 9.5. The concentrations were determined with NanoDrop ND-1000 (NanoDrop Technologies, Wilmington, DE, USA) with an OD 260/280 between 1.95 and 2.03.

2.3. Target Prediction

Target prediction tools used to predict the targets of differentially expressed miRNA in this study were, e.g., TargetScan, Release 6.2: June 2012 (http://www.targetscan.org/), PicTar (http://pictar.mdc-berlin.de/) [29], and the miRanda algorithm [30,31].

2.4. Quantification by Real-Time PCR

Twenty nanograms of total RNA from each time point was reverse transcribed using TaqMan MicroRNA Reverse Transcription Kit (Applied Biosystems, Foster City, CA, USA), according to the manufacturer's instructions, enabling miRNA specific cDNA synthesis for the miR-203 and miR-320 human miRNAs and 1 miRNA control.

Following the RT step, TaqMan MicroRNA Assays (Applied Biosystems) were performed using specific TaqMan miRNA probes hsa-miR-320b (ID 002844) and hsa-miR-203 (ID 000507) (Applied Biosystems) according to the manufacturer's instructions. PCR cycling began with AmpliTaq Gold enzyme activation at 95 °C for 10 min, then 40 cycles of 95 °C for 15 s, and 60 °C for 60 s performed on a 7500 Fast Real-Time PCR System (Applied Biosystems).

For the target osteogenic genes cDNA synthesis was performed in triplicate using total RNA reverse transcribed using High Capacity cDNA reverse transcription kit (Applied Biosystems), according to the manufacturer's instructions, with no template control added to ensure a lack of signal in assay background. The real-time PCR reactions were carried out with 10 µL of 2x TaqMan® Universal PCR Master Mix, no AmpErase® UNG (Applied Biosystems), 9 µL diluted cDNA, and 1 µL of TaqMan gene specific assay mix in a 20 µL final reaction volume. Reference gene beta-actin (*ACTB*) and *GAPDH* (Applied Biosystems) was selected as control for normalization of TaqMan data. Probes specific for *Runx2* (Hs00231692_m1), *Dlx5* (Hs00193291_m1), *Osx* (Hs01866874_s1), *ACTB* (Hs01060665_g1) and *GAPDH* (Hs02758991_g1) were purchased from Applied Biosystems. The amplification was carried out using the 7500 Fast Real-Time PCR System (Applied Biosystems) using a 40-cycle program. The 7500 software automatically calculates raw Ct (cycle threshold) values.

The comparative quantitation $2^{-\Delta\Delta CT}$ method (also called the $\Delta\Delta CT$ method) [32] was used to compare differences in cycle number thresholds for samples normalized for endogenous controls.

2.5. Western Blot Analysis

The total cell lysates were prepared at seven different time points using RIPA lysis buffer (50 mM Tris-HCl, pH 8.0, with 150 mM NaCl, 1.0% Igepal CA-630 (NP-40), 0.5% sodiumdeoxycholate, 0.1% SDS, supplemented with 1.0% protease inhibitor cocktail (SIGMA-ALDRICH®, St. Louis, MO, USA). Lysates were centrifuged at 10,000-RPM for 20 min to collect supernatant. Coomassie Plus—The Better Bradford Assay™ Reagent (Thermo Scientific, Waltham, MA, USA) was used to quantify the proteins. Twenty micrograms of soluble protein was subjected to SDS-PAGE and the separated proteins were transferred to polyvinylidene difluoride membrane (Millipore). The primary antibodies were used to probe the protein bands against Dlx5 (1:500 dilution; SIGMA-ALDRICH®), Runx2 (1:250 dilution, SIGMA-ALDRICH®), Osx (1:500 dilution; SIGMA-ALDRICH®) and ACTB (1:1000 dilution; Cell Signaling Technology®, Danvers, MA, USA). Anti-Rabbit-HRP conjugated secondary antibodies (1:3000 dilution; R&D Systems®, Minneapolis, MN, USA) were used to detect the primary antibodies, followed by the target protein visualization with EMD Millipore Immobilon™ Western Chemiluminescent HRP Substrate (ECL). Images were acquired using LI-COR Odyssey® Fc Dual-Mode Imaging system (LI-COR® Biosciences, Lincoln, NE, USA) and Image Studio Software (LI-COR® Biosciences).

2.6. Luciferase Reporter Assay

After in silico target prediction as described above, the target region for miR-203 and miR-320 in *Dlx5* was verified using Dual-Luciferase Reporter Assay. The psiCHECK-2 vector, a dual-luciferase plasmid, has both the synthetic Firefly Luciferase (Fluc) gene and the synthetic *Renilla Luciferase* (*hRluc*) gene incorporated, each possessing its own promoter and poly (A)-addition sites. Luciferase reporter plasmids were constructed by inserting a perfectly complementary (Wild type) 3′ UTR fragment of *Dlx5* between the XhoI–NotI restriction sites in the multiple cloning regions in the hRluc gene in the psiCHECK-2 vector (Promega, Madison, WI, USA) by Generay Company (Shanghai, China). We also constructed 3 mutant types by replacing 6–8 base pairs at the 3′-UTR of the seed sequence. DNA sequencing was employed confirming the nucleotide sequences of the constructed plasmids.

Luciferase assays were conducted in 96 well plates cells, were HOBs were co-transfected with 100 ng/well of the wild reporter plasmid or the 3 mutant reporter plasmid and 40 nM of mirVana hsa-miR-203 mimic (Ambion, Catalog No. 4464066), mirVana hsa-miR-203 inhibitor (Ambion, Catalog No. 4464084), mirVana hsa-miR-320b mimic (Ambion, Catalog No. 4464066), mirVana hsa-miR-320b inhibitor (Ambion, Catalog No. 4464084) or with mirVana miRNA inhibitor negative control #1 (Ambion, Catalog No. 4464077) and mirVana miRNA mimic negative control #1 (Ambion, Catalog No. 4464061) negative control (NC) using 0.2 µL/well DharmaFECT™ Duo, Dharmacon™. Forty-eight hours post transfection, cells were lysed using passive lysis buffer. *Firefly* and *Renilla luciferase* activity were measured consecutively using the Dual-Luciferase® Reporter Assay System (Promega, Cat. #E1910) by using Lumat LB 9507 luminometer (Berthold Technologies) per manufacturer's instructions. *Renilla luciferase* activity was normalized to that of *Firefly luciferase*. All experiments were performed in triplicates. The relative luciferase activity was expressed as a ratio to the negative control miRNA.

2.7. Statistical Analysis

The statistical difference between pairs of groups was determined by student's t-test. Two-way analysis of variance (ANOVA) was used to evaluate the statistical significance for comparisons within groups. $p < 0.05$ was considered as statistically significant. Data were analyzed using Statistica v12 software (Stat Soft Inc., Tulsa, OK, USA,) and GraphPad Prism software package (version 6.0, GraphPad Software, Inc., La Jolla, CA, USA).

3. Results

3.1. miR-203 and miR-320b Targets Dlx5 as Shown by in Silico Analyses

In a BMP-2 induced osteoblast differentiation model, we aimed to determine target genes for miR-203 and miR-320b, that we previously had identified as miRNAs important in modulating osteoblast differentiation [33]. The putative binding sites for miR-203 and miR-320b were predicted using TargetScan, PicTar and miRanda, and *Dlx5* was selected as one of the candidate target genes (Figure 1A). The seed sequences of these miRNAs were conserved across species (Figure 1B).

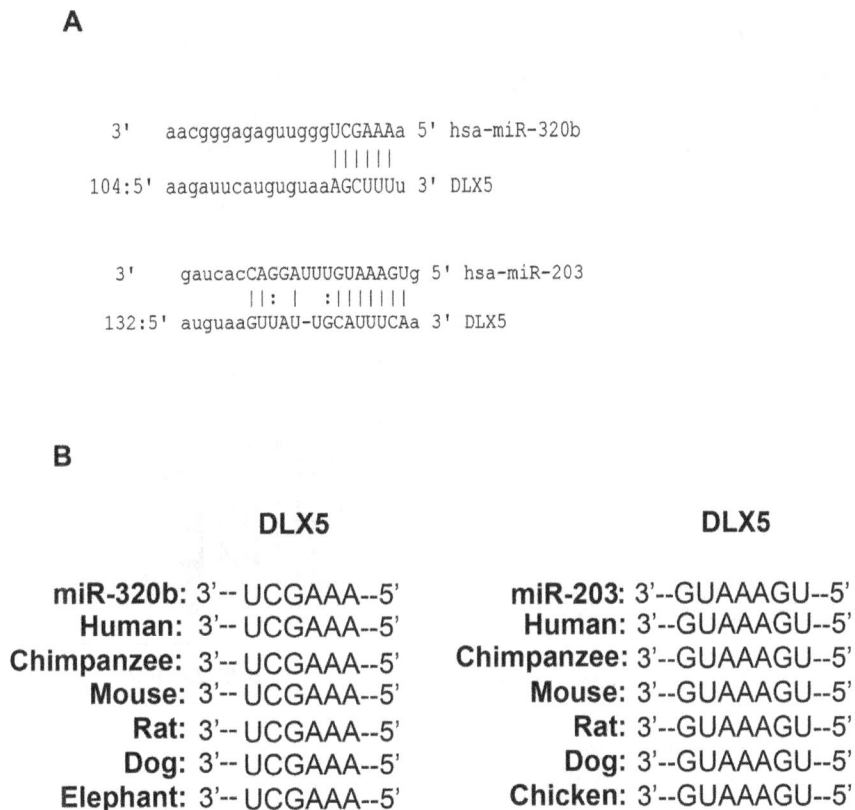

A

```
    3'    aacgggagaguugggUCGAAAa 5' hsa-miR-320b
                          ||||||
104:5' aagauucauguguaaAGCUUUu 3' DLX5


    3'    gaucacCAGGAUUUGUAAAGUg 5' hsa-miR-203
                ||: |  :||||||||
132:5' auguaaGUUAU-UGCAUUUCAa 3' DLX5
```

B

DLX5		**DLX5**	
miR-320b:	3'-- UCGAAA--5'	**miR-203:**	3'--GUAAAGU--5'
Human:	3'-- UCGAAA--5'	**Human:**	3'--GUAAAGU--5'
Chimpanzee:	3'-- UCGAAA--5'	**Chimpanzee:**	3'--GUAAAGU--5'
Mouse:	3'-- UCGAAA--5'	**Mouse:**	3'--GUAAAGU--5'
Rat:	3'-- UCGAAA--5'	**Rat:**	3'--GUAAAGU--5'
Dog:	3'-- UCGAAA--5'	**Dog:**	3'--GUAAAGU--5'
Elephant:	3'-- UCGAAA--5'	**Chicken:**	3'--GUAAAGU--5'

Figure 1. Putative binding sites and interspecies conservation: (**A**) the putative binding sites for miR-320b and miR-203 in the *Dlx5* 3′ UTR as predicted by TargetScan, PicTar and miRanda; and (**B**) seed sequences for miR-320b and miR-203 show conservation between species as depicted in the TargetScan alignment.

3.2. BMP-2 Stimulates Dlx5, Runx2 and Osx Expression

We examined the expression pattern of *Dlx5*, *Runx2* and *Osx* in human osteoblast cells when stimulated with BMP-2 at different time intervals. In the absence of BMP-2, the expressions of these genes are low, but under BMP-2 stimulation, the HOBs displayed a gradual increase in expression of *Dlx5*, *Runx2* and *Osx* mRNA up to 120 h (Figure 2A–C). The expression of miR-203 and miR-320b in BMP-2 stimulated cells showed a significant down regulation up to 120 h compared to cells untreated cells (Figure 2D).

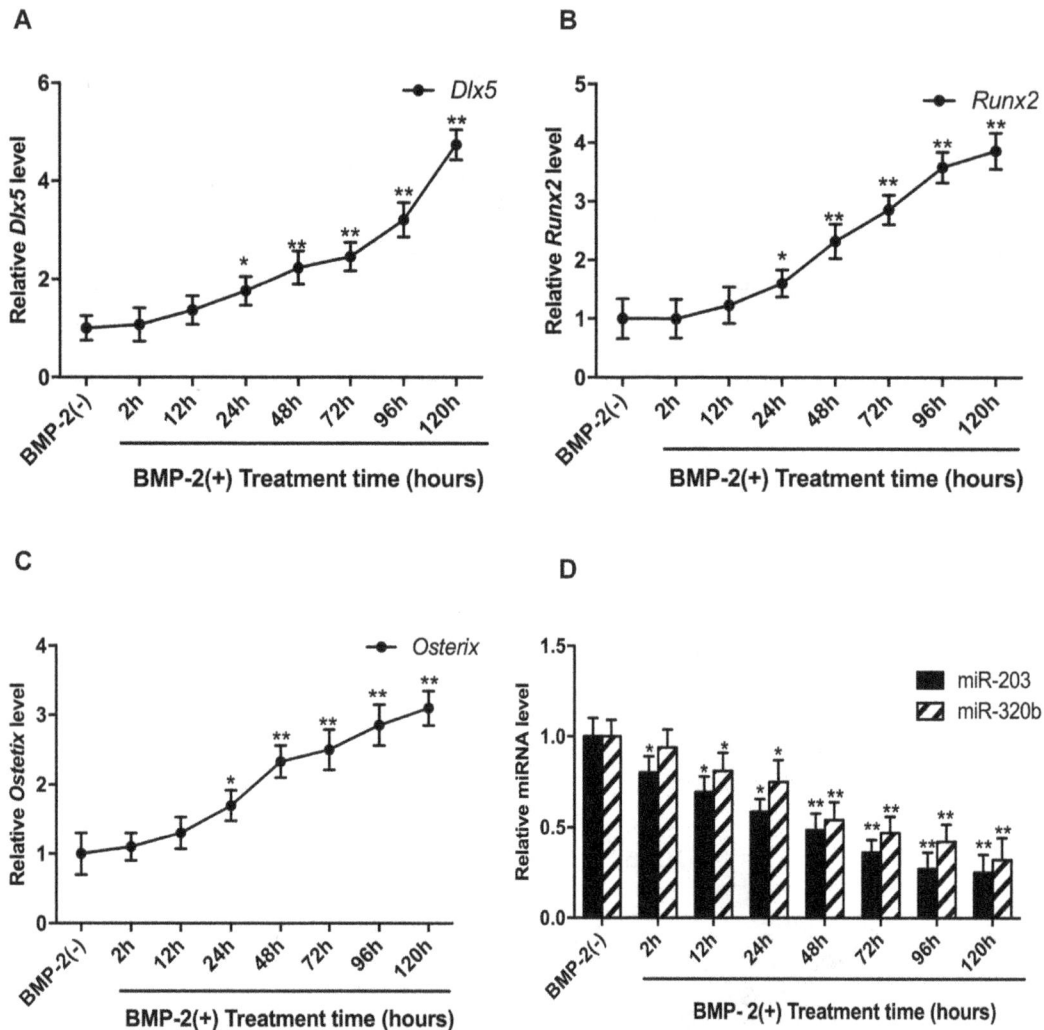

Figure 2. Time course for the expression of *Dlx5*, *Runx2*, *Osx* and miR-203 and miR-320b. The time course shows an increase in the relative expression levels of: *Dlx5* (**A**); *Runx2* (**B**); and *Osx* (**C**) at the mRNA level in BMP-2 induced cells. A concomitant decrease in the miR-203 and miR-320b expression levels is seen in the cells treated with BMP-2 (**D**). Expression levels in cells untreated with BMP-2 were set to 1. All values are represented as mean ± S.D. of three independent experiments. Statistical significance is indicated as * $p < 0.05$ and ** $p < 0.001$.

3.3. miR-203 and miR-320b Regulate the Expression of Osteogenic Transcription Factor Dlx5

To verify experimentally if the putative binding sites of miR-203 and miR-320b were functional and to evaluate the role of miR-203 and miR-320b on osteoblast differentiation stimulated by BMP-2, HOBs were transfected with miR-203 and miR-320b mimic, anti-miR and NC. mRNA and protein levels were assessed at set time points using qPCR and Western blotting. miRNAs were overexpressed by transfecting HOBs with either miR-203 or miR-320b mimics or both, showing that endogenously expressed miRNA significantly reduced the expression of *Dlx5* (Figure 3A,B). A concurrent decrease in the levels of *Runx2* (Figure 3C,D) and *Osx* (Figure 3E,F) at the mRNA and protein levels up to 120 h when compared to cells treated with mimic NC was also observed.

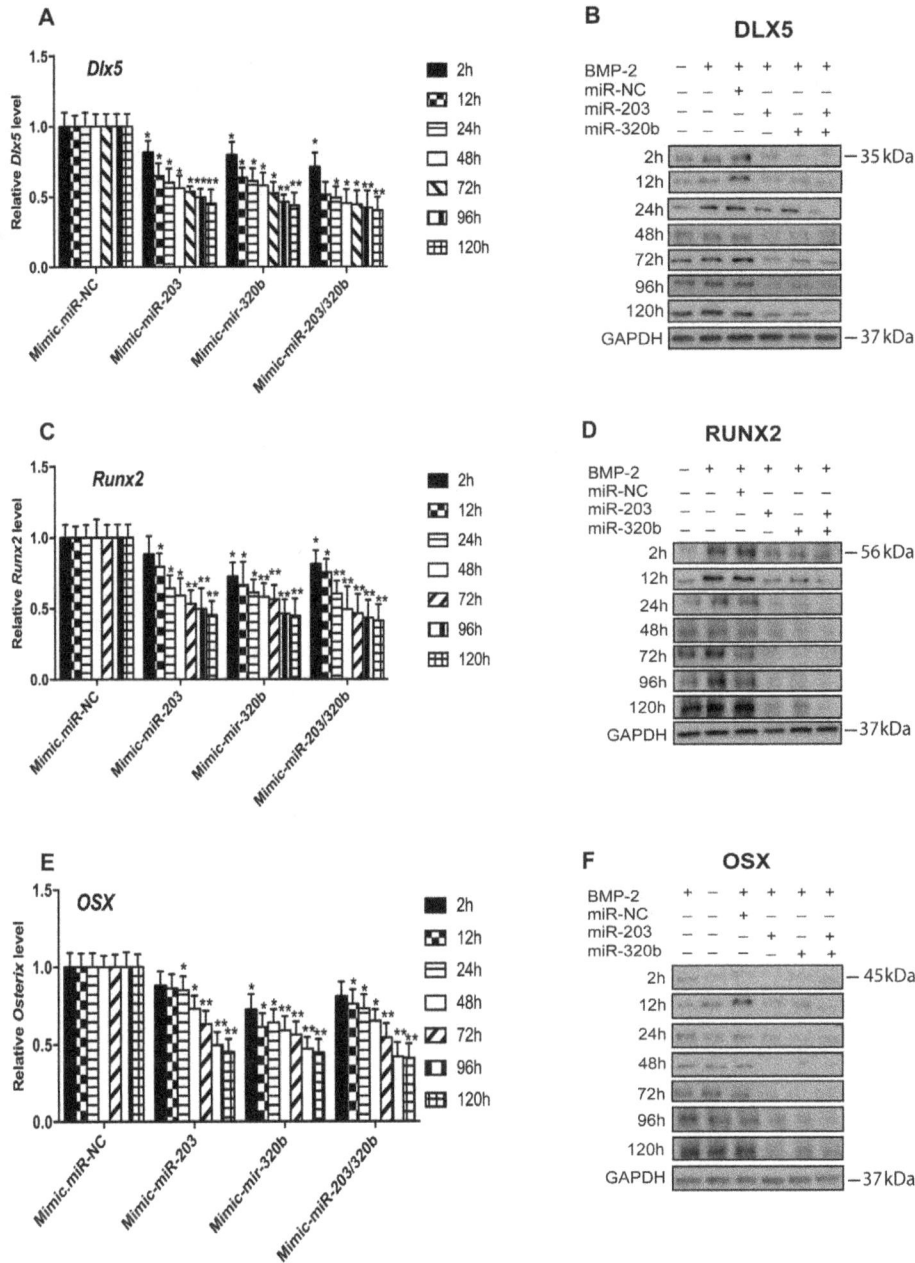

Figure 3. Overexpression of miR-203 and miR-320b negatively regulates *Dlx5*, *Runx2* and *Osx* in osteoblast differentiation. Primary human bone cells were transfected with mimic-miRNA-203 and mimic-miR-320b and RNA and protein isolated at different time points. q-PCR analysis of: *Dlx5* (**A**); *Runx2* (**C**); and *Osx* (**E**) showed down regulation when compared to the mimic-miRNA negative control. mRNA levels were normalized to *GAPDH*. Down regulation at the protein levels of: (**B**) Dlx5, Lane 1, BMP-2 (−); lane 2, BMP-2 (+); (**D**) Runx2, Lane 1, BMP-2 (−); lane 2, BMP-2 (+); and (**F**) Osx, Lane 1, BMP-2 (+); lane 2, BMP-2 (−); lane 3, BMP-2 (+) and mimic-miRNA negative control; lane 4, BMP-2 (+) and miR-203; lane 5, BMP-2 (+) and miR-320b; lane 6, BMP-2 (+) and miR-203 and miR-320b; with GAPDH as a loading control. All values are represented as mean ± S.D. of three independent experiments. Statistical significance is indicated as * $p < 0.05$ and ** $p < 0.001$.

Conversely, the expression of either miR-203 or miR-320b or both was repressed using anti-miR, to assess if miRNA repression regulated osteoblast differentiation. *Dlx5* expression was significantly increased (Figure 4A,B), and also observed was a concurrent increase in the levels of Runx2

(Figure 4C,D) and Osx (Figure 4E,F) at the mRNA and protein levels up to 120 h when compared to cells treated with anti-miR NC. GAPDH was used as an internal control for the Western blots. GAPDH for only one time point (120 h) is shown as the representative blot for all the time points. GAPDH levels for the rest of the time points are similar. All these results suggest that miR-203 and miR-320b negatively regulates *Dlx5* expression at translational level in BMP-2 stimulated osteoblast differentiation, and in turn regulates *Runx2* and *Osx* expression.

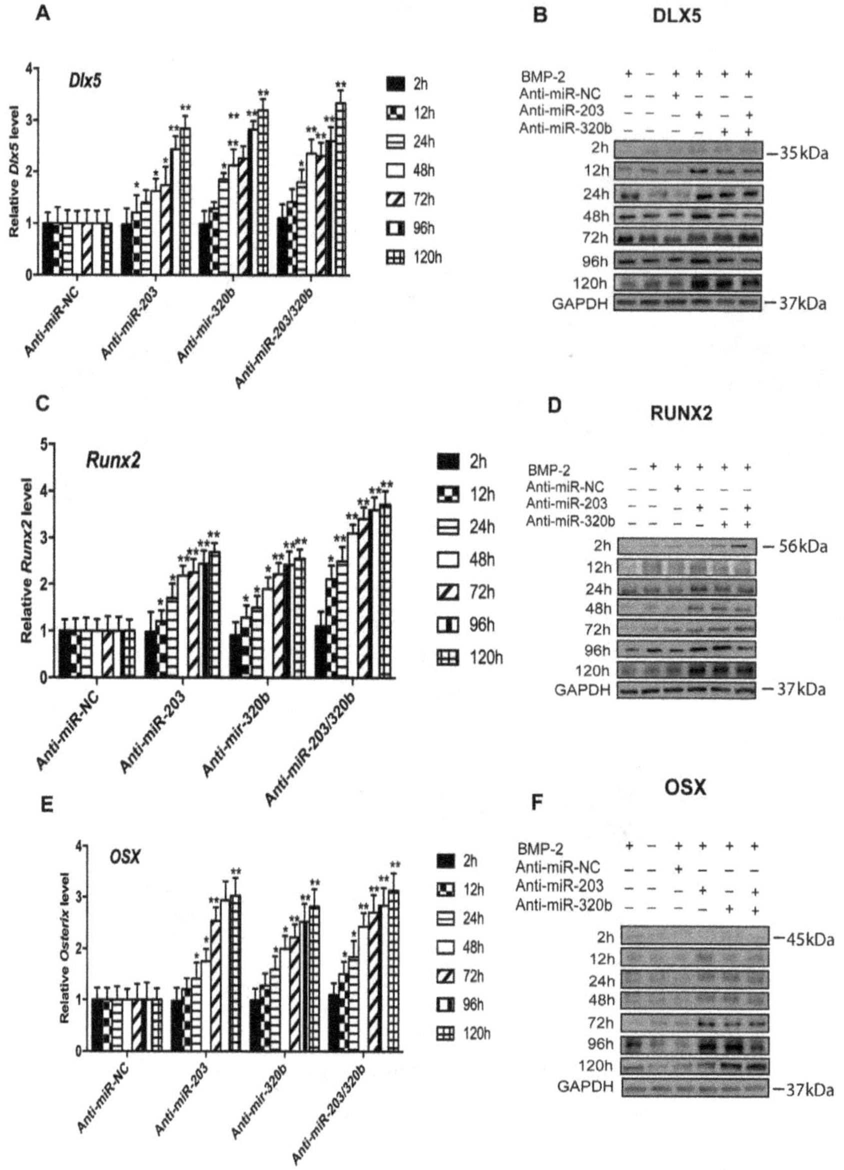

Figure 4. Knockdown of miR-203 and miR-320b up regulates *Dlx5*, *Runx2* and *Osx* in osteoblast differentiation. Primary human bone cells were transfected with pre-miRNA-203 and pre-miR-320b and RNA and protein isolated at different time points. q-PCR analysis of: *Dlx5* (**A**); *Runx2* (**C**); and *Osx* (**E**) showed up regulation when compared to the pre-miRNA negative control. mRNA levels were normalized to GAPDH. Up regulation at the protein levels of: (**B**) Dlx5, Lane 1, BMP-2 (+); lane 2, BMP-2 (−); (**D**) Runx2, Lane 1, BMP-2 (−); lane 2, BMP-2 (+); and (**F**) Osx, Lane 1, BMP-2 (+); lane 2, BMP-2 (−); lane 3, BMP-2 (+) and anti-miRNA negative control; lane 4, BMP-2 (+) and anti-miR-203; lane 5, BMP-2 (+) and anti-miR-320b; lane 6, BMP-2 (+) and anti-miR-203 and -320b; with GAPDH as a loading control. All values are represented as mean $\pm$ S.D. of three independent experiments. Statistical significance is indicated as * $p < 0.05$ and ** $p < 0.001$.

3.4. miR-203 and miR-320b Regulates BMP-2 Stimulated Human Osteoblast Differentiation

To study the impact of miR-203 and miR-320b on BMP-2 stimulated differentiation of human osteoblast, HOBs were transfected with either pre-miR or anti-miR. The osteoblast differentiation marker ALP expression was significantly enhanced with the inhibition of miR-203 and miR-320b after BMP-2 treatment compared to no BMP-2 treatment and NC, and in vitro matrix mineralization was visualized by Alizarin red staining. In contrast, HOBs transfected with pre-miR miR-203 and miR-320b, the ALP activity and matrix mineralization was reduced (Figure 5A,B). Taken together, these results suggest that miR-203 and miR-320b act as negative regulators of osteoblast differentiation.

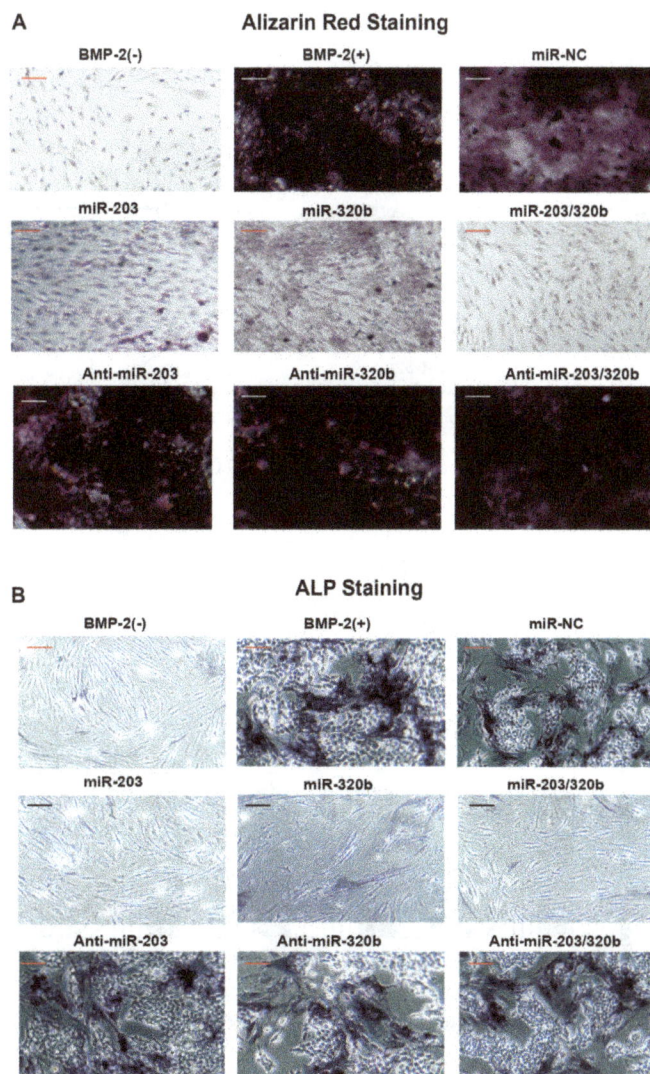

Figure 5. Effects of miR-203 and miR-320b on nodule formation and ALP activity. Human osteoblast cells were cultured and were transfected with miR-203, miR-320b miRs or Anti-miRs and with negative control (NC), and cells were stimulated with BMP-2 for 12 days. The deposition of calcium was detected by: (**A**) Alizarin Red staining; and (**B**) an ALP staining assay. Experiments were performed in triplicate and representative micrographs are shown with scale bar = 200 μm.

3.5. miR-203 and miR-320b Directly Target Dlx5

Target prediction algorithms TargetScan, PicTar and miRanda were utilized identifying *Dlx5* as a potential target gene for miR-203 and miR-320b. To determine whether miR-203 and miR-320b directly target *Dlx5*, a *Renilla* luciferase reporter was constructed in the psiCHECK-2 vector containing the

3′ UTR (Untranslated region) of *Dlx5* (wild-type). In addition, three types of mutant luciferase reporter plasmids were constructed, containing mutation in the 3′ UTR of *Dlx5* (Figure 6A). The wild-type and the three mutant plasmids were co-transfected with pre-miR-203 and pre-miR-320b in HOBs and the level of firefly and *Renilla* luciferase activity was measured consecutively. Luciferase activity was suppressed in cells transfected with the wild-type plasmid overexpressed with either miR-203 or miR-320b or both compared to NC. Mutations at the binding site of both the miRNA abolished this decrease in luciferase activity (Figure 6B). In all experiments, the psiCHECK-2 vector constitutively expressing firefly luciferase activity served as a normalization control for transfection efficiency. These results strongly indicate that miR-203 and miR-320b specifically target *Dlx5*.

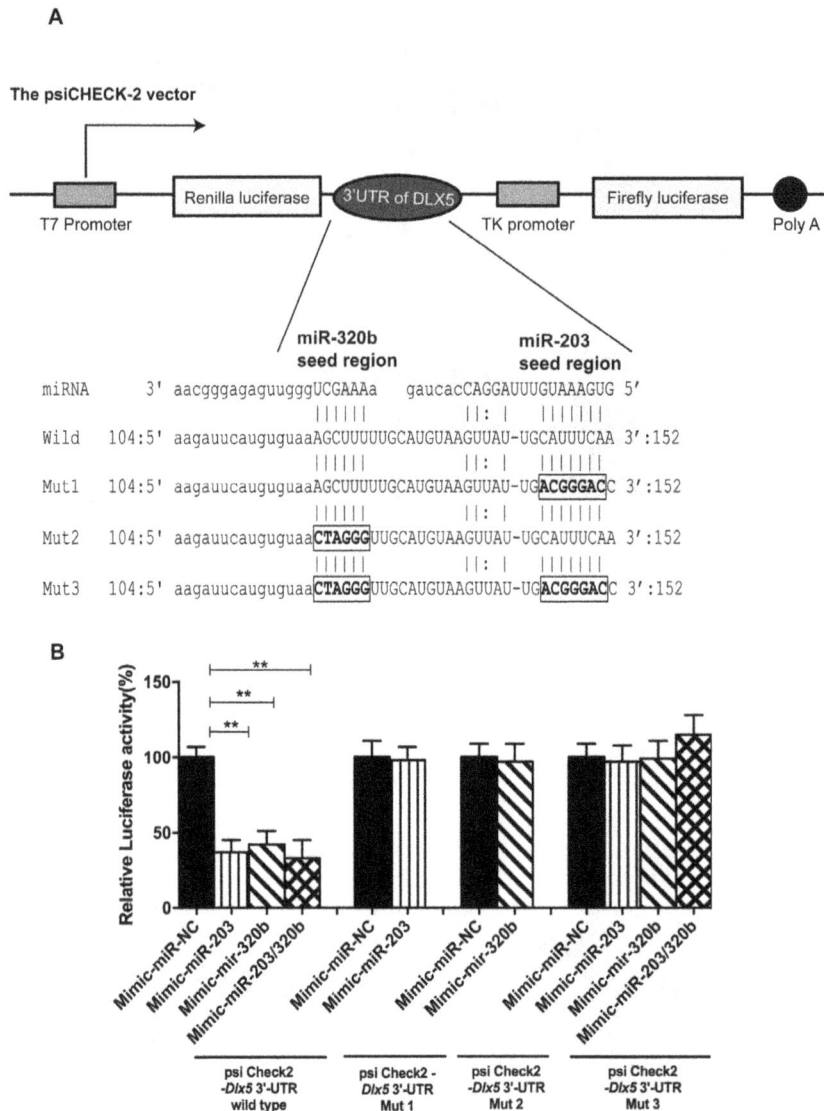

Figure 6. Characterization and functional analyses of the *Dlx5* 3′ UTR. (**A**) Schematic representation of dual luciferase reporter plasmid psiCHECK-2 vector containing the 3′ UTR of *Dlx5* inserted downstream of Renilla luciferase gene. Four types of dual luciferase reporter plasmids were constructed, one wild type and three types of mutations were constructed within the seed region of the miR-203 and miR-320b binding site. (**B**) Effect of miR-203 and miR-320b overexpression on the plasmid containing the 3′ UTR of *Dlx5* with the mutations was analyzed. Firefly and luciferase activity of each construct were measured in cell lysates. The luciferase activity was normalized to the miRNA negative control (NC). All values are represented as mean ± S.D. of three independent experiments. Statistical significance is indicated as and ** $p < 0.001$.

4. Discussion

BMP-2 is a highly efficient inducer of osteogenesis, regulating osteoblast differentiation by binding to and thereby stimulating the BMP receptors (BMPR) BMPR-I and BMPR-II. The downstream regulators Smad proteins play an important role in relaying BMP signals from receptors to activate the expression of three osteogenic master transcription factors *Dlx5*, *Runx2* and *Osx* in the nucleus [34]. *Runx2* in turn regulates the expression of many osteoblastic genes, e.g., *Col1A1* (Collagen type I AI), alkaline phosphatase (*ALP*), Osteopontin (*OPN*) and osteonectin (*ON*) [35,36] (Figure 7). Many bone regulating miRNAs have been recently identified exerting their effects at different stages of osteoblast differentiation when stimulated with BMP-2. Studies show that Smad proteins may bind to miRNA promoter genes and control their transcription and regulate miRNA biogenesis and processing [37–40]. Furthermore, during BMP-2 induced osteogenesis, the expressions of many osteomiRs are downregulated.

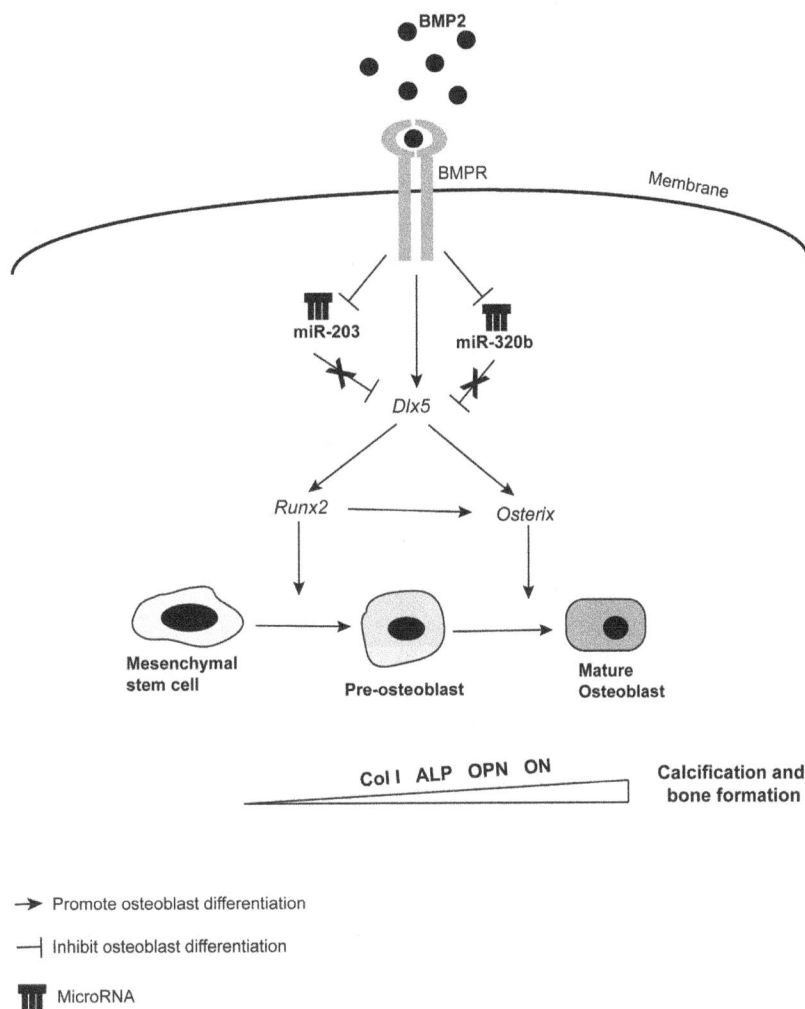

Figure 7. Schematic outline highlighting the complex regulation of BMP-2-induced osteoblastogenesis. BMP-2 binding to the BMP-2 Receptor (BMPR) activates the canonical BMP signaling pathway that regulates osteoblastic differentiation. Our present results suggest that BMP-2 downregulates the expression of miR-203 and miR-320b, in turn upregulating the transcription factor *Dlx5* (Distal-less Homeobox 5). *Dlx5* up regulation activates both the transcription factors *Runx2* and *Osx*, promoting osteoblast differentiation. This in turn results in the upregulation of genes important for bone formation and calcification, e.g., collagen type I (*COL1A1*), alkaline phosphatase (ALP), Osteopontin (OPN) and osteonectin (ON).

To our best knowledge, this is the first study that relates miR-203 and miR-320b to *Dlx5* in a BMP-2-induced human osteoblast differentiation. Previous studies have shown that miR-141 and miR-200a regulate BMP-2 mediated osteoblast differentiation by targeting *Dlx5* [22], miR-208 target *Ets1* and regulate BMP-2 induced mouse preosteoblast differentiation [41]. miR-203 has previously been shown to target *Runx2* in osteolytic bone disease [42,43]. Our study was conducted in primary human osteoblast cells, and we used BMP-2 to induce osteoblast differentiation. Based on previous studies conducted in our group [33], we identified miR-203 and miR-320b as miRNAs important during osteoblast differentiation. Our results indicate that miR-203 and miR-320b, at least in part, regulate osteoblast differentiation by downregulating a common target, *Dlx5*. Studies have shown that several miRNAs may bind to their target mRNAs and cooperatively fine-tune the degree of protein translation of a single mRNA [44].The regulation of common target genes in which upstream regulators also have an impact on each other, is reminiscent of feed-forward loops established within a transcription factors-based hierarchy. *Dlx5*. We validated the *Dlx5* as a target of miR-203 and miR-320b using Luciferase assays. BMP-2 downregulates miR-203 and miR-320b in a time dependent manner.

Studies by Ulsamer and Holleville et al [45,46] have shown that *Dlx5* regulates the expression of *Runx2* and *Osx* and their expression is upregulated when induced by BMP-2. To assess the effect of miR-203 and miR-320b on *Dlx5* in regulating osteoblast differentiation, we overexpressed miR-203 and miR-320b and observed that osteoblast differentiation was suppressed by the decrease in the expression of *Dlx5, Runx2* and *Osx*. Reciprocally, we also observed that when miR-203 and miR-320b are repressed, this results in an increased expression of the all three transcription factors. Although there was a clear effect by knockdown of miR-203 and miR-320b increasing the levels of Dlx5, Runx2 and Osx by 70%–80%, effects of other modulating factors including other osteogenic miRNAs cannot be ruled out. ALP activity was also diminished and mineral nodule formation was suppressed with miR-203 and miR-320b overexpression. In this paper, we demonstrate for the first time that miRNA 203 and miR-320b regulate *Dlx5* in a human model of BMP-2 induced osteoblast differentiation.

In summary, we have identified two osteoblast related miRNAs that are crucial for regulating osteoblast differentiation. We indicate in our study that miR-203 and miR-320b actively inhibit osteogenesis and BMP-2 addition down regulated the miRNAs and promoted osteoblast differentiation. We observed that miR-203 and miR-320b together mediated repression of the transcription factor *Dlx5*, which is upstream of and activator of *Runx2* and *Osx*, and they are important for the regulation of osteoblast differentiation. In view of these results, it can be suggested that downregulation of miR-203 and miR-320b by BMP-2 augmented the expression of *Dlx5*, hence bone formation. Taken altogether, we describe a novel function of miR-203 and miR-320b in negatively regulating BMP-2-induced osteoblast differentiation by suppressing *Dlx5*, which in turn suppresses the downstream osteogenic master transcription factor *Runx2* and *Osx*. We also show that inhibition of miR-203 and miR-320b can increase osteoblast differentiation. Further studies are needed, investigating the role of miR-203 and miR-320b in modulating bone metabolism in vivo.

5. Conclusions

In summary, our study reveals a novel function of miR-203 and miR-320b in negatively regulating BMP-2-induced osteoblast differentiation by suppressing *Dlx5*, a bone-inducing transcription factor, which in turn suppresses the downstream osteogenic master transcription factor *Runx2* and *Osx* and together they suppress osteoblast differentiation. Taken together, we propose that miR-203 and miR-320b suppresses BMP-induced osteogenic differentiation by suppressing *Dlx5* and its downstream signaling.

Acknowledgments: We thank Elin Carlsson for skillful assistance during qPCR experiments and ALP and Alizarin staining experiments. Funding: This work was supported by Vetenskapsrådet (2009-2852); and Avtal om Läkarutbildning och Forskning (ALF-grants) (46945).

Author Contributions: All authors contributed to the writing of the manuscript. Study design: N.L. and A.K. Study conduct: N.L., H.M., O.N. and A.K. N.L. Performed the experiments. Data interpretation: N.L. and A.K. Revising manuscript content: N.L. and A.K. Approving final version of manuscript: N.L., H.M., O.N. and A.K.

Abbreviations

The following abbreviations are used in this manuscript:

ALP	alkaline phosphatase
BMP-2	Bone morphogenetic protein 2
Col1A1	Collagen type I AI
Dlx5	Distal-less Homeobox 5
HOBs	primary human osteoblasts
miRNA or miR	microRNA
mRNA	messenger RNA
Osx	Osterix
OPN	Osteopontin
ON	osteonectin
PPARγ	Peroxisome proliferator-activated receptor gamma
Runx2	Runt-related transcription factor 2
TGFβ	transforming growth factor-β
UTR	Untranslated region

References

1. Groeneveld, E.H.; Burger, E.H. Bone morphogenetic proteins in human bone regeneration. *Eur. J. Endocrinol.* **2000**, *142*, 9–21. [CrossRef] [PubMed]

2. Chen, G.; Deng, C.; Li, Y.P. TGF-beta and BMP signaling in osteoblast differentiation and bone formation. *Int. J. Biol. Sci.* **2012**, *8*, 272–288. [CrossRef] [PubMed]

3. Akiyama, S.; Katagiri, T.; Namiki, M.; Yamaji, N.; Yamamoto, N.; Miyama, K.; Shibuya, H.; Ueno, N.; Wozney, J.M.; Suda, T. Constitutively active BMP type I receptors transduce BMP-2 signals without the ligand in C2C12 myoblasts. *Exp. Cell. Res.* **1997**, *235*, 362–369. [CrossRef] [PubMed]

4. Yamamoto, N.; Akiyama, S.; Katagiri, T.; Namiki, M.; Kurokawa, T.; Suda, T. Smad1 and smad5 act downstream of intracellular signalings of BMP-2 that inhibits myogenic differentiation and induces osteoblast differentiation in C2C12 myoblasts. *Biochem. Biophys. Res. Commun.* **1997**, *238*, 574–580. [CrossRef] [PubMed]

5. Heldin, C.H.; Miyazono, K.; Dijke, P. TGF-[beta] signalling from cell membrane to nucleus through SMAD proteins. *Nature* **1997**, *390*, 465–471. [CrossRef] [PubMed]

6. Miyama, K.; Yamada, G.; Yamamoto, T.S.; Takagi, C.; Miyado, K.; Sakai, M.; Ueno, N.; Shibuya, H. A BMP-inducible gene, dlx5, regulates osteoblast differentiation and mesoderm induction. *Dev. Biol.* **1999**, *208*, 123–133. [CrossRef] [PubMed]

7. Luo, T.; Matsuo-Takasaki, M.; Lim, J.H.; Sargent, T.D. Differential regulation of Dlx gene expression by a BMP morphogenetic gradient. *Int. J. Dev. Biol.* **2001**, *45*, 681–684. [PubMed]

8. Lee, M.H.; Kim, Y.J.; Kim, H.J.; Park, H.D.; Kang, A.R.; Kyung, H.M.; Sung, J.H.; Wozney, J.M.; Kim, H.J.; Ryoo, H.M. BMP-2-induced Runx2 expression is mediated by Dlx5, and TGF-beta 1 opposes the BMP-2-induced osteoblast differentiation by suppression of Dlx5 expression. *J. Biol. Chem.* **2003**, *278*, 34387–34394. [CrossRef] [PubMed]

9. Lee, K.S.; Kim, H.J.; Li, Q.L.; Chi, X.Z.; Ueta, C.; Komori, T.; Wozney, J.M.; Kim, E.G.; Choi, J.Y.; Ryoo, H.M.; et al. Runx2 is a common target of transforming growth factor beta1 and bone morphogenetic protein 2, and cooperation between Runx2 and Smad5 induces osteoblast-specific gene expression in the pluripotent mesenchymal precursor cell line C2C12. *Mol. Cell. Biol.* **2000**, *20*, 8783–8792. [CrossRef] [PubMed]

10. Nakashima, K.; Zhou, X.; Kunkel, G.; Zhang, Z.P.; Deng, J.M.; Behringer, R.R.; de Crombrugghe, B. The novel zinc finger-containing transcription factor osterix is required for osteoblast differentiation and bone formation. *Cell* **2002**, *108*, 17–29. [CrossRef]

11. Ryoo, H.M.; Hoffmann, H.M.; Beumer, T.; Frenkel, B.; Towler, D.A.; Stein, G.S.; Stein, J.L.; van Wijnen, A.J.; Lian, J.B. Stage-specific expression of Dlx-5 during osteoblast differentiation: Involvement in regulation of osteocalcin gene expression. *Mol. Endocrinol.* **1997**, *11*, 1681–1694. [CrossRef] [PubMed]

12. Lee, M.H.; Kim, Y.J.; Yoon, W.J.; Kim, J.I.; Kim, B.G.; Hwang, Y.S.; Wozney, J.M.; Chi, X.Z.; Bae, S.C.; Choi, K.Y.; et al. Dlx5 specifically regulates Runx2 type II expression by binding to homeodomain-response elements in the Runx2 distal promoter. *J. Biol. Chem.* **2005**, *280*, 35579–35587. [CrossRef] [PubMed]

13. Lee, H.L.; Kim, Y.J.; Yoon, W.J.; Kim, J.I.; Kim, B.G.; Hwang, Y.S.; Wozney, J.M.; Chi, X.Z.; Bae, S.C.; Choi, K.Y.; et al. Dlx5 inhibits adipogenic differentiation through down-regulation of PPAR? *J. Cell. Physiol.* **2013**, *228*, 87–98. [CrossRef] [PubMed]

14. Nishimura, R.; Hata, K.; Harris, S.E.; Ikeda, F.; Yoneda, T. Core-binding factor alpha 1 (Cbfa1) induces osteoblastic differentiation of C2C12 cells without interactions with Smad1 and Smad5. *Bone* **2002**, *31*, 303–312. [CrossRef]

15. Stefani, G.; Slack, F.J. Small non-coding RNAs in animal development. *Nat. Rev. Mol. Cell. Biol.* **2008**, *9*, 219–230. [CrossRef] [PubMed]

16. Bartel, D.P. MicroRNAs: Genomics, biogenesis, mechanism, and function. *Cell* **2004**, *116*, 281–297. [CrossRef]

17. Doench, J.G.; Sharp, P.A. Specificity of microRNA target selection in translational repression. *Genes Dev.* **2004**, *18*, 504–511. [CrossRef] [PubMed]

18. Zhao, X.; Xu, D.; Li, Y.; Zhang, J.Y.; Liu, T.T.; Ji, Y.L.; Wang, J.F.; Zhou, G.M.; Xie, X.X. MicroRNAs regulate bone metabolism. *J. Bone Miner. Metab.* **2014**, *32*, 221–231. [CrossRef] [PubMed]

19. Eguchi, T.; Watanabe, K.; Hara, E.S.; Ono, M.; Kuboki, T.; Calderwood, S.K. OstemiR: A novel panel of microRNA biomarkers in osteoblastic and osteocytic differentiation from mesencymal stem cells. *PLoS ONE* **2013**, *8*, e58796. [CrossRef] [PubMed]

20. Hu, R.; Li, H.; Liu, W.; Yang, L.; Tan, F.Y.; Luo, X.H. Targeting miRNAs in osteoblast differentiation and bone formation. *Exp. Opin. Ther. Targets* **2010**, *14*, 1109–1120. [CrossRef] [PubMed]

21. Taipaleenmaki, H.; Hokland, L.B.; Chen, L.; Kauppinen, S.; Kassem, M. Mechanisms in endocrinology: micro-RNAs: Targets for enhancing osteoblast differentiation and bone formation. *Eur. J. Endocrinol.* **2012**, *166*, 359–371. [CrossRef] [PubMed]

22. Itoh, T.; Nozawa, Y.; Akao, Y. MicroRNA-141 and -200a are involved in bone morphogenetic protein-2-induced mouse pre-osteoblast differentiation by targeting distal-less homeobox 5. *J. Biol. Chem.* **2009**, *284*, 19272–19279. [CrossRef] [PubMed]

23. Li, Z.; Hassan, M.Q.; Volinia, S.; van Wijnen, A.J.; Stein, J.L.; Croce, C.M.; Lian, J.B.; Stein, G.S. A microRNA signature for a BMP2-induced osteoblast lineage commitment program. *Proc. Natl. Acad. Sci. USA* **2008**, *105*, 13906–13911. [CrossRef] [PubMed]

24. Yang, L.; Cheng, P.; Chen, C.; He, H.B.; Xie, G.Q.; Zhou, H.D.; Xie, H.; Wu, X.P.; Luo, X.H. miR-93/Sp7 function loop mediates osteoblast mineralization. *J. Bone Miner. Res.* **2012**, *27*, 1598–1606. [CrossRef] [PubMed]

25. Inose, H.; Ochi, H.; Kimura, A.; Fujita, K.; Xu, R.; Sato, S.; Iwasaki, M.; Sunamura, S.; Takeuchi, Y.; Fukumoto, S.; et al. A microRNA regulatory mechanism of osteoblast differentiation. *Proc. Natl. Acad. Sci. USA* **2009**, *106*, 20794–20799. [CrossRef] [PubMed]

26. Grundberg, E.; Adoue, V.; Kwan, T.; Ge, B.; Duan, Q.L.; Lam, K.C.L.; Koka, V.; Kindmark, K.; Weiss, S.T.; Tantisira, K.; et al. Global analysis of the impact of environmental perturbation on cis-regulation of gene expression. *PLoS Genet.* **2011**, *7*, e1001279. [CrossRef] [PubMed]

27. Grundberg, E.; Kwan, T.; Ge, B.; Duan, Q.L.; Lam, K.C.L.; Koka, V.; Kindmark, A.; Mallmin, H.; Dias, J.; Verlaan, D.J.; et al. Population genomics in a disease targeted primary cell model. *Genome. Res.* **2009**, *19*, 1942–1952. [CrossRef] [PubMed]

28. Laxman, N.; Rubin, G.J.; Mallmin, H.; Nilsson, O.; Pastinen, T.; Grundberg, E.; Kindmark, A. Global miRNA expression and correlation with mRNA levels in primary human bone cells. *RNA* **2015**, *21*, 1433–1443. [CrossRef] [PubMed]

29. Krek, A.; Grün, D.; Poy, M.N.; Wolf, R.; Rosenberg, L.; Epstein, E.J.; MacMenamin, P.; da Piedade, I.; Gunsalus, K.C.; Stoffel, M.; et al. Combinatorial microRNA target predictions. *Nat. Genet.* **2005**, *37*, 495–500. [CrossRef] [PubMed]

30. Enright, A.J.; John, B.; Gaul, U.; Tuschl, T.; Sander, C.; Marks, D.S. MicroRNA targets in *Drosophila*. *Genome. Biol.* **2003**. [CrossRef]

31. John, B.; Enright, A.J.; Aravin, A.; Tuschl, T.; Sander, C.; Marks, D.S. Human MicroRNA targets. *PLoS Biol.* **2004**, *2*, e363. [CrossRef] [PubMed]

32. Livak, K.J.; Schmittgen, T.D. Analysis of relative gene expression data using real-time quantitative PCR and the 2(-Delta Delta C(T)) Method. *Methods* **2001**, *25*, 402–408. [CrossRef] [PubMed]

33. Laxman, N.; Rubin, C.J.; Mallmin, H.; Nilsson, O.; Tellgren-Roth, C.; Kindmark, A. Second generation sequencing of microRNA in human bone cells treated with parathyroid hormone or dexamethasone. *Bone* **2016**, *24*, 181–188. [CrossRef] [PubMed]

34. Nohe, A.; Hassel, S.; Ehrlich, M.; Neubauer, F.; Sebald, W.; Henis, Y.I.; Knaus, P. The mode of bone morphogenetic protein (BMP) receptor oligomerization determines different BMP-2 signaling pathways. *J. Biol. Chem.* **2002**, *277*, 5330–5338. [CrossRef] [PubMed]

35. Ducy, P.; Karsenty, G. Two distinct osteoblast-specific cis-acting elements control expression of a mouse osteocalcin gene. *Mol. Cell. Biol.* **1995**, *15*, 1858–1869. [CrossRef] [PubMed]

36. Javed, A.; Barnes, G.L.; Jasanya, B.O.; Stein, J.L.; Gerstenfeld, L.; Lian, J.B.; Stei, G.S. Runt homology domain transcription factors (Runx, Cbfa, and AML) mediate repression of the bone sialoprotein promoter: Evidence for promoter context-dependent activity of Cbfa proteins. *Mol. Cell. Biol.* **2001**, *21*, 2891–2905. [CrossRef] [PubMed]

37. Blahna, M.T.; Hata, A. Smad-mediated regulation of microRNA biosynthesis. *FEBS Lett.* **2012**, *586*, 1906–1912. [CrossRef] [PubMed]

38. Davis, B.N.; Hilyard, A.C.; Lagna, G.; Hata, A. SMAD proteins control DROSHA-mediated microRNA maturation. *Nature* **2008**, *454*, 56–61. [CrossRef] [PubMed]

39. Davis, B.N.; Hilyard, A.C.; Nguyen, P.H.; Lagna, G.; Hata, A. Smad proteins bind a conserved RNA sequence to promote microRNA maturation by Drosha. *Mol. Cell* **2010**, *39*, 373–384. [CrossRef] [PubMed]

40. Butz, H.; Rácz, K.; Hunyady, L.; Patócs, A. Crosstalk between TGF-β signaling and the microRNA machinery. *Trends Pharmacol. Sci.* **2012**, *33*, 382–393. [CrossRef] [PubMed]

41. Itoh, T.; Takeda, S.; Akao, Y. MicroRNA-208 modulates BMP-2-stimulated mouse preosteoblast differentiation by directly targeting V-ets erythroblastosis virus E26 oncogene homolog 1. *J. Biol. Chem.* **2010**, *285*, 27745–27752. [CrossRef] [PubMed]

42. Taipaleenmäki, H.; Browne, G.; Akech, J.; Zustin, J.; van Wijnen, A.J.; Stein, J.L.; Hesse, E.; Stein, G.S.; Lian, J.B. Targeting of Runx2 by miR-135 and miR-203 impairs progression of breast cancer and metastatic bone disease. *Cancer Res.* **2015**, *75*, 1433–1444. [CrossRef] [PubMed]

43. Saini, S.; Majid, S.; Yamamura, S.; Tabatabai, L.; Suh, S.O.; Shahryari, V.; Chen, Y.; Deng, G.; Tanaka, Y.; Dahiya, R. Regulatory Role of mir-203 in prostate cancer progression and metastasis. *Clin. Cancer Res.* **2011**, *17*, 5287–5298. [CrossRef] [PubMed]

44. Bartel, D.P.; Chen, C.Z. Micromanagers of gene expression: The potentially widespread influence of metazoan microRNAs. *Nat. Rev. Genet.* **2004**, *5*, 396–400. [CrossRef] [PubMed]

45. Ulsamer, A.; Ortuño, M.J.; Ruiz, S.; Susperregui, A.R.G.; Osses, N.; Rosa, J.L.; Ventura, F. BMP-2 induces Osterix expression through up-regulation of Dlx5 and its phosphorylation by p38. *J. Biol. Chem.* **2008**, *283*, 3816–3826. [CrossRef] [PubMed]

46. Holleville, N.; Matéosa, S.; Bontouxc, M.; Bollerotd, K.; Monsoro-Burq, A.H. Dlx5 drives Runx2 expression and osteogenic differentiation in developing cranial suture mesenchyme. *Dev. Biol.* **2007**, *304*, 860–874. [CrossRef] [PubMed]

2

Contribution of the RgfD Quorum Sensing Peptide to *rgf* Regulation and Host Cell Association in Group B *Streptococcus*

Robert E. Parker [1], David Knupp [1], Rim Al Safadi [1], Agnès Rosenau [2] and Shannon D. Manning [1,*

[1]   Department of Microbiology and Molecular Genetics, Michigan State University, East Lansing, MI 48824, USA; parke274@msu.edu (R.E.P.); knuppdav@msu.edu (D.K.); rimalsafadi@gmail.com (R.A.S.)

[2]   Infectiologie et Santé Publique ISP, Institut National de la Recherche Agronomique, Université de Tours, Equipe Bactéries et Risque Materno-fœtal, UMR1282 Tours, France; rosenau@univ-tours.fr

*   Correspondence: mannin71@msu.edu

Academic Editor: Helen J. Wing

Abstract: *Streptococcus agalactiae* (group B *Streptococcus*; GBS) is a common inhabitant of the genitourinary and/or gastrointestinal tract in up to 40% of healthy adults; however, this opportunistic pathogen is able to breach restrictive host barriers to cause disease and persist in harsh and changing conditions. This study sought to identify a role for quorum sensing, a form of cell to cell communication, in the regulation of the fibrinogen-binding (*rgfBDAC*) two-component system and the ability to associate with decidualized endometrial cells in vitro. To do this, we created a deletion in *rgfD*, which encodes the putative autoinducing peptide, in a GBS strain belonging to multilocus sequence type (ST)-17 and made comparisons to the wild type. Sequence variation in the *rgf* operon was detected in 40 clinical strains and a non-synonymous single nucleotide polymorphism was detected in *rgfD* in all of the ST-17 genomes that resulted in a truncation. Using qPCR, expression of *rgf* operon genes was significantly decreased in the ST-17 Δ*rgfD* mutant during exponential growth with the biggest difference (3.3-fold) occurring at higher cell densities. Association with decidualized endometrial cells was decreased 1.3-fold in the mutant relative to the wild type and *rgfC* expression was reduced 22-fold in Δ*rgfD* following exposure to the endometrial cells. Collectively, these data suggest that this putative quorum sensing molecule is important for attachment to human tissues and demonstrate a role for RgfD in GBS pathogenesis through regulation of *rgfC*.

Keywords: *Streptococcus agalactiae*; quorum sensing; colonization; group B *Streptococcus*

---

1. Introduction

Streptococcus agalactiae, or group B *Streptococcus* (GBS), resides as a commensal in the gastrointestinal and/or urogenital tracts in up to 40% of healthy men and women but is an opportunistic pathogen presenting a threat to newborns, pregnant women, the chronically ill, and the elderly [1]. In neonates, GBS is a leading cause of meningitis and sepsis. Although there has been a reduction in the incidence of neonatal early onset disease (EOD) over the past 30 years [2], GBS is still a major concern in both industrialized and developing nations, and there remain significant gaps in our understanding of the molecular mechanisms of pathogenesis. The identification of features that allow one GBS strain to become more invasive than another is incomplete. Several studies utilizing multilocus sequence typing (MLST), a method targeting seven conserved housekeeping genes [3], have shown that most strains belong to one of four clonal complexes (CCs): 1, 17, 19, and 23. Strains belonging to CC-17, however, have been shown to cause an increased frequency of neonatal infections [3,4] and were suggested to be more virulent with unique features that impact disease development and progression [5–8].

While GBS is well adapted to survival in the host, crossing restrictive barriers like the extraplacental membranes and blood-brain barrier presents a challenge to the bacterium as disease progression requires the complex regulation of multiple virulence factors [9,10]. The ability to respond to environmental cues occurs through transcriptome remodeling, which facilitates adaptation and survival in distinct niches [11]. Indeed, remodeling of the GBS transcriptome has been observed in response to growth temperature and exposure to other host-specific environments [12–14]. For most bacterial pathogens, the ability to recognize extracellular stimuli and respond occurs via signal transduction systems (STS), with the most common being two-component systems (TCSs) [15]. Most TCSs are composed of a membrane-bound sensor kinase, which reacts to an extracellular stimulus by phosphorylating and activating a response regulator that serves as a transcription factor driving downstream behavioral changes [16]. The number of TCSs in bacterial chromosomes have been shown to correlate with the genome size at a rate of ~2.3 TCSs per 1 Mb for genomes up to 5 Mb [17]. GBS has a disproportionately high number of TCSs with 17–20 predicted for the 2.2 Mb genome [18]. Several of these systems have been shown to play a role in pathogenesis, including the controller of virulence (CovR/S) [19], the regulator of D-Alanyl-lipotechoic acid biosynthesis (DltR/S) [20], the competence and β-lactam-resistance promoting system (CiaR/C) [21], and the regulator of fibrinogen binding (RgfA/C) [22].

The Rgf system, encoded by the *rgfBDAC* operon, was identified as a polycistronically transcribed system that promotes binding of host cell components through the regulation of cell surface proteins including the fibronectin binding protein, *scpB*, and two fibrinogen binding proteins, *fbsA* and *fbsB* [22,23]. Importantly, the *rgf* operon is homologous to the accessory gene regulator operon (*agrBDCA*), a TCS found in staphylococci [22]. The *agr* TCS is a well-studied quorum sensing circuit important for virulence via the regulation of secreted virulence factors and surface proteins [24,25]. Regulation of this operon, however, is complex and has been linked to multiple factors [24–27]. Similar to the *agr* system, the *rgf* operon is composed of a putative ABC transporter, *rgfB*, a putative quorum sensing protein, *rgfD*, and the TCS *rgfA/C* [22].

Genetic variation has been described in both the *S. aureus agr* and GBS *rgf* system. Mutations conferring a non-hemolytic, non-invasive phenotype have been detected in the *agr* operon from strains recovered from patients [24,28]. For the *rgf* operon, one study identified a truncation in the gene encoding the response regulator, *rgfA*, in several clinical strains and deletion of both *rgfC*, the sensor histidine kinase, and *rgfA* resulted in increased virulence in a mouse and rat model [29]. This increase was possibly due to other virulence mechanisms including increased sialic acid production and capsule operon transcription, which were both altered in the deletion mutant. The same study also found that extrachromosomal *rgfC* expression altered the transcriptome, indicating that regulation of the sensor histidine kinase may be important for GBS pathogenesis [29]. Separate analyses of the genome of NEM316, a serotype III ST-23 GBS strain isolated from a fatal case of neonatal septicemia, also identified a large deletion within *rgfD*, which encodes a putative auto-inducing peptide, and part of *rgfC* [30]. The role of *rgfD* and different *rgfD* mutations on the regulation of the *rgf* operon in GBS, however, has not been examined nor has the impact of both on phenotypes relevant for pathogenesis. We therefore sought to investigate the contribution of *rgfD* to biofilm production, host cell association, and operon regulation in distinct growth stages and following exposure to decidualized endometrial cells.

2. Materials and Methods

2.1. Bacterial Strains, Growth Conditions, and rgf Sequence Analysis

GBS was cultured in Todd-Hewitt broth (THB) or agar (THA) or trypticase soy agar plus 5% sheep's blood (Becton Dickinson, Franklin Lakes, NJ, USA) at 37 °C with 5% CO_2. Growth curves were performed in THB using the same conditions with samples taken for determination of OD_{595}

at different times. Three serotype III, CC-17 strains (GB00451, GB00546, and GB00097) were used to quantify *rgf* transcription by growth phase. Mutagenesis was performed in GB00451 (ST-17) and GB00012 (ST-1).

To examine sequence variation, additional *rgf* operon sequences were extracted from 40 draft genomes sequenced by the J. Craig Venter Institute (Table 1) using the Basic Local Alignment Search Tool (BLAST) available in the National Center for Biotechnology Information (NCBI) with strain O90R as the *rgf* reference sequence (AF390107.1) [22]. All base locations in the 40 genomes are named relative to the 3320 bp O90R sequence, which begins 94 bp prior to the start of *rgfB* in the *rgf* operon. Multiple alignments were performed using the ClustalW algorithm in MegAlign and a Neighbor joining phylogeny based on p-distance was generated using MEGA6 with bootstrapping [31]. The 40 clinical strains, which were recovered from colonized mothers or young adults, were previously characterized by MLST [7,32]. Although biofilm production was performed previously using OD_{595} values ≥ 1.8 as the cutoff for strong biofilms [33], this study sought to examine the relationship between biofilm level and *rgf* sequence variation, which was not examined initially.

Table 1. Strains examined in the study with sequence accession numbers.

Strain	Accession Number	Strain	Accession Number
rgf reference sequence	AF390107.1	GB00557	GCA_000290235.1
GB00002	GCA_000289475.1	GB00614	GCA_000290335.1
GB00012	GCA_000288135.1	GB00651	GCA_000290375.1
GB00013	GCA_000288095.1	GB00654	GCA_000290395.1
GB00020	GCA_000288235.1	GB00663	GCA_000290435.1
GB00082	GCA_000288215.1	GB00679	GCA_000290475.1
GB00083	GCA_000288255.1	GB00865	GCA_000290495.1
GB00092	GCA_000290055.1	GB00867	GCA_000289595.1
GB00097	GCA_000289495.1	GB00874	GCA_000289615.1
GB00111	GCA_000290075.1	GB00884	GCA_000289635.1
GB00112	GCA_000291585.1	GB00887	GCA_000289655.1
GB00115	GCA_000290095.1	GB00891	GCA_000290215.1
GB00190	GCA_000290135.1	GB00904	GCA_000288375.1
GB00206	GCA_000289535.1	GB00923	GCA_000288475.1
GB00226	GCA_000288195.1	GB00929	GCA_000288515.1
GB00241	GCA_000288175.1	GB00932	GCA_000288535.1
GB00245	GCA_000288335.1	GB00959	GCA_000288615.1
GB00279	GCA_000288355.1	GB00984	GCA_000288655.1
GB00300	GCA_000289575.1	GB00986	GCA_000289715.1
GB00555	GCA_000290235.1	GB00992	GCA_000289735.1

Accession numbers were assigned by the European Nucleotide Archive (http://www.ebi.ac.uk/ena). Sequences are also available at www.pathogenportal.org/portal/portal/PathPort/Data.

2.2. rgfD Mutagenesis and Complementation

Mutagenesis was performed using a double-homologous recombination strategy with the pG+host5 thermosensitive plasmid [34] for the deletion of *rgfD* as described [35]. Flanking regions were amplified by PCR using primers rgfD_del 1 and 2 and rgfD_del 3 and 4 (Table 2). An assembly PCR resulting in a single product was accomplished using equal amounts of the flanking products with the primers rgfD_del1 and rgfD_del4. Restriction digestion using *BamHI* and *KpnI* (New England Biolabs, Ipswich, MA, USA) of the resulting product and the plasmid pG+Host5 were performed followed by ligation and electroporation into Max Efficiency DH5α *Escherichia coli* electrocompetent cells (Thermo Fisher Scientific, Waltham, MA, USA) using a Micropulser (Bio-Rad, Hercules, CA, USA). The plasmid was confirmed to be present by PCR amplification with primers PGhost 4630 and PGhost 5117 and sequencing of the resulting product followed by electroporation into GB00451 and growth at 28 °C with erythromycin (2 μg/mL). Chromosomal integration of pG+host:Δ*rgfD* was selected for by growth on agar at 40 °C in the presence of erythromycin. Excision and loss of the plasmid was stimulated by growth at 28 °C without antibiotics in broth for six generations followed by dilution and plating. Single colonies were tested for erythromycin susceptibility to ensure plasmid loss and PCR was performed using primers rgfD_del 5 and 6 to identify a mutant with a gene deletion (GB00451Δ*rgfD*). Complementation of *rgfD* was completed using the pLZ12 plasmid with a constitutive *rofA* promoter sequence regulating transcription [36]. For construction, *rgfD* was amplified from GB00012 with Plz:rgfD F and R, digested with *Pst*I and *Bam*HI enzymes, and ligated into the pLZ12 plasmid. The constructed plasmid was transformed into the DH5α MAX Efficiency Chemically-Competent Cells by Invitrogen™ (Thermo Fisher Scientific, Waltham, MA, USA) and chloramphenicol resistant transformants were identified. The plasmid was extracted and electroporated into GB00451Δ*rgfD* competent cells, and transformants were selected for growth on THA and chloramphenicol (3 µg/mL).

2.3. Association Assays

Telomerase-immortalized human endometrial stromal cells (T-HESCs) were decidualized and grown to approximately 50% confluence followed by treatment with 0.5 mM 8-bromo-cyclic adenosine monophosphate (Sigma-Aldrich, St. Louis, MO, USA) for 3–6 days as described [37]. Decidualization was confirmed by examining the expression of prolactin and insulin-like growth factor-binding protein 1. Assays were performed in triplicate at least three times when cells reached 100% confluence. GBS was washed with phosphate-buffered saline (PBS) and resuspended in infection medium (HESC medium with 2% charcoal-treated fetal bovine serum, insulin, human transferrin, and selenous acid without antibiotics) following overnight growth in THB. Host cells were washed three times with PBS and infected with GBS at a multiplicity of infection (MOI) of one bacterial cell per host cell. After 2 h at 37 °C with 5% CO_2, samples were taken, diluted, and plated to quantify bacteria (CFU/mL). Each well was washed three times with PBS to remove non-adherent bacteria, and host cells were lysed with 0.1% Triton X-100 (Sigma-Aldrich) for 30 min at 37 °C and mixed to liberate intracellular bacteria. After serial dilution, lysates were plated on THA, incubated overnight at 37 °C, and quantified (CFU/mL). All data were expressed as percentages of the total number of bacteria per well after 2 h.

2.4. RNA Extraction, Preparation, and Quantitation

RNA was extracted, cDNA was synthesized and transcripts were quantified as previously described [37]. For collection, samples were added to two volumes of RNA Protect (Qiagen, Germantown, MD, USA) and pelleted followed by RNA extraction using the RNeasy Kit (Qiagen). DNA was removed with TURBO™ DNase (Thermo Fisher Scientific) and purified RNA was quantified. For samples exposed to host cells, total RNA was precipitated following Turbo DNase treatment and bacterial RNA was separated using the MICROB*Enrich*™ Kit by Ambion (Thermo Fisher Scientific). Following purification, 1 µg of RNA was used for reverse-transcription with the iScript Reverse Transcription Kit (Bio-Rad), while the iQ SYBR Supermix (Bio-Rad) was used for quantitative RT-PCR (qRT-PCR) in 15 µL reactions with 10 µM (each) of gene-specific primers (Table 2). Products were amplified and quantified using a CFX384 Touch™ Real-Time PCR detection system (Bio-Rad) under the following conditions: 1 cycle of 3 min at 95 °C and 39 cycles of 95 °C for 10 s and 60 °C for 30 s. Relative transcript quantities were calculated using the comparative threshold cycle (C_T) method ($2^{-\Delta CT}$) [38] with *gyrA* as the internal control gene.

2.5. Statistical Analysis

Data shown were either pooled from or were representative of at least three independent experiments performed in triplicate. The *t*-test was used to compare differences in expression levels across groups of strains, while the paired ratio *t*-test was used to compare percent association to host cells. The likelihood Chi-square test was used to examine differences in categorical variables. Analyses were performed in GraphPad Prism (version 6.0; GraphPad Software, Inc., La Jolla, CA, USA) and Epi Info™ (CDC, Atlanta, GA, USA). $p \leq 0.05$ was considered significant.

Table 2. Oligonucleotide primers used in this study.

Primer/Gene	Forward Primer (5′ to 3′)	Reverse Primer (5′ to 3′)
	Mutagenesis	
rgfD_del 1 & 2	CCGCGGATCCCCACTTTTACTCATGGGTGACTT	CCCATCCACTAAACTTAAACAGCATTCCAAACTTTGTAAGGAGTC
rgfD_del 3 & 4	TGTTTAAGTTTAGTGGATGGGTTTTATTCAACAGGCACGTTTAG	GGGGGTACCAAAACTTCTTCAATCCTTCTGCT
rgfD_del 5 & 6	TCATACTCGTCGTCCTCGG	CAACTCTATGTGACCTTAATGACG
plz12:rgfD	CGCGGATCCAGGAGGACAGCTATGCGAAGTTTGGAATGCATGAG	AAAACTGCAGTTCTCTCTAAACGTGCCTGTTG
	qPCR Detection	
gyrA	CGGGACACGTACAGGCTACT	CGATACGAGAAGCTCCCACA
rgfC	GCGAAGTAGTGAAGTTTCGCCCAT	CCGGTCTAAACTGGCTATTGCTCC
rgfB	GCAAGTACCATGAAGGGGTAGCG	TCAGCTACCAGAGCACGACGAGT
fbsB	GCGATTGTTGAATAGAATGAGTG	ACAGAAGCGGCGATTTCATT

Underline designates restriction enzyme sites, *Italic* designates complementary sequence, **Bold** designates ribosomal-binding sequence.

3. Results

3.1. Allelic Variation in rgf among Diverse GBS Lineages

Because sequence variation within the *rgf* operon has been observed [23,39], we compared the O90R *rgf* reference sequence [22] to 40 *rgf* sequences from clinical strains representing 14 STs. In all, 39 strains were classified as belonging to five CCs including CC-1 ($n = 10$), CC-12 ($n = 2$), CC-17 ($n = 7$), CC-19 ($n = 10$), and CC-23 ($n = 10$); two strains were singletons. Phylogenetic analysis of the complete 3320 bp *rgf* operon extracted from NCBI resulted in two *rgf* clusters (Figure 1), which differed based on the presence of an 881 bp deletion within *rgfC* at position 2328 as well as multiple single nucleotide polymorphisms (SNPs) within both *rgfA* and *rgfC*. A total of 21 (52.5%) strains contained the complete *rgf* operon with an intact *rgfC*, while the remaining 19 (47.5%) strains contained the 881 bp *rgfC* deletion.

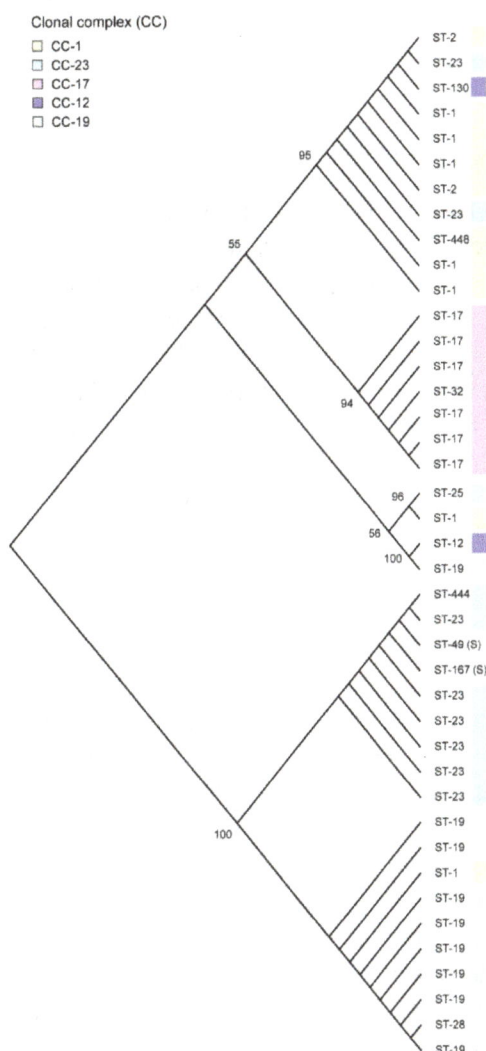

Figure 1. Neighbor joining phylogeny of *rgf* operon alleles by multilocus sequence type (ST). The evolutionary distances between *rgf* operon sequences (3320 bp) for 41 strains of different STs were calculated using the p-distance method, which is represented as the number of base differences per site. The bootstrap test (1000 replicates) values are represented at the nodes. The *rgfC* sequence, which was classified as complete or with an 881 bp deletion, contributed to the clustering observed in the phylogeny. S = singleton.

When stratified by ST, strains of the same ST were more likely to cluster together. The 20 ST-19 and ST-23 strains, for example, clustered together on the tree depending on whether they harbored the complete ($n = 4$) or deleted ($n = 16$) version of *rgfC*. The same was true for CC-1 strains, though one ST-1 strain had the *rgfC* deletion and clustered separately from the others. By contrast, strains belonging to ST-17 were homogeneous with only three detectable SNPs among all seven ST-17 genomes. These SNPs were located within *rgfD* (T1115G), *rgfA* (A1848T), and *rgfC* (G2338A) and each mutation was exclusive to one of the three different ST-17 strains. Relative to the other STs, two unique non-synonymous SNPs were detected in all of the ST-17 strains. The first SNP (C246A) is located in *rgfB* and the second (A1131T) is located 54 bp into *rgfD*. Importantly, the *rgfD* SNP results in a truncated coding sequence after 17 amino acids due to the introduction of a stop codon. This finding suggests that *rgfD* may function differently in ST-17 strains relative to strains belonging to other lineages. Mutations within *rgfD* also resulted in the separate clustering of a ST-12 and ST-19 strain with a complete *rgfC* near the bottom of the top branch of the phylogeny. Both strains had three unique SNPs, G1044A, C1048T, and A1054G, located 21 bp, 25 bp, and 31 bp into *rgfD*, respectively, as well as an additional SNP (A3018T) in *rgfC* that was shared only with the seven ST-17 strains. Only C1048T and A1054G in *rgfD* represent non-synonymous mutations.

3.2. Association between rgf Variation and Biofilm Production

Since allelic variation in the *agr* system has previously been related to biofilm production in *S. aureus* [40], we assessed the importance of *rgf* allelic variation on biofilm phenotypes. Of the 40 clinical strains examined, 13 (32.5%) were previously classified as strong biofilm producers and 27 (67.5%) were weak. Those strains possessing a complete *rgfC* were not more likely to produce a strong biofilm relative to the strains containing the *rgfC* deletion ($p = 0.83$). Among the 21 strains with a complete *rgfC*, 28.6% ($n = 6$) were strong biofilm producers relative to 36.8% ($n = 7$) of strains with the *rgfC* deletion. It is notable that all but one of the seven CC-17 strains containing the *rgfD* truncation were classified as weak biofilm producers. Although the strains containing a complete *rgfD* were 3.4 times more likely to form a strong biofilm, the association was not statistically significant (95% confidence interval: 0.42, 85.59; Fisher's exact $p = 0.39$), which may be due to the small sample size.

3.3. rgfD-Dependent Expression of the rgf Operon

Since quorum-sensing controlled systems are characterized by increased expression when the extracellular inducer reaches a specific concentration, we sought to quantify *rgf* expression in a subset of GBS strains. The *rgf* operon was previously shown to be transcribed polycistronically [22]; therefore, we examined expression of *rgfC*, the gene encoding the sensor histidine kinase (*rgfC*), in three ST-17 clinical strains over time. No difference in relative *rgfC* transcript quantity was observed between the three clinical strains at any of the time points. It is important to note that all three strains contained complete *rgf* operons with a complete *rgfC* and genetically identical *rgfB* and *rgfD* genes. Next, we deleted *rgfD* in one of the three CC-17 strains, GB00451 (wild type; WT), in order to compare *rgfC* expression along the growth curve to the same strain lacking *rgfD*. Samples were also subcultured to ensure that the bacterial densities were similar at each OD_{595} value; no difference in colony forming units (CFU) was observed between the WT and GB00451Δ*rgfD* mutant (data not shown). Significantly reduced relative *rgfC* transcript quantity was observed at all growth points in the GB00451Δ*rgfD* mutant relative to the WT (Figure 2). The largest difference was observed in lag phase (OD_{595} = ~0.2) with relative expression values of 0.16 ± 0.03 and 0.02 ± 0.01 for WT and mutant, respectively. In early log phase (OD_{595} = ~0.4), *rgfC* expression was reduced from 0.13 ± 0.04 to 0.06 ± 0.01 in the WT versus mutant ($p = 0.04$) and at mid-log phase (OD_{595} = ~0.6), expression values of 0.13 ± 0.05 and 0.05 ± 0.008 were observed for the WT and mutant ($p = 0.05$). At late log phase, the WT had a significantly higher level of *rgfC* expression (0.17 ± 0.02) compared to the GB00451Δ*rgfD* mutant (0.08 ± 0.03; $p = 0.01$). Although expression of *rgfC* was highest for both strains at stationary phase, the level of expression in the mutant (0.44 ± 0.06) was still significantly lower than in the WT (0.72 ± 0.12;

$p = 0.03$). Complementation of GB00451$\Delta rgfD$ with the pLZ12 plasmid containing the truncated version of $rgfD$ from GB00451 (WT) was not capable of restoring $rgfC$ expression. To determine whether this result was partly due to the $rgfD$ mutation in the WT strain, complementation with pLZ12-$rgfD$ from GB00012, a ST-1 strain lacking the $rgfD$ truncation, was performed. Importantly, complementation of GB00451$\Delta rgfD$ with pLZ12-$rgfD$ from GB00012 (Figure 3) was sufficient to restore relative expression of $rgfC$ to 0.15 ± 0.02 at OD$_{595}$ = 0.4 compared to the empty vector control (0.08 ± 0.1); t-test $p < 0.01$.

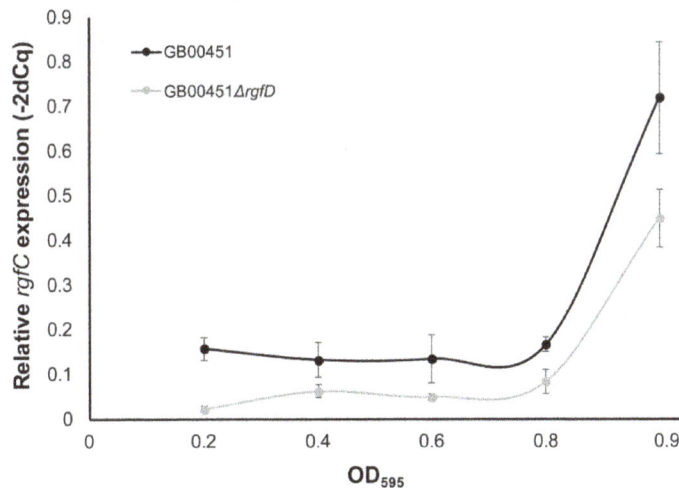

Figure 2. Expression of $rgfC$ increased over the growth phase in sequence type (ST)-17 strains. $rgfC$ expression was assessed in three clinical ST-17 strains including GB00451 (shown here), which was compared to GB00451$\Delta rgfD$. The relative $rgfC$ transcript quantity is represented as the optical density (OD)$_{595}$ increases. Error bars represent the standard deviation between the strains at a given OD$_{595}$.

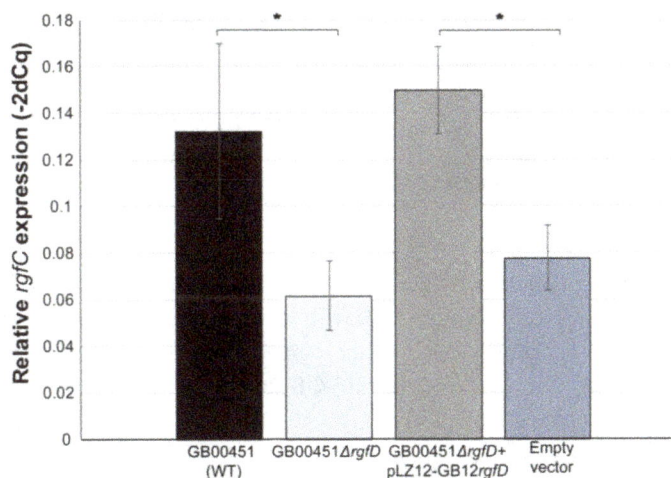

Figure 3. Expression of $rgfD$ is necessary for $rgfC$ expression. Comparison of the relative $rgfC$ transcript quantity between the GB00451 wild-type (WT) and GB00451$\Delta rgfD$ mutant during early mid-log (OD$_{595}$ = 0.4) growth. The GB00451$\Delta rgfD$ mutant complemented with GB0012$rgfD$ on the pLZ12 plasmid (pLZ12-GB12$rgfD$) and complementation with pLZ12 alone (empty vector) are also shown. Bars represent the standard deviation of four biological replicates. * t-test p-value < 0.05.

To determine whether RgfD alters expression of other genes that were suggested to be regulated by the rgf operon, the WT and GB00411$\Delta rgfD$ mutant were examined for expression changes in the gene encoding the fibrinogen binding surface protein ($fbsB$), which is activated by $rgfA/C$ [23]. Notably, relative $fbsB$ transcript quantity was similar for the WT (0.010 ± 0.003) and GB00451$\Delta rgfD$ (0.008 ± 0.003) mutant at OD$_{595}$ = 0.4 as well as OD$_{595}$ = 0.6 (0.012 ± 0.005 for the WT versus

0.012 ± 0.006 for GB00451ΔrgfD). Expression levels in stationary phase (OD$_{595}$ = 0.8) were slightly more variable, though there was still no significant difference between the WT (0.0007 ± 0.0004) and mutant (0.0009 ± 0.0008).

3.4. Role of rgfD in Association with Host Cells and Biofilm Production

Since the *rgf* operon has been shown to promote binding to host cell components like fibrinogen [23], the ability to associate with T-HESCs was investigated. Interestingly, the GB00451ΔrgfD mutant had an average 1.3-fold decrease in the ability to associate with the decidualized endometrial cells compared to the WT. Association with T-HESCs was $0.40\% \pm 0.03\%$ for GB00451ΔrgfD compared to $0.55\% \pm 0.07\%$ for the WT (ratio *t*-test $p < 0.03$) (Figure 4). The empty vector control had an average 1.6-fold reduction in the level of association with T-HESCs compared to GB00451ΔrgfD complemented with pLZ12 containing *rgfD* from GB00012. The association level for the complemented mutant was $0.53\% \pm 0.11\%$ versus $0.37\% \pm 0.07\%$ for the empty vector (ratio *t*-test $p = 0.002$). It is important to note that even though the trend remained consistent across biological replicates, association percentages varied between experiments. When biofilms were examined, no difference in biofilm production was observed between the WT, GB00451ΔrgfD mutant, both complemented mutants, or empty vector control.

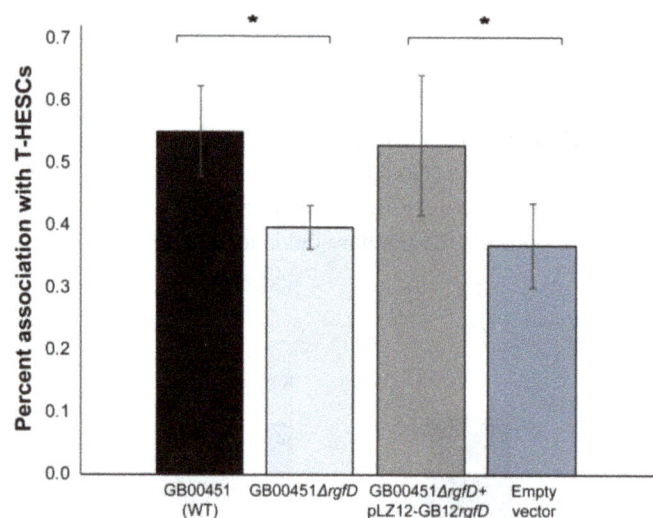

Figure 4. *rgfD* plays a role in association with decidualized endometrial stromal cells. Association percentages for GB00451 (WT) and GB00451ΔrgfD with telomerase-immortalized human endometrial stromal cells (T-HESCs) are shown as well as the percentages for GB00451ΔrgfD complemented with pLZ12 containing *rgfD* from GB00012 (pLZ12-GB12*rgfD*) and the empty vector (pLZ12 only). The histogram represents a single biological replicate with three technical replicates and error bars representing the standard deviation between technical replicates; the assay was performed four times in triplicate with identical trends per assay. * paired ratio *t*-test *p*-value < 0.05.

Because the association assays were performed in different conditions than the *rgfC* expression analysis, we also sought to compare *rgfC* expression in the WT, *rgfD* mutant, complemented *rgfD* mutant, and empty vector control to determine whether differential regulation of the operon was detectable following host cell exposure. Notably, a 22.8-fold reduction in *rgfC* expression was observed in the GB00451ΔrgfD mutant compared to the WT following a 2 h exposure to decidualized T-HESCs; relative transcript levels were 0.0019 ± 0014 and 0.043 ± 0.019, respectively (Figure 5). No difference in *rgfC* expression was observed, however, between the complemented and empty vector controls with relative transcription values of 0.029 ± 0.01 and 0.031 ± 0.01, respectively, following exposure to T-HESCs.

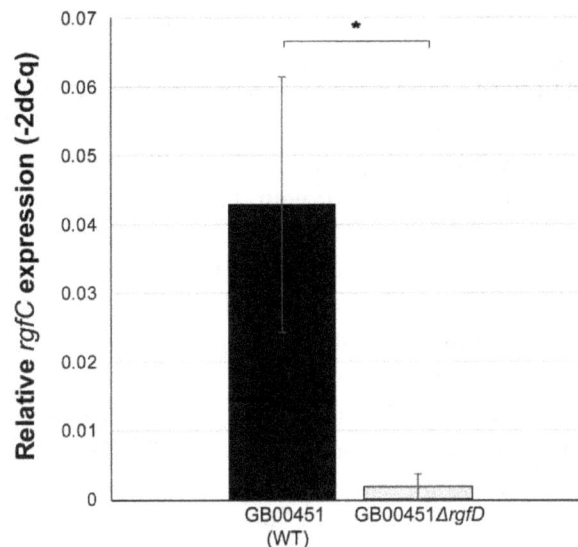

Figure 5. *rgfC* is upregulated by *rgfD* following exposure to decidualized endometrial stromal cells. Comparison of the relative *rgfC* transcript quantity between the GB00451*rgfD* wild-type (WT) and GB00451Δ*rgfD* mutant following 2 h exposure to telomerase-immortalized human endometrial stromal cells (T-HESCs). Bars represent the standard deviation of three biological replicates. * *t*-test *p*-value < 0.05.

4. Discussion

Because the *rgf* operon was found to facilitate binding to host cell components and impact virulence in vivo, [22,23,29] we sought to better understand the role of the putative autoinducing peptide, RgfD, in phenotypes important for colonization. Similar to prior studies [29,39], we have demonstrated *rgf*-dependent expression of the *rgf* operon and have identified genetic variation in the *rgf* operon genes among a diverse set of GBS strains. The large 881 bp deletion within *rgfC* is notable given that it was present in almost half of the 40 strains recovered from women with asymptomatic GBS colonization. Since these strains represented multiple STs and were collected from patient populations in different geographic locations and time periods, the presence of this mutation suggests parallel evolution in the *rgfC* locus. Evidence for gene loss as well as lateral transfer and gene duplication have been described for genes involved in other quorum sensing systems (e.g., *Pseudomonas* [41]).

The identification of a point mutation within *rgfD* that was exclusive to the seven clinical strains belonging to CC-17, the lineage most commonly associated with neonatal disease, [4] is also noteworthy. The A1131T mutation results in transcription of a premature stop codon within the *rgfD* open reading frame to encode a truncated protein. Because of this mutation, it is possible that *rgfD* functions differently in ST-17 strains versus strains of other lineages with a complete *rgfD*. In *S. aureus*, sequence variation has been observed within the auto-inducing peptide gene (*agrD*) and was shown to influence activation of the two component system [42,43]. The AgrD peptide was also found to be post-translationally modified and reduced to a functional eight amino acid peptide [44]. In all phases of growth, we found that the truncated Δ*rgfD* mutant had decreased expression of *rgfC*, which encodes the sensor histidine kinase and represents the last of the four genes in the polycistronically transcribed operon [22]. These data suggest that this truncated RgfD protein is functional in this CC-17 strain and likely serves as an auto-inducing peptide in conjunction with other factors. This hypothesis is in agreement with *agr* regulation in *S. aureus* in which there are several factors affecting expression besides AgrD [26,27]. Furthermore, maximum *rgfC* expression was observed as the cells entered stationary phase in both the WT as well as the GB00451Δ*rgfD* mutant, suggesting that other factors can impact transcription of this operon even in the absence of a functional version of RgfD. In the future, additional studies should focus on clarifying the specific regulatory role of *rgfD* during each growth phase and identifying other factors that contribute to transcription. It is important to note that quorum quenching has also been described to occur following entry into stationary phase in several bacterial species [45,46]. Since complementation with the truncated *rgfD* from the WT did not restore *rgfC* expression relative to complementation with a complete *rgfD* from a different strain (genotype), we further hypothesize that *rgfB* may be needed for *rgfD* processing in CC-17 strains containing a truncated RgfD protein. Indeed, the *agrBD* complex was found to be responsible for activation of *agr* in *S. aureus* [42] and hence, future work is required to determine whether extrachromosomal transcription of *rgfBD* from a CC-17 strain can restore *rgfC* activity and if certain mutations within *rgf* can result in altered protein function. Little is known about the structure of the mature RgfD peptide in GBS and virtually nothing is known about how that structure varies across diverse strain types with different *rgfD* alleles.

The Δ*rgfD* mutant also had a significant decrease in *rgfC* expression following exposure to decidualized T-HESCs and in its ability to associate with the T-HESCs; the latter could be restored following complementation with a complete *rgfD* from the GB00012 strain. Although the decrease in association was modest and the biological relevance is not clear, it was consistently observed and statistically significant. Because we have previously shown that GBS strains of different genetic backgrounds vary in their ability to attach to A549 lung epithelial cells and T-HESCs [37], it is possible that the *rgf* operon plays a role in regulating distinct adherence factors. These factors may be needed to colonize different tissues inside the host and likely vary across the GBS genotypes. Because *rgf* was previously found to activate *fbsB*, the gene encoding one of two fibrinogen binding proteins [23], it is possible that the reduction in host cell association in the truncated Δ*rgfD* CC-17 mutant is due to a decreased ability to bind fibrinogen or other cell components via the lack of *rgf* activation. Similar findings were observed when both *rgfA* and *rgfC* were interrupted in a prior study [23]. Since our prior study demonstrated that only a small fraction (<1%) of associated bacteria invaded host cells [37]; however, the association reduction observed in the GB00451Δ*rgfD* mutant cannot be explained solely by *fbsB* activation or through reduced invasion. Support for this hypothesis also comes from the observation that *fbsB* expression was not significantly different in the WT and GB00451Δ*rgfD* mutant across the growth phase despite the observed differences in relative transcript levels of *rgfC*. Nonetheless, it is also possible that host cell exposure does not represent the optimal conditions for *rgf* activation given the higher level of *rgfC* expression that we observed during growth at $OD_{595} = 0.4$. Another possibility affecting host cell association is that the CC-17 strains containing the truncated *rgfD* are less likely to form biofilms due to altered activation of genes important for adherence. Reduced biofilm production has been demonstrated with deletion of the *agr* operon in *S. aureus* [47] and our prior study of biofilms in 293 GBS strains showed that strains belonging to ST-17, which more commonly possessed the truncated *rgfD* in this study, were significantly more likely to form weak biofilms relative to strains from other lineages [33]. Because we also observed an association with weak biofilm production in ST-19 strains, a more comprehensive comparative genomics analysis is warranted. In the present study, most of the ST-19 strains possessed the large deletion within *rgfC* and hence, it is possible that altered transcription of *rgfC* combined with a complete *rgfD* can also impact biofilms. Although there was no association between the *rgfC* deletion and biofilm production overall, it is notable that only one of the nine CC-19 strains formed a strong biofilm and possessed a complete *rgfC*; the remaining eight CC-19 strains had the *rgfC* deletion and formed weak biofilms. Further testing of the different *rgf* mutants is therefore warranted, particularly those clinical strains with natural mutations, which can enhance understanding of the relationship between sequence variation, *rgf* activation and colonization.

The only other verified quorum sensing system in GBS involves RovS, an Rgg-type transcriptional regulator and its activator, a short hydrophobic peptide (SHP), has been found in many specis of *Streptococcus* [48,49]. Similar to the *rgf* system, SHP is post-translationally modified by one or more peptidases and secreted extracellularly [50,51]. Rather than indirectly affecting downstream gene regulation through extracellular recognition, however, the SHP interacts directly with RovS following importation by the Ami oligopeptide transporter [49]. Similar to the *rgf* operon, this system is autoregulating and affects expression of fibrinogen-binding proteins and host-cell attachment [51]; hence, differential expression of the RovS system could have an impact on our findings and warrants further investigation. Interestingly, the SHP has been linked to persistence [51], while inactivation of *rgfC* has specifically been shown to induce a disseminating and invasive phenotype [29,39]. Because the SHP system has also been shown to function differently in different mediums [51], further work should also focus on identifying the optimal conditions for *rgf* expression and *rgf*-associated regulatory networks, particularly during the course of an infection. Additional studies that aim to isolate the mature RgfD peptide from cell-free supernatant cultures and assess its impact on *rgf* expression over time are also needed.

5. Conclusions

Because GBS disease progression can involve transcriptional remodeling in response to changing host environments, quorum sensing offers a potential explanation for the variation in pathogenicity that has been observed between strains belonging to different phylogenetic lineages. Although quorum sensing has been demonstrated to affect pathogenesis for many bacterial species, there are few studies specific to GBS. The work described herein adds to the knowledge of quorum sensing systems in GBS and better defines the role of *rgfD*, the gene encoding the putative auto-inducing peptide, as a regulator of *rgfC* and stimulus for host-cell association.

Acknowledgments: The authors wish to thank Dr. Melody Neely for sharing the pLZ12 plasmid. We would also like to thank Pallavi Singh and Michelle Korir for productive scientific conversations aiding this project. This study was supported in part by the Global Alliance to Prevent Prematurity and Stillbirth (GAPPS) in collaboration with the Bill and Melinda Gates Foundation (project N015615, SDM), while salary support was provided by the USDA NIFA (grant #2011-67005-30004, SDM). Graduate student support was provided by the Thomas S. Whittam Graduate Fellowship, the Rudolph Hugh Graduate Fellowship and the Graduate School at Michigan State University.

Author Contributions: Robert E. Parker, Rim Al Safadi, Agnès Rosenau and Shannon D. Manning conceived the study and contributed materials; Robert E. Parker and Shannon D. Manning planned the study and experiments; Robert E. Parker, Rim Al Safadi and David Knupp performed the experiments; Robert E. Parker and Shannon D. Manning analyzed the data and drafted the paper; all authors approved the final version.

Abbreviations

GBS	Group B *Streptococcus*
qPCR	quantitative polymerase chain reaction
EOD	early onset disease
MLST	multilocus sequence type
STS	signal transduction system
TCS	two component system
ST	sequence type
CC	clonal complex
THB	Todd-Hewitt broth
THG	Todd-Hewitt broth plus 1% glucose
THA	Todd-Hewitt agar
PBS	phosphate-buffered saline
T-HESC	Telomerase-immortalized human endometrial stromal cells
MOI	multiplicity of infection
SNP	single nucleotide polymorphism

References

1. Schuchat, A.; Wenger, J.D. Epidemiology of group B streptococcal disease. Risk factors, prevention strategies, and vaccine development. *Epidemiol. Rev.* **1994**, *16*, 374–402. [PubMed]

2. Phares, C.R.; Lynfield, R.; Farley, M.M.; Mohle-boetani, J.; Harrison, L.H.; Petit, S.; Craig, A.S.; Schaffner, W.; Gershman, K.; Stefonek, K.R.; et al. Epidemiology of invasive group B streptococcal disease in the United States, 1999–2005. *J. Am. Med. Assoc.* **2008**, *299*, 2056–2065. [CrossRef] [PubMed]

3. Jones, N.; Bohnsack, J.F.; Takahashi, S.; Oliver, K.A.; Chan, M.-S.; Kunst, F.; Glaser, P.; Rusniok, C.; Crook, D.W.M.; Harding, R.M. Multilocus sequence typing system for group B *Streptococcus*. *J. Clin. Microbiol.* **2003**, *41*, 2530–2536. [CrossRef] [PubMed]

4. Manning, S.D.; Springman, A.C.; Lehotzky, E.; Lewis, M.A.; Whittam, T.S.; Davies, H.D. Multilocus sequence types associated with neonatal group B streptococcal sepsis and meningitis in canada. *J. Clin. Microbiol.* **2009**, *47*, 1143–1148. [CrossRef] [PubMed]

5. Brochet, M.; Couve, E.; Zouine, M.; Vallaeys, T.; Rusniok, C.; Lamy, M.C.; Buchrieser, C.; Trieu-Cuot, P.; Kunst, F.; Poyart, C.; et al. Genomic diversity and evolution within the species *Streptococcus agalactiae*. *Microbes Infect.* **2006**, *8*, 1227–1243. [CrossRef] [PubMed]

6. Bisharat, N.; Crook, D.W.; Leigh, J.; Harding, R.M.; Ward, P.N.; Coffey, T.J.; Maiden, M.C.; Peto, T.; Jones, N. Hyperinvasive neonatal group B *Streptococcus* has arisen from a bovine ancestor. *J. Clin. Microbiol.* **2004**, *42*, 2161–2167. [CrossRef] [PubMed]

7. Springman, A.C.; Lacher, D.W.; Waymire, E.A.; Wengert, S.L.; Singh, P.; Zadoks, R.N.; Davies, H.D.; Manning, S.D. Pilus distribution among lineages of group B *Streptococcus*: An evolutionary and clinical perspective. *BMC Microbiol.* **2014**. [CrossRef] [PubMed]

8. Springman, A.C.; Lacher, D.W.; Wu, G.; Milton, N.; Whittam, T.S.; Davies, H.D.; Manning, S.D. Selection, recombination, and virulence gene diversity among group B streptococcal genotypes. *J. Bacteriol.* **2009**, *191*, 5419–5427. [CrossRef] [PubMed]

9. Doran, K.S.; Nizet, V. Molecular pathogenesis of neonatal group B streptococcal infection: No longer in its infancy. *Mol. Microbiol.* **2004**, *54*, 23–31. [CrossRef] [PubMed]

10. Huang, S.-H.; Stins, M.F.; Kim, K.S. Bacterial penetration across the blood-brain barrier during the development of neonatal meningitis. *Microbes Infect.* **2000**, *2*, 1237–1244. [CrossRef]

11. Winzer, K.; Williams, P. Quorum sensing and the regulation of virulence gene expression in pathogenic bacteria. *Int. J. Med. Microbiol.* **2001**, *291*, 131–143. [CrossRef] [PubMed]

12. Mereghetti, L.; Sitkiewicz, I.; Green, N.M.; Musser, J.M. Remodeling of the *Streptococcus agalactiae* transcriptome in response to growth temperature. *PLoS ONE* **2008**, *3*, e2785. [CrossRef] [PubMed]

13. Mereghetti, L.; Sitkiewicz, I.; Green, N.M.; Musser, J.M. Extensive adaptive changes occur in the transcriptome of *Streptococcus agalactiae* (group B *Streptococcus*) in response to incubation with human blood. *PLoS ONE* **2008**, *3*, e3143. [CrossRef] [PubMed]

14. Sitkiewicz, I.; Green, N.M.; Guo, N.; Bongiovanni, A.M.; Witkin, S.S.; Musser, J.M. Transcriptome adaptation of group B *Streptococcus* to growth in human amniotic fluid. *PLoS ONE* **2009**, *4*, e6114. [CrossRef] [PubMed]

15. Hoch, J.A. Two-component and phosphorelay signal transduction. *Curr. Opin. Microbiol.* **2000**, *3*, 165–170. [CrossRef]

16. Kleerebezem, M.; Quadri, L.E.N.; Kuipers, O.P.; de Vos, W.M. Quorum sensing by peptide pheromones and two-component signal-transduction systems in Gram-positive bacteria. *Mol. Microbiol.* **1997**, *24*, 895–904. [CrossRef] [PubMed]

17. Ulrich, L.E.; Koonin, E.V.; Zhulin, I.B. One-component systems dominate signal transduction in prokaryotes. *Trends Microbiol.* **2005**, *13*, 52–56. [CrossRef] [PubMed]

18. Tettelin, H.; Masignani, V.; Cieslewicz, M.J.; Donati, C.; Medini, D.; Ward, N.L.; Angiuoli, S.V.; Crabtree, J.; Jones, A.L.; Durkin, A.S.; et al. Genome analysis of multiple pathogenic isolates of *Streptococcus agalactiae*: Implications for the microbial "pan-genome". *Proc. Natl. Acad. Sci. USA* **2005**, *102*, 13950–13955. [CrossRef] [PubMed]

19. Lamy, M.-C.; Zouine, M.; Fert, J.; Vergassola, M.; Couve, E.; Pellegrini, E.; Glaser, P.; Kunst, F.; Msadek, T.; Trieu-Cuot, P.; et al. CovS/CovR of group B *Streptococcus*: A two-component global regulatory system involved in virulence. *Mol. Microbiol.* **2004**, *54*, 1250–1268. [CrossRef] [PubMed]

20.	Poyart, C.; Pellegrini, E.; Marceau, M.; Baptista, M.; Jaubert, F.; Lamy, M.-C.; Trieu-Cuot, P. Attenuated virulence of *Streptococcus agalactiae* deficient in d-alanyl-lipoteichoic acid is due to an increased susceptibility to defensins and phagocytic cells. *Mol. Microbiol.* **2003**, *49*, 1615–1625. [CrossRef] [PubMed]

21.	Quach, D.; van Sorge, N.M.; Kristian, S.a.; Bryan, J.D.; Shelver, D.W.; Doran, K.S. The ciar response regulator in group B *Streptococcus* promotes intracellular survival and resistance to innate immune defenses. *J. Bacteriol.* **2009**, *191*, 2023–2032. [CrossRef] [PubMed]

22.	Spellerberg, B.; Rozdzinski, E.; Martin, S.; Weber-Heynemann, J.; Lütticken, R. *rgf* encodes a novel two-component signal transduction system of *Streptococcus agalactiae*. *Infect. Immun.* **2002**, *70*, 2434–2440. [CrossRef] [PubMed]

23.	Al Safadi, R.; Mereghetti, L.; Salloum, M.; Lartigue, M.-F.; Virlogeux-Payant, I.; Quentin, R.; Rosenau, A. Two-component system RgfA/C activates the *fbsB* gene encoding major fibrinogen-binding protein in highly virulent CC-17 clone group B *Streptococcus*. *PLoS ONE* **2011**, *6*, e14658. [CrossRef] [PubMed]

24.	Traber, K.E.; Lee, E.; Benson, S.; Corrigan, R.; Cantera, M.; Shopsin, B.; Novick, R.P. *agr* function in clinical *Staphylococcus aureus* isolates. *Microbiology* **2008**, *154*, 2265–2274. [CrossRef] [PubMed]

25.	Novick, R.P.; Projan, S.J.; Kornblum, J.; Ross, H.F.; Ji, G.; Kreiswirth, B.; Vandenesch, F.; Moghazeh, S. The *agr* p2 operon: An autocatalytic sensory transduction system in *Staphylococcus aureus*. *Mol. Gen. Genet.* **1995**, *248*, 446–458. [CrossRef] [PubMed]

26.	Roux, A.; Todd, D.A.; Velázquez, J.V.; Cech, N.B.; Sonenshein, A.L. Cody-mediated regulation of the *Staphylococcus aureus agr* system integrates nutritional and population density signals. *J. Bacteriol.* **2014**, *196*, 1184–1196. [CrossRef] [PubMed]

27.	Chien, Y.-T.; Cheung, A.L. Molecular interactions between two global regulators, *sar* and *agr*, in *Staphylococcus aureus*. *J. Biol. Chem.* **1998**, *273*, 2645–2652. [CrossRef] [PubMed]

28.	Wright, J.S.; Traber, K.E.; Corrigan, R.; Benson, S.A.; Musser, J.M.; Novick, R.P. The *agr* radiation: An early event in the evolution of staphylococci. *J. Bacteriol.* **2005**, *187*, 5585–5594. [CrossRef] [PubMed]

29.	Gendrin, C.; Lembo, A.; Whidbey, C.; Burnside, K.; Berry, J.; Ngo, L.; Banerjee, A.; Xue, L.; Arrington, J.; Doran, K.S.; et al. The sensor histidine kinase RgfC affects group B streptococcal virulence factor expression independent of its response regulator RgfA. *Infect. Immun.* **2015**, *83*, 1078–1088. [CrossRef] [PubMed]

30.	Glaser, P.; Rusniok, C.; Buchrieser, C.; Chevalier, F.; Frangeul, L.; Msadek, T.; Zouine, M.; Couvé, E.; Lalioui, L.; Poyart, C.; et al. Genome sequence of *Streptococcus agalactiae*, a pathogen causing invasive neonatal disease. *Mol. Microbiol.* **2002**, *45*, 1499–1513. [CrossRef] [PubMed]

31.	Tamura, K.; Stecher, G.; Peterson, D.; Filipski, A.; Kumar, S. MEGA6: Molecular evolutionary genetics analysis version 6.0. *Mol. Biol. Evol.* **2013**, *30*, 2725–2729. [CrossRef] [PubMed]

32.	Manning, S.D.; Schaeffer, K.E.; Springman, A.C.; Lehotzky, E.; Lewis, M.A.; Ouellette, L.M.; Wu, G.; Moorer, G.M.; Whittam, T.S.; Davies, H.D. Genetic diversity and antimicrobial resistance in group B *Streptococcus* colonizing young, nonpregnant women. *Clin. Infect. Dis.* **2008**, *47*, 388–390. [CrossRef] [PubMed]

33.	Parker, R.E.; Laut, C.; Gaddy, J.A.; Zadoks, R.N.; Davies, H.D.; Manning, S.D. Association between genotypic diversity and biofilm production in group B *Streptococcus*. *BMC Microbiol.* **2016**. [CrossRef] [PubMed]

34.	Biswas, I.; Gruss, A.; Ehrlich, S.D.; Maguin, E. High-efficiency gene inactivation and replacement system for gram-positive bacteria. *J. Bacteriol.* **1993**, *175*, 3628–3635. [CrossRef] [PubMed]

35.	Schubert, A.; Zakikhany, K.; Schreiner, M.; Frank, R.; Spellerberg, B.; Eikmanns, B.J.; Reinscheid, D.J. A fibrinogen receptor from group B *Streptococcus* interacts with fibrinogen by repetitive units with novel ligand binding sites. *Mol. Microbiol.* **2002**, *46*, 557–569. [CrossRef] [PubMed]

36.	Hanson, B.R.; Lowe, B.A.; Neely, M.N. Membrane topology and DNA-binding ability of the streptococcal cpsa protein. *J. Bacteriol.* **2011**, *193*, 411–420. [CrossRef] [PubMed]

37.	Korir, M.L.; Knupp, D.; LeMerise, K.; Boldenow, E.; Loch-Caruso, R.; Aronoff, D.M.; Manning, S.D. Association and virulence gene expression vary among serotype III group B *Streptococcus* isolates following exposure to decidual and lung epithelial cells. *Infect. Immun.* **2014**, *82*, 4587–4595. [CrossRef] [PubMed]

38.	Schmittgen, T.D.; Livak, K.J. Analyzing real-time PCR data by the comparative Ct method. *Nat. Protoc.* **2008**, *3*, 1101–1108. [CrossRef] [PubMed]

39.	Faralla, C.; Metruccio, M.M.; de Chiara, M.; Mu, R.; Patras, K.A.; Muzzi, A.; Grandi, G.; Margarit, I.; Doran, K.S.; Janulczyk, R. Analysis of two-component systems in group B *Streptococcus* shows that RgfAC and the novel FspSR modulate virulence and bacterial fitness. *mBio* **2014**. [CrossRef] [PubMed]

40. Cafiso, V.; Bertuccio, T.; Santagati, M.; Demelio, V.; Spina, D.; Nicoletti, G.; Stefani, S. *agr*-genotyping and transcriptional analysis of biofilm-producing *Staphylococcus aureus*. *FEMS Immunol. Med.Microbiol.* **2007**, *51*, 220–227. [CrossRef] [PubMed]

41. Lerat, E.; Moran, N.A. The evolutionary history of quorum-sensing systems in bacteria. *Mol. Biol. Evol.* **2004**, *21*, 903–913. [CrossRef] [PubMed]

42. Ji, G.; Beavis, R.; Novick, R.P. Bacterial interference caused by autoinducing peptide variants. *Science* **1997**, *276*, 2027–2030. [CrossRef] [PubMed]

43. Takeuchi, S.; Maeda, T.; Hashimoto, N.; Imaizumi, K.; Kaidoh, T.; Hayakawa, Y. Variation of the agr locus in *Staphylococcus aureus* isolates from cows with mastitis. *Vet. Microbiol.* **2001**, *79*, 267–274. [CrossRef]

44. Ji, G.; Beavis, R.C.; Novick, R.P. Cell density control of staphylococcal virulence mediated by an octapeptide pheromone. *Proc. Natl. Acad. Sci. USA* **1995**, *92*, 12055–12059. [CrossRef] [PubMed]

45. Barber, C.E.; Tang, J.L.; Feng, J.X.; Pan, M.Q.; Wilson, T.J.G.; Slater, H.; Dow, J.M.; Williams, P.; Daniels, M.J. A novel regulatory system required for pathogenicity of *Xanthomonas campestris* is mediated by a small diffusible signal molecule. *Mol. Microbiol.* **1997**, *24*, 555–566. [CrossRef] [PubMed]

46. Surette, M.G.; Bassler, B.L. Quorum sensing in *Escherichia coli* and *Salmonella typhimurium*. *Proc. Natl. Acad. Sci. USA* **1998**, *95*, 7046–7050. [CrossRef] [PubMed]

47. Vuong, C.; Saenz, H.L.; Götz, F.; Otto, M. Impact of the *agr* quorum-sensing system on adherence to polystyrene in *Staphylococcus aureus*. *J. Infect. Dis.* **2000**, *182*, 1688–1693. [CrossRef] [PubMed]

48. Ibrahim, M.; Nicolas, P.; Bessières, P.; Bolotin, A.; Monnet, V.; Gardan, R. A genome-wide survey of short coding sequences in streptococci. *Microbiology* **2007**, *153*, 3631–3644. [CrossRef] [PubMed]

49. Fleuchot, B.; Gitton, C.; Guillot, A.; Vidic, J.; Nicolas, P.; Besset, C.; Fontaine, L.; Hols, P.; Leblond-Bourget, N.; Monnet, V.; et al. Rgg proteins associated with internalized small hydrophobic peptides: A new quorum-sensing mechanism in streptococci. *Mol. Microbiol.* **2011**, *80*, 1102–1119. [CrossRef] [PubMed]

50. Ibrahim, M.; Guillot, A.; Wessner, F.; Algaron, F.; Besset, C.; Courtin, P.; Gardan, R.; Monnet, V. Control of the transcription of a short gene encoding a cyclic peptide in *Streptococcus thermophilus*: A new quorum-sensing system? *J. Bacteriol.* **2007**, *189*, 8844–8854. [CrossRef] [PubMed]

51. Pérez-Pascual, D.; Gaudu, P.; Fleuchot, B.; Besset, C.; Rosinski-Chupin, I.; Guillot, A.; Monnet, V.; Gardan, R. RovS and its associated signaling peptide form a cell-to-cell communication system required for *Streptococcus agalactiae* pathogenesis. *mBio* **2015**. [CrossRef] [PubMed]

3

Identification of the Ovine Keratin-Associated Protein 22-1 (KAP22-1) Gene and its Effect on Wool Traits

Shaobin Li [1,2], **Huitong Zhou** [1,2,3], **Hua Gong** [2,3], **Fangfang Zhao** [1], **Jiqing Wang** [1,2], **Xiu Liu** [1,2], **Yuzhu Luo** [1,2,*] **and Jon G. H. Hickford** [2,3,*]

[1] Gansu Key Laboratory of Herbivorous Animal Biotechnology, Faculty of Animal Science and Technology, Gansu Agricultural University, Lanzhou 730070, China; lisb2008@hotmail.com (S.L.); Zhou@lincoln.ac.nz (H.Z.); zhaoFangfang@gsau.edu.cn (F.Z.); wangjq@gsau.edu.cn (J.W.); liuxiu@gsau.edu.cn (X.L.)

[2] International Wool Research Institute, Gansu Agricultural University, Lanzhou 730070, China; Hua.Gong@lincolnuni.ac.nz

[3] Gene-marker Laboratory, Faculty of Agricultural and Life Sciences, Lincoln University, Lincoln 7647, New Zealand

* Correspondence: luoyz@gsau.edu.cn (Y.L.); Jon.hickford@lincoln.ac.nz (J.G.H.H.)

Academic Editor: Paolo Cinelli

Abstract: Keratin-associated proteins (KAPs) are structural components of wool and hair fibers. To date, eight high glycine/tyrosine KAP (HGT-KAP) families have been identified in humans, but only three have been identified in sheep. In this study, the putative ovine homolog of the human KAP22-1 gene (*KRTAP22-1*) was amplified using primers designed based on a human *KRTAP22-1* sequence. Polymerase chain reaction-single stranded conformational polymorphism (PCR-SSCP) was used to screen for variation in *KRTAP22-1* in 390 Merino × Southdown-cross lambs and 75 New Zealand (NZ) Romney sheep. Three PCR-SSCP banding patterns were detected and DNA sequencing revealed that the banding patterns represented three different nucleotide sequences (*A–C*). Two single nucleotide polymorphisms (SNPs) were identified in these sequences. Variant *B* was most common with a frequency of 81.3% in NZ Romney sheep, while in the Merino × Southdown-cross lambs, *A* was more common with a frequency of 51.8%. The presence of *B* was found to be associated with increased wool yield and decreased mean fiber curvature (MFC). Sheep of genotype *BB* or *AB* had a higher wool yield than those of genotype *AA*. These results suggest that ovine *KRTAP22-1* variation may be useful when developing breeding programs based on increasing wool yield, or decreasing wool curvature.

Keywords: Keratin-associated protein KAP22-1; variation; mean fiber curvature (MFC); wool yield; sheep

1. Introduction

Keratin-associated proteins (KAPs) and keratins are the main structural proteins of wool and hair fibers. The former create a semi-rigid matrix with the keratin intermediate filaments (IFs) [1] and they play an important role in defining the physico-mechanical properties of the fibers. KAPs are a complex class of proteins and typically possess a high cysteine content. The KAPs have been classified into three broad groups according to their amino acid composition: the high sulfur (HS; ≤30 mol% cysteine), the ultra-high sulfur (UHS; >30 mol% cysteine) and the high glycine-tyrosine (HGT; 35–60 mol% glycine and tyrosine) groups [2]. More than 100 KAP genes have been identified across species and they have been divided into 27 KAP families [3]. Of these KAP families: 1–3, 10–16 and 23–27 are HS-KAPs; 4, 5, 9 and 17 are UHS-KAPs and 6–8 and 18–22 are HGT-KAPs [3,4].

Wool varies in HGT-KAP content ranging from less than 1% in Lincoln wool, to between 4% and 12% in Merino wool [5]. The wide range in the proportional content of HGT-KAPs in different wools and the extensive variation in the genes for the HGT-KAP genes [6–8] suggests that the HGT-KAPs may have important function in the wool fiber.

To date, three HGT-KAP gene families have been reported in sheep: KAP6, KAP7 and KAP8. There has been no report of the presence of other HGT-KAPs. The KAP22-1 gene (*KRTAP22-1*) has been identified in humans [9], but it has not been described in sheep. In this study, we describe the identification of a sequence encoding the putative ovine *KRTAP22-1*, report variation in this gene detected using polymerase chain reaction-single stranded conformational polymorphism (PCR-SSCP), and reveal associations between this genetic variation and variation in wool traits.

2. Materials and Methods

2.1. Sheep Blood and Wool Samples

Three hundred ninety lambs, produced over three years from crosses of Merino ewes × Southdown (n = 4; 188, 75, 59 and 68 progeny per ram), and seventy-five New Zealand (NZ) Romney lambs (n = 75, sourced from five farms) were used to search for variation in *KRTAP22-1*. The 390 Merino × Southdown lambs were subsequently used for the association study. Blood samples from all these sheep were collected onto FTA cards (Whatman BioScience, Middlesex, UK) and genomic DNA was purified using a two-step procedure described by Zhou et al. [10].

Wool samples were collected at 12 months of age (first shearing) from the mid-side of the Merino × Southdown-cross lambs. Greasy fleece weight (GFW) was measured at shearing, and other wool traits were measured by the New Zealand Wool Testing Authority Ltd (Ahuriri, Napier, NZ), including mean fiber diameter (MFD), fiber diameter standard deviation (FDSD), coefficient of variation of fiber diameter (CVFD), mean staple length (MSL), mean fiber curvature (MFC), mean staple strength (MSS) and prickle factor (PF). Wool yield (%) was measured and used to calculate the clean fleece weight (CFW).

2.2. Search for an Ovine Homolog of the Human KRTAP22-1 Gene in the Sheep Genome Sequence

The coding sequence of a human *KRTAP22-1* sequence (AP001708) was used to BLAST search the Ovine Genome Assembly v4.0 (http://blast.ncbi.nlm.nih.gov/Blast.cgi). The genome sequences that showed highest homology with the human *KRTAP22-1* sequence were presumed to be ovine *KRTAP22-1* sequences. These sequences were used to design PCR primers for amplifying the entire coding region of this gene from sheep genomic DNA.

2.3. PCR Primers and Amplification of Sheep Genomic DNA

The sequences of the PCR primers designed were: 5′-TATGAGTGCAACAGTGACTG-3′ and 5′-CCATGTTTTGAATAGACAAGC-3′. They were synthesized by Integrated DNA Technologies (Coralville, IA, USA). PCR amplification was performed in a 15-μL reaction containing the genomic DNA on one 1.2-mm punch of FTA paper, 0.25 μM of each primer, 150 μM of each dNTP (Bioline, London, UK), 2.5 mM of Mg^{2+}, 0.5 U of *Taq* DNA polymerase (Qiagen, Hilden, Germany) and 1× reaction buffer supplied with the enzyme. The thermal profile consisted of 2 min at 94 °C, followed by 35 cycles of 30 s at 94 °C, 30 s at 61 °C and 30 s at 72 °C, with a final extension of 5 min at 72 °C. Amplification was carried out in S1000 thermal cyclers (Bio-Rad, Hercules, CA, USA).

Amplicons were visualized by electrophoresis in 1% agarose gels (Quantum Scientific, Queensland, Australia), using 1 × TBE buffer (89 mM Tris, 89 mM boric acid, 2 mM Na_2EDTA) containing 200 ng/mL of ethidium bromide.

2.4. Screening for Variation in KRTAP22-1

The PCR amplicons were screened for sequence variation using SSCP analysis. A 0.7-μL aliquot of each amplicon was mixed with 7 μL of loading dye (98% formamide, 10 mM EDTA, 0.025%

bromophenol blue, 0.025% xylene-cyanol). After denaturation at 95 °C for 5 min, the samples were rapidly cooled on wet ice and then loaded on 16 cm × 18 cm, 14% acrylamide: bisacrylamide (37.5:1) (Bio-Rad) gels. Electrophoresis was performed using Protean II xi cells (Bio-Rad) in 0.5× TBE buffer, under the electrophoretic conditions 18 °C, 300 V for 16 h. Gels were silver-stained according to the method of Byun et al. [11].

2.5. Sequencing of Allelic Variants and Sequence Analysis

PCR amplicons representing different banding patterns from sheep that appeared to be homozygous were sequenced in both directions at the Lincoln University DNA sequencing facility, New Zealand. Alleles that were only found in heterozygous sheep were sequenced using an approach described by Gong et al. [12]. Briefly, a band corresponding to the allele was excised as a gel slice form the polyacrylamide gel, macerated, and then used as a template for re-amplification with the original primers. This second amplicon was then sequenced. Sequence alignments, translations and phylogenetic analysis were carried out using DNAMAN (version 5.2.10, Lynnon BioSoft, Vaudreuil, Canada). Phylogenetic tree was constructed using Observed Divergency method with 1000 bootstrap trials based on the predicated amino acid sequence.

2.6. Statistical Analyses

Statistical analyses were performed using Minitab version 16 (Minitab Inc., State College, PA, USA). General linear models (GLMs) were used to assess the effect of the presence/absence of the *KRTAP22-1* variants on various wool traits for the 390 Merino × Southdown lambs. For genotypes with a frequency >5% (and thus that had an adequate sample size), GLMs were used to compare the various wool traits among these genotypes and with a Bonferroni correction being applied to reduce the chances of obtaining false positive results during the multiple comparisons. Sire was found to affect ($p < 0.05$) all the wool traits and was included in the models as a random factor. Gender was found to affect ($p < 0.05$), or potentially affect ($p < 0.20$), wool traits, and was therefore fitted as a fixed factor into the models. Birth rank was not found to affect or potentially affect wool traits, and was not factored into the models.

3. Results

3.1. Identification of KRTAP22-1 in the Sheep Genome

A BLAST search of the Ovine Genome Assembly v4.0 using the human *KRTAP22-1* coding sequence (AP001708) revealed a homologous region on sheep chromosome 1. Analysis of the sequence in this homologous region led to the identification of a 144-bp open reading frame at OAR1: 123213996–123214139. Five previously identified ovine KAP genes were also found near this open reading frame and these from centromere to telomere were *KRTAP6-1*, *KRTAP6-3*, *KRTAP6-4*, *KRTAP6-2* and *KRTAP6-5* (Figure 1).

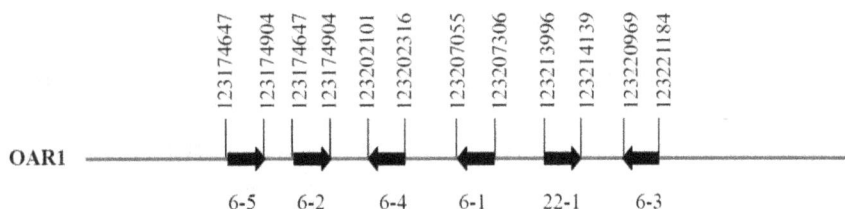

Figure 1. Location of sheep genome region that is homologous to *KRTAP22-1*, together with five other previously identified *KRTAP*s on sheep chromosome 1. Horizontal arrow bars represent the coding regions of *KRTAP*s and the arrowheads indicate the direction of transcription. The numbers below the horizontal arrow bars indicate the name of the respective KAP gene (e.g., 6-5 represents *KRTAP6-5*). The nucleotide positions refer to NC_019458.2.

The open reading frames identified were translated into amino acid sequences, and sequence comparison with known sheep KAPs together with human KAP22-1, revealed that this region was clustered with the human KAP22-1 sequence and formed a group that is distinct to other sheep KAP families (Figure 2). It suggested the presence of sheep *KRTAP22-1*.

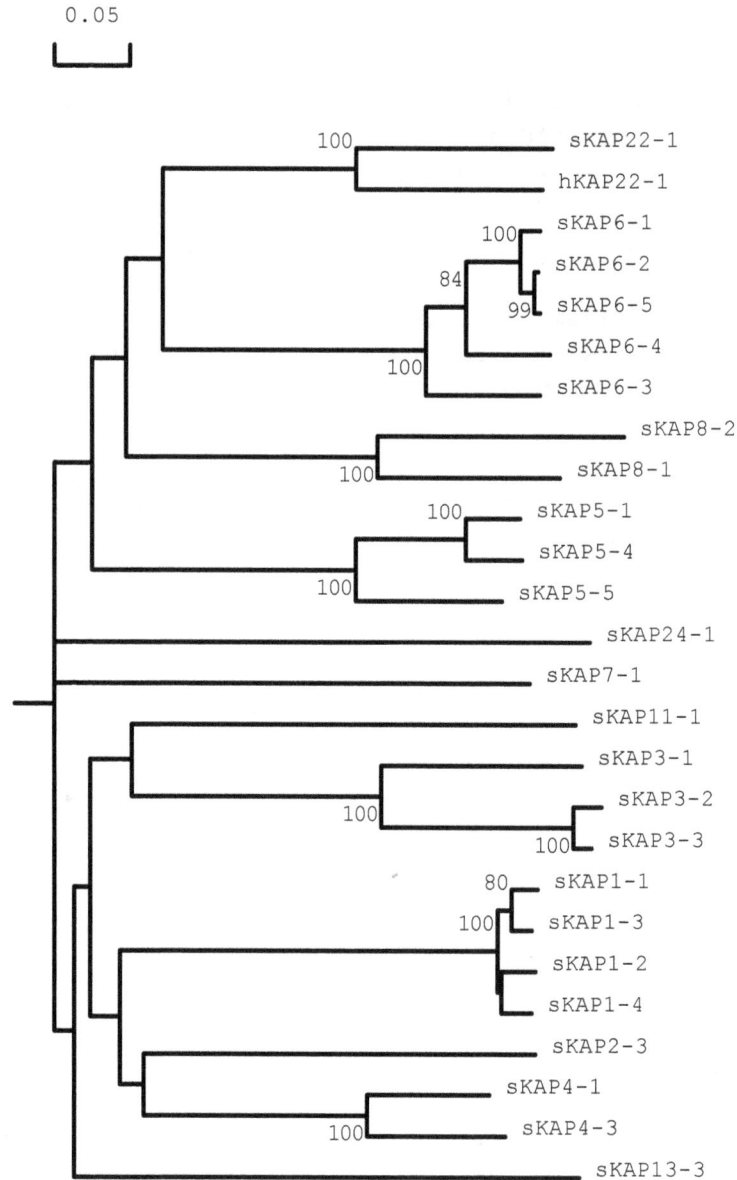

Figure 2. Phylogenetic tree of the sheep genomic regions identified, together with human KAP22-1. The tree was constructed using the predicted amino acid sequences. The ovine KAPs are indicated with the prefix "s" and the human KAP has the prefix "h". The numbers at the forks indicate the bootstrap confidence values and only those equal to or higher than 70% are shown. The GenBank accession numbers for the human KAP22-1 is AP001708. The GenBank accession numbers for the sheep KAPs are X01610, HQ897973, X02925, X01610, U60024, M21099, M21100, M21103, X73462, EU239778, X55294, X73434, X73435, M95719, KT725827, KT725833, KT725838, KT725841, X05638, X05639, KF220646, HQ595347, JN377429 and JX112014 (for sKAP1-1, sKAP1-2, sKAP1-3, sKAP1-4, sKAP2-3, sKAP3-1, sKAP3-2, sKAP3-3, sKAP4-1, sKAP4-3, sKAP5-1, sKAP5-4, sKAP5-5, sKAP6-1, sKAP6-2, sKAP6-3, sKAP6-4, sKAP6-5, sKAP7-1, sKAP8-1, sKAP8-2, sKAP11-1, sKAP13-3 and sKAP24-1, respectively).

3.2. Detection of Variation in Ovine KRTAP22-1

There were three PCR-SSCP banding patterns detected for ovine *KRTAP22-1*, with either one or a combination of two banding patterns observed for each sheep (Figure 3). DNA sequencing revealed that these PCR-SSCP patterns represented three different nucleotide sequences (*A, B* and *C*) (Table 1). These sequences have been deposited into GenBank with accession numbers KX377616, KX377617 and KX377618. Two single nucleotide polymorphisms (SNPs) were identified among the three sequences. One SNP was located 28 bp upstream of the nominal TATA box sequence and the other SNP was located in the coding region. The coding region SNP was synonymous.

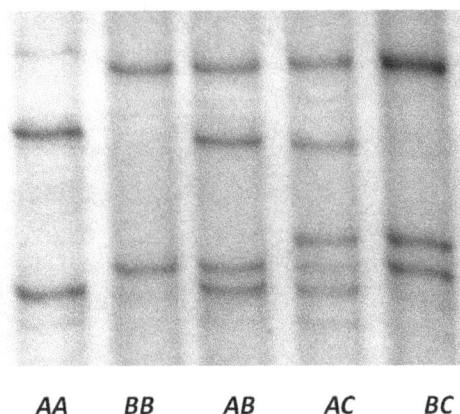

AA   BB   AB   AC   BC

Figure 3. PCR–single-stranded conformational polymorphism (PCR–SSCP) of the ovine KAP22-1 gene.

Table 1. SNPs and alleles of the ovine *KRTAP22-1*.

SNP	Allele			Amino Acid Change
	A	*B*	*C*	
C.-100C/T	C	C	T	5'UTR
C.45T/C	T	C	C	No change

3.3. Amino Acid Composition of Ovine KAP22-1

The ovine *KRTAP22-1* sequences would encode a polypeptide of 47 amino acid residues. This would have a high content of glycine and tyrosine (51.1 mol%), and moderate levels of cysteine (14.9 mol%) and serine (8.3 mol%). The putative KAP22-1 protein would therefore be basic, with a predicted isoeletric point (pI) value of 7.65.

3.4. Genotypes and Allele Frequencies in NZ Romney and Merino-Cross Sheep

Five genotypes were detected in Merino × Southdown-cross lambs, and they were as follows: *AA, AB, AC, BB* and *BC*. Genotypes *AC* and *BC* were not detected in the NZ Romney sheep.

The frequencies of the *KRTAP22-1* variants in the NZ Romney sheep were: *A*: 18.7 % and *B*: 81.3%; while those in Merino × Southdown-cross lambs were: *A*: 51.8%, *B*: 47.2% and *C*: 1.0%. Variant *B* was very common in NZ Romney lambs, while, in Merino × Southdown-cross lambs, *A* was more common. The four sires of the Merino × Southdown-cross lambs were all of genotype *AB*.

3.5. Effect of Variation in KRTAP22-1 on Wool Traits

Of the three variants detected in Merino × Southdown-cross lambs, variant *C* occurred at a very low frequency (<5%) and accordingly its association with wool traits was not analyzed (sheep with these genotypes were discarded from the genotype analyses). In the presence/absence models, the presence of *B* was found to be associated with increased wool yield and decreased MFC (Table 2).

With three genotypes (*AA*, *BB* and *AB*) that occurred at a frequency >5% in Merino × Southdown-cross lambs, an effect of genotype was detected for wool yield (Table 3). Sheep of genotype *AB* or *BB* had a higher wool yield than those of genotype *AA*.

Table 2. Association of *KRTAP22-1* variants with various wool traits.

Trait [1]	Variant	n		Mean ± SE [2]		p [3]
		Absent	Present	Absent	Present	
GFW (kg)	A	79	311	2.30 ± 0.07	2.33 ± 0.06	0.628
	B	99	291	2.37 ± 0.07	2.31 ± 0.06	0.191
CFW (kg)	A	79	311	1.69 ± 0.06	1.69 ± 0.05	0.931
	B	99	291	1.70 ± 0.05	1.69 ± 0.05	0.844
Yield (%)	A	79	311	72.9 ± 0.99	72.0 ± 0.79	0.244
	B	99	291	70.8 ± 0.91	72.6 ± 0.79	**0.008**
MFD (μm)	A	79	311	19.6 ± 0.31	19.5 ± 0.25	0.705
	B	99	291	19.4 ± 0.30	19.6 ± 0.25	0.547
FDSD (μm)	A	79	311	4.28 ± 0.11	4.16 ± 0.09	0.139
	B	99	291	4.17 ± 0.10	4.19 ± 0.30	0.828
CVFD (%)	A	79	311	22.0 ± 0.36	21.7 ± 0.30	0.281
	B	99	291	21.9 ± 0.35	21.7 ± 0.30	0.503
MSL (mm)	A	79	311	84.2 ± 1.99	84.6 ± 1.63	0.796
	B	99	291	85.5 ± 1.93	84.3 ± 1.62	0.368
MSS (N/ktex)	A	79	311	21.1 ± 1.25	23.4 ± 1.02	0.192
	B	99	291	21.7 ± 1.21	22.2 ± 1.02	0.526
MFC (°/mm)	A	79	311	86.5 ± 2.43	89.4 ± 1.98	0.128
	B	99	291	91.6 ± 2.34	87.9 ± 1.97	**0.032**
PF (%)	A	79	311	2.12 ± 0.36	2.05 ± 0.29	0.816
	B	99	291	2.13 ± 0.35	2.05 ± 0.28	0.762

[1] GFW—Greasy Fleece Weight; CFW—Clean Fleece Weight; MFD—Mean Fiber Diameter; FDSD—Fiber Diameter Standard Deviation; CVFD—Coefficient of Variation of Fiber Diameter; MSL—Mean Staple Length; MSS—Mean Staple Strength; MFC—Mean Fiber Curvature; PF—Prickle Factor (percentage of fibers over 30 μm). [2] Predicted marginal means and standard errors derived from GLMs with variant absent/ present, sire (random effect) and gender (fixed effect) being factored into the models. [3] $p < 0.05$ are in bold.

Table 3. The effect of *KRTAP22-1* genotype on various wool traits.

Trait [1]	Mean ± SE [2]			p
	AA (n = 93)	*AB* (n = 212)	*BB* (n = 77)	
GFW (kg)	2.38 ± 0.07	2.31 ± 0.06	2.29 ± 0.07	0.341
CFW (kg)	1.71 ± 0.06	1.69 ± 0.05	1.70 ± 0.06	0.918
Yield (%)	70.9 ± 0.97 [b]	72.6 ± 0.83 [a]	73.0 ± 1.00 [a]	**0.044**
MFD (μm)	19.5 ± 0.31	19.6 ± 0.27	19.6 ± 0.32	0.838
FDSD (μm)	4.20 ± 0.10	4.16 ± 0.09	4.27 ± 0.11	0.385
CVFD (%)	21.9 ± 0.36	21.6 ± 0.31	21.9 ± 0.37	0.321
MSL (mm)	85.9 ± 1.98	84.1 ± 1.71	84.1 ± 2.03	0.488
MSS (N/ktex)	21.8 ± 1.24	22.7 ± 1.08	21.2 ± 1.28	0.306
MFC (°/mm)	91.7 ± 2.45	89.0 ± 2.12	86.8 ± 2.51	0.120
PF (%)	2.46 ± 0.53	2.45 ± 0.46	3.04 ± 0.55	0.387

[1] GFW—Greasy Fleece Weight; CFW—Clean Fleece Weight; MFD—Mean Fiber Diameter; FDSD—Fiber Diameter Standard Deviation; CVFD—Coefficient of Variation of Fiber Diameter; MSL—Mean Staple Length; MSS—Mean Staple Strength; MFC—Mean Fiber Curvature; PF—Prickle Factor (percentage of fibers over 30 μm). [2] Predicted marginal means and standard errors derived from the GLMs with genotype, sire (random effect) and gender (fixed effect) being factored into the models with a Bonferroni correction to adjust for repetitive testing. Means within rows that do not share a superscript letter are significantly ($p < 0.05$) different and bolded.

4. Discussion

This study describes the identification of a new ovine HGT-KAP. The putative ovine *KRTAP22-1* was clustered with several previously described KAP genes and displayed a lower sequence similarity to any known ovine KAP gene, when compared to *KRTAP22-1* from humans. This gene was located between *KRTAP* 6-1 and *KRTAP* 6-3, and this is consistent with the location of human *KRTAP22-1* [9]. This suggests that the gene represents the ovine *KRTAP22-1* sequence. The identification of ovine *KRTAP22-1* brings the total number of HGT-KAPs identified in sheep, from eight to nine.

The peptide encoded for by the ovine *KRTAP22-1* sequence was predicted to contain 47 amino acids and more than half of these amino acids were glycine and tyrosine. This is consistent with other HGT-KAPs; however, the number of repeated occurrences of glycine-tyrosine and glycine-tyrosine-glycine in KAP22-1 was small when compared to other ovine HGT-KAPs (KAP6–KAP8). Until now, nearly all the HGT-KAPs are basic (except KAP8-2) [2], and this was also the case for KAP22-1.

Two SNPs were detected in ovine *KRTAP22-1* and these produced three unique variant sequences. Both of the SNPs were T/C transitions. The coding region SNP was synonymous. Among the three variants, *B* was very common in Romney sheep, while in Merino × Southdown-cross lambs *A* was more common.

Variation in *KRTAP22-1* was found to be associated with two wool traits, MFC and wool yield. However, the most enduring effect by *KRTAP22-1* appeared to be on the wool yield, for which a sizeable difference in mean among common genotypes was detected and a difference that reinforced the conclusions drawn from the variant absence/presence models. Sheep with *B* have a higher wool yield, but a lower MFC. This is consistent with the correlation that has been reported between these two traits, with a moderate negative correlation ($0.3 < |r| \leq 0.7$) found between MFC and wool yield ($r = -0.518$) [13]. The wool yield for the Merino × Southdown-cross lambs was lower at 73% (Table 3), than in Romney sheep (80%) and this is consistent with other reports [14].

Lambs with *B* had lower fiber curvature. The correlation between MFC and fiber crimp has been reported to be high at 0.85 [15], so curvature is a reasonable proxy for the crimp of fiber. "Low-curvature" wool generally has a curvature less than 50°/mm, while "medium curvature" wool is from 60 to 90°/mm and "high curvature" wool is greater than 100°/mm, this being associated typically with a high crimp frequency. The Merino × Southdown-cross lambs in this study had a medium curvature wool (Table 3), while the Romney sheep had low-curvature wool (38°/mm) [16].

The results are consistent with the predicted function of HGT-KAPs. It has been reported that the content of HGT-KAPs is decreased in Merino felting luster (FL) mutant wool that loses crimp [17], and that the helical angle of IFs in the orthocortex is associated with fiber curvature [18]. The HGT-KAPs are predominantly present in the orthocortex and are thought to have some associations with the crimp of wool fiber [19]. The results from this research confirm that variation in *KRTAP22-1* may affect those traits.

Although the variation in the SNP in the coding region of *KARTAP22-1* is synonymous, and would not result in any amino acid substitution, it may affect the expression or structure of the protein. It has been reported that silent mutations may affect mRNA translation rates and thus potentially change the way that protein folds [20]. It is also possible that the effects observed for *KRTAP22-1* are be due to its linkage to other *KRTAPs* or *KRTs* on the same chromosome. The location of *KRTAP22-1* is interesting in that it is found within KAP6 family members and the biological significance of this needs more study.

5. Conclusions

These findings confirm that ovine *KRTAP22-1* is variable and suggests that variation in the gene may need to be considered when developing breeding programs based on improving wool curvature or wool yield.

Acknowledgments: Financial support from the China Scholarship Council, International S & T Cooperation Program of China (2011DFG33310) and the Lincoln University Gene-Marker Laboratory are acknowledged.

We acknowledge the support of the AGMARDT Postdoctoral Fellowship to Hua Gong. We thank Qian Fang and Seung OK Byun for technical assistance.

Author Contributions: Shaobin Li, Huitong Zhou, Yuzhu Luo and Jon G. H. Hickford conceived and designed the project. Shaobin Li, Hua Gong and Fangfang Zhao performed the experiments. Huitong Zhou, Jiqing Wang and Xiu Liu analyzed the data. Shaobin Li, Huitong Zhou, Yuzhu Luo and Jon G. H. Hickford wrote the manuscript. All authors reviewed and commented on the manuscript.

References

1. Powell, B.C.; Rogers, G.E. The role of keratin proteins and their genes in the growth, structure and properties of hair. In *Formation and Structure of Human Hair*; Jollès, P., Zahn, H., Höcker, H., Eds.; Birkhäuser Verlag: Basel, Switzerland, 1997; pp. 59–148.
2. Gong, H.; Zhou, H.; Forrest, R.H.J.; Li, S.; Wang, J.; Dyer, J.M.; Luo, Y.; Hickford, J.G. Wool keratin-associated protein genes in Sheep—A Review. *Genes* **2016**. [CrossRef] [PubMed]
3. Gong, H.; Zhou, H.; McKenzie, G.W.; Yu, Z.D.; Clerens, S.; Dyer, J.M.; Plowman, J.E.; Wright, M.W.; Arora, R.; Bawden, C.S.; et al. An updated nomenclature for keratin-associated proteins (KAPs). *Int. J. Bio. Sci.* **2012**, *8*, 258–264. [CrossRef] [PubMed]
4. Rogers, M.A.; Langbein, L.; Praetzel-Wunder, S.; Winter, H.; Schweizer, J. Human hair keratin-associated proteins (KAPs). *Int. Rev. Cytol.* **2006**, *251*, 209–263. [PubMed]
5. Gillespie, J.M. The proteins of hair and other hard α-keratins. In *Cellular and Molecular Biology of Intermediate Filaments*; Goldman, R.D.S., Peter, M., Eds.; Plenum press: New York, NY, USA, 1990; pp. 95–128.
6. Gong, H.; Zhou, H.; Hickford, J.G.H. Diversity of the glycine/tyrosine-rich keratin-associated protein 6 gene (KAP6) family in sheep. *Mol. Biol. Rep.* **2011**, *38*, 31–35. [CrossRef] [PubMed]
7. Gong, H.; Zhou, H.; Plowman, J.E.; Dyer, J.M.; Hickford, J.G.H. Search for variation in the ovine *KAP7-1* and *KAP8-1* genes using polymerase chain reaction–single-stranded conformational polymorphism screening. *DNA Cell B* **2012**, *31*, 367–370. [CrossRef] [PubMed]
8. Zhou, H.; Gong, H.; Wang, J.Q.; Dyer, J.M.; Luo, Y.Z.; Hickford, J.G.H. Identification of four new gene members of the *KAP6* gene family in sheep. *Sci. Rep.* **2016**. [CrossRef] [PubMed]
9. Rogers, M.A.; Langbein, L.; Winter, H.; Ehmann, C.; Praetzel, S.; Schweizer, J. Characterization of a first domain of human high glycine-tyrosine and high sulfur keratin-associated protein (KAP) genes on chromosome 21q22.1. *J. Biol. Chem.* **2002**, *277*, 48993–49002. [CrossRef] [PubMed]
10. Zhou, H.; Hickford, J.G.H.; Fang, Q. A two-step procedure for extracting genomic DNA from dried blood spots on filter paper for polymerase chainreaction amplification. *Anal. Biochem.* **2006**, *354*, 159–161. [CrossRef] [PubMed]
11. Byun, S.O.; Fang, Q.; Zhou, H.; Hickford, J.G.H. An effective method for silver-staining DNA in large numbers of polyacrylamide gels. *Anal. Biochem.* **2009**, *385*, 174–175. [CrossRef] [PubMed]
12. Gong, H.; Zhou, H.; Dyer, J.M.; Hickford, J.G. Identification of the ovine KAP11-1 gene (*KRTAP11-1*) and genetic variation in its coding sequence. *Mol. Biol. Rep.* **2011**, *38*, 5429–5433. [CrossRef] [PubMed]
13. Gong, H.; Zhou, H.; Hodge, S.; Dyer, J.M.; Hickford, J.G. Association of wool traits with variation in the ovine *KAP1-2* gene in Merino cross lambs. *Small Rum. Res.* **2015**, *124*, 24–29. [CrossRef]
14. Elliott, K.H.; Bigham, M.L.; Sumner, R.M.W.; Dalton, D.C. Wool production of yealing ewes of different breeds on hill country. *N. Z. J. Agric. Res.* **1978**, *21*, 179–186. [CrossRef]
15. Nimbs, M.A.; Hygate, L.; Behrendt, R. The relationship between fibre curvature, crimp frequence and other wool traits. *Anim. Prod. Aust.* **1998**, *22*, 396.
16. Sumner, R.M.W.; Young, S.R.; Upsdell, M.P. Wool yellowing and pH within Merino and Romney fleeces. *Proc. N. Z. Soc. Anim. Prod.* **2003**, *63*, 155–159.
17. Li, S.W.; Ouyang, H.S.; Rogers, G.E.; Bawden, C.S. Characterization of the structural and molecular defectsin fibres and follicles of the merino felting lustre mutant. *Exp. Dermatol.* **2009**, *18*, 134–142. [CrossRef] [PubMed]
18. Caldwell, J.P.; Mastronarde, D.N.; Woods, J.L.; Bryson, W.G. The three-dimensional arrangement of intermediate filaments in Romney wool cortical cells. *J. Struct. Biol.* **2005**, *151*, 298–305. [CrossRef] [PubMed]

19. Powell, B.C.; Rogers, G.E. Hard keratin IF and associatedproteins. In *Cellular and Molecular Biology of Intermediate Filaments*; Goldman, R.D., Steinert, P.M., Eds.; Plenum Press: New York, NY, USA, 1990; pp. 267–300.

20. Hurst, L.D. Molecular genetics: The sound of silence. *Nature* **2011**, *471*, 582–583. [CrossRef] [PubMed]

4

Characterization of the Transcriptome and Gene Expression of Brain Tissue in Sevenband Grouper (*Hyporthodus septemfasciatus*) in Response to NNV Infection

Jong-Oh Kim, Jae-Ok Kim, Wi-Sik Kim and Myung-Joo Oh *

Department of Aqualife Medicine, College of Fisheries and Ocean Science, Chonnam National University, Yeosu 550-749, Korea; jongoh.kim77@gmail.com (J.-O.K.); hoy0924@naver.com (J.-O.K.); wisky@jnu.ac.kr (W.-S.K.)
* Correspondence: ohmj@jnu.ac.kr

Academic Editor: J. Peter W. Young

Abstract: Grouper is one of the favorite sea food resources in Southeast Asia. However, the outbreaks of the viral nervous necrosis (VNN) disease due to nervous necrosis virus (NNV) infection have caused mass mortality of grouper larvae. Many aqua-farms have suffered substantial financial loss due to the occurrence of VNN. To better understand the infection mechanism of NNV, we performed the transcriptome analysis of sevenband grouper brain tissue, the main target of NNV infection. After artificial NNV challenge, transcriptome of brain tissues of sevenband grouper was subjected to next generation sequencing (NGS) using an Illumina Hi-seq 2500 system. Both mRNAs from pooled samples of mock and NNV-infected sevenband grouper brains were sequenced. Clean reads of mock and NNV-infected samples were de novo assembled and obtained 104,348 unigenes. In addition, 628 differentially expressed genes (DEGs) in response to NNV infection were identified. This result could provide critical information not only for the identification of genes involved in NNV infection, but for the understanding of the response of sevenband groupers to NNV infection.

Keywords: nervous necrosis virus (NNV); sevenband grouper; transcriptome; next generation sequencing (NGS); differential expressed genes (DEGs)

---

1. Introduction

Grouper is one of the highest valued marine fish and has become an important species in the aquaculture industry of various Asian countries. In Korea, sevenband grouper (*Hyporthodus septemfasciatus*) is one the favorite grouper fish consumed. Its production rate is increasing. However, viral nervous necrosis (VNN) causes high mortality, especially at the larval and juvenile stage of sevenband groupers during the summer season, which has caused vast economic losses [1].

Viral Nervous Necrosis is a serious disease in the world aquaculture industry [2–4]. Firstly, it was reported in bigeye trevally (*Caranx sexfasciatus*) in the 1980s and since then it has been reported in over twenty species [2–4]. The infected fish are usually swimming abnormally and having vacuolization and necrosis of the central nervous system in the brain [3]. In Korea, mass mortalities caused by VNN have been reported from various cultured marine fish such as sevenband grouper (*Hyporthodus septemfasciatus*), rock bream (*Oplegnathus fasciatus*), red drum (*Sciaenops ocellatus*) and olive flounder (*Paralichthys olivaceus*) since 1990 [5–7].

Nervous necrosis virus (NNV), the causative agent of VNN, has non-enveloped icosahedral structure and belongs to the family *Nodaviridae* (genus *Betanodavirus*). Its genome contains two single-stranded positive senses RNA: RNA1 (approximately 3.1 kb in length) encodes an RNA-dependent RNA polymerase for viral replication, whereas RNA2 (1.4 kb) encrypts a protein α. During RNA replication, a sub-genomic RNA3 was produced from the 3' terminus of RNA1 that encodes non-structural proteins B2 [8]. Betanodaviruses have five genogroups based on the T4 region sequence of RNA2 as barfin flounder nervous necrosis virus (BFNNV), red-spotted grouper nervous necrosis virus (RGNNV), striped jack nervous necrosis virus (SJNNV), tiger puffer nervous necrosis virus (TPNNV) [9], and turbot nervous necrosis virus [10].

Although virological and genetic characterizations of NNV have been reported, its infection mechanisms and disease outbreak mechanisms remain unclear. Therefore, systematic approaches are needed to determine its infection mechanisms.

Recently, rapid progress has been made in next generation sequencing (NGS) technology and bioinformatics. They are powerful tools for transcriptome analysis. RNA-Seq by NGS is one of the most useful methods to survey the landscape of a transcriptome since it produces millions of data on gene expression. Numerous novel genes and unraveling expression profiles related phenotypic changes, such as development stages, were identified using RNA-Seq [11,12]. Several studies using RNA-Seq have reported the immune relevant genes of fish after pathogen challenges. For example, zebrafish (*Danio rerio*) [13], orange-spotted grouper (*Epinephelus coioides*) [14], large yellow croaker (*Pseudosciaena crocea*) [15], turbot (*Scophthalmus maximus*) [16], Japanese sea bass (*Lateolabrax japonicus*) [17] and common carp (*Cyprinus carpio*) [18] in response to pathogen challenges have been analyzed.

In this study, we analyzed the brain transcriptome of sevenband grouper in response to NNV infection. We annotated the transcripts using Gene Ontology (GO) and non-redundant (nr) databases of GenBank. In addition, we obtained differential expression genes (DEGs) between sevenband grouper brain infected by NNV and non-infected sevenband grouper brain. To the best of our knowledge, this is the first study that reports the transcriptome of brain tissue of NNV-infected sevenband grouper by RNA-Seq. Our transcriptome analysis data will provide valuable genomic information to determine the functional roles of these genes related to NNV infection and VNN outbreak in the future.

2. Materials and Methods

2.1. Ethics Statment

The experiments using sevenband grouper were carried out in strict accordance with the recommendations of the Institutional Animal Care and Use Committee of Chonnam National University (Permit Number: CNU IACUC-YS-2015-4).

2.2. Experimental Fish and Virus

Sevenband groupers were purchased from an aqua-farm without history of VNN occurrence. Prior to experiment, brain tissues of 10 fish were randomly sampled from the aqua-farm and subjected to reverse transcription polymerase chain reaction (RT-PCR) analysis to determine betanodavirus infection according to a previous report [9]. The NNV (SGYeosu08) [19] isolated from sevenband grouper aqua-farm in Korea was propagated in the striped snakehead (SSN-1) cell line. SSN-1 cells were grown at 25 °C in Leibovitz L-15 medium (Sigma Aldrich, St. Louis, MO, USA) containing 10% fetal bovine serum (FBS, Gibco, Waltham, MA, USA), 100 μg/mL streptomycin and 150 U/mL penicillin G. NNV was inoculated onto SSN-1 cell monolayer and incubated at 25 °C for the virus propagation. After the cells were completely lysed, virus titer was calculated by the Reed and Muench method [20]. Viral samples were aliquoted into small volumes and stored at −80 °C until use.

2.3. Virus Challenge

Twenty fish (mean body weight, 20 g) were reared in two 40 L tanks ($n = 10$/tank) at 25 °C. Ten fish were intramuscularly injected with NNV at doses of $10^{3.8}$ TCID50/fish. The remaining 10 fish were injected with L15 medium as a control. The challenged fish were observed daily. The NNV infected fish died from day 3 after infection and showed 100% of cumulative mortality after 1 week. The moribund fish at days 3 and 4 were selected for sampling. Brain tissues of three of ten challenged fish from mock and the virus-challenge group were collected and pooled for NGS analysis, respectively.

2.4. Next Generation Sequencing of Transcriptome

To obtain high-throughput transcriptome data of sevenband grouper, complementary DNA (cDNA) libraries were prepared for 100 bp paired-end sequencing using a TruSeq RNA Sample Preparation Kit (Illumina, San Diego, CA, USA) according to the manufacturer's protocols. They were then paired-end (2 × 100 bp) sequenced using an Illumina HiSeq2500 system (Illumina, San Diego, CA, USA).

2.5. Transcriptome Assembly and Functional Annotation

Prior to de novo assembly, paired-end sequences were filtered and cleaned using an NGS QC toolkit [21] to remove low quality reads (Q < 20) and adapter sequences. In addition, bases of both ends less than Q20 of filtered reads were removed additionally. This process is to enhance the quality of reads due to mRNA degradation in both ends of it as time goes on [22]. Only high quality reads were used for de novo assembly performed by Trinity (version 20130225) using default values [23]. To remove the redundant sequences, CD-HIT-EST [24] was used. NCBI Blast (version 2.2.28) was applied for the homology search to predict the function of unigenes. The function of unigenes was predicted by Blastx to search all possible proteins against the NCBI Non-redundant (NR) database (accessed on 17 July 2013). The criterion regarding significance of the similarity was set at Expect-value less than 1×10^{-5}.

2.6. Differentially Expressed Genes Analysis

After obtaining the assembled transcriptome data using Trinity, gene expression level was measured with RNA-Seq by Expectation Maximization (RSEM), a tool for measuring the expression level of transcripts without any information on its reference [25]. The TCC package was used for DEG analysis through the iterative DEGES/DEseq method [26]. Normalization was progressed three times to search meaningful DEGs between comparable samples [27]. The DEGs were identified based on the p-value of less than 0.05.

2.7. GO Enrichment of Differentially Expressed Genes

The GO database classifies genes according to the three categories of Biological Process (BP), Cellular Component (CC) and Molecular Function (MF) and provides information on the function of genes. To characterize the identified genes from DEG analysis, a GO based trend test was carried out through the Fisher's exact test. Selected genes with p-values of less than 1×10^{-5} were regarded as statistically significant.

2.8. Data Deposition

All the raw read files were submitted to the sequence reads archive (SRA), NCBI database (accession number—SRR5091816).

3. Results

3.1. Sequence Analysis of the Transcriptome

Sequencing of the two libraries (mock and NNV-infected brain tissue) using the Illumina Hiseq 2500 platform generated a total of 45,101,102 (5,682,738,852 bases) and 34,715,846 (4,374,196,596 bases) raw reads, respectively (Table 1). After the cleansing step with an NGS QC Toolkit and removal of low quality (Q < 20) reads, 39,932,160 (5,006,434,933 bases) and 31,353,144 (3,932,946,324 bases) remained as clean reads, respectively (Table 1). The percentages of clean reads were 88.1% and

89.9%, respectively (Table 1). All the clean reads were submitted to the Trinity for de novo assembly. Unigenes were identified after removing redundant sequences from assembled transcripts. The number of unigenes was 104,348, the total length and the average length of the unigenes were 88,123,224 bp and 845 bp, respectively (Table 1). The length distribution of unigenes is presented in Figure 1. Among these unigenes, 66,204 unigenes (63.4%) were no more than 500 bp. A total of 15,382 unigenes (14.7%) were 501–1000 bp, 6991 unigenes (6.7%) were 1001–1500 bp, 4727 unigenes (4.5%) were 1501–2000 bp, 3332 unigenes (3.2%) were 2001–2500 bp, and 7712 unigenes (7.4%) were longer than 2500 bp.

Table 1. Sequencing, assembly and annotation of transcriptome.

A. Sequencing and Preprocessing		
Sample Type	Not Infected (Mock)	NNV Infected
Number of raw reads	45,101,102	34,715,846
Total number of raw bases	5,682,738,852	4,374,196,596
Number of clean reads	39,932,160 (88.5%)	31,353,144 (90.3%)
Total number of clean bases	5,006,434,933 (88.1%)	3,932,946,324 (89.9%)
B. De novo Assembly and Annotation		
Number of unigenes		104,348
Total bases		88,123,224
Average length of unigenes		845 bases
Annotation by BLAST		43,280 (41.5%)

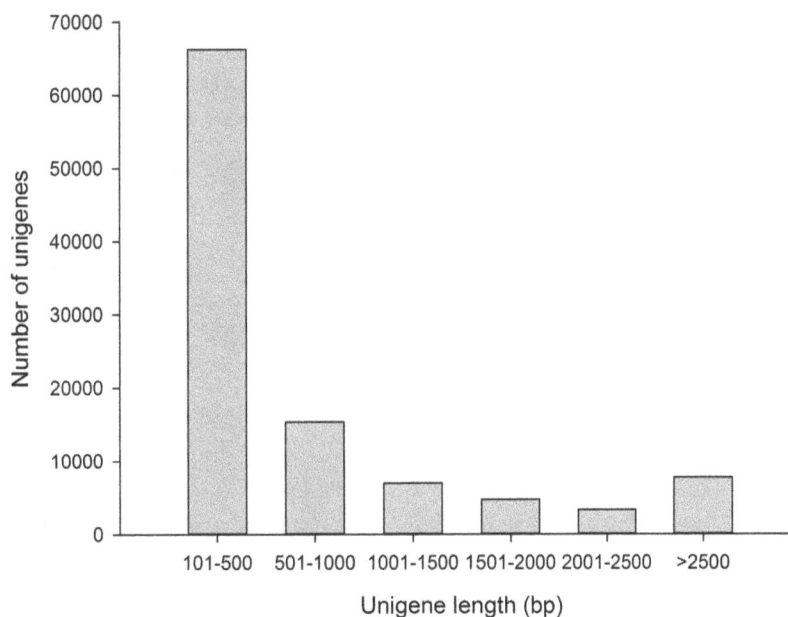

Figure 1. Length distribution of unigenes obtained from transcriptome analysis.

3.2. The Most Abundantly Expressed Gene in the Transcriptome Profile

To estimate gene expression levels, we calculated the abundance of reads in the transcriptome. The top 20 most highly expressed transcripts are shown in Table 2. Commonly, the most abundant genes in both the mock and NNV-infected groups were ribosomal proteins, such as ribosomal protein (RPS) 15a, RPL39, RPS28, RPS14, RPLS2, RPS27a, RPL21 and RPL32 essential for biological metabolism. Ubiquitin-like protein 4a (UBL4a), C-C motif chemokine 2 (CCL2), lysozyme g (LYG_EPICO) and two novel genes (ID: SGU016297, SGU008676) were highly expressed in the NNV-infected group compared to that in the mock group. Of them, Ubiquitin-like protein 4a was the most abundant gene in the NNV-infected group.

Table 2. The top 20 most abundant unigenes.

No	Mock			Infected		
	ID	Name	Description	ID	Name	Description
1	SGU067144	-	Hypothetical protein	SGU020577	UBL4a	Ubiquitin-like protein 4a
2	SGU067145	-	Hypothetical protein	SGU005369	-	Hyphothetical protein
3	SGU004764	EPD1	Ependymin-1	SGU067144	-	Hyphothetical protein
4	SGU023233	MT-CO1	Cytochrome c oxidase subunit 1	SGU067145	-	Hyphothetical protein
5	SGU005369	-	Hypothetical protein	SGU051992	RPS15a	40S ribosomal protein S15a
6	SGU023234	MT-ND5	NADH dehydrogenase subunit 5	SGU028675	RPL38	60S ribosomal protein L38
7	SGU051992	RPS15a	40S ribosomal protein S15a	SGU004764	EPD1	Ependymin-1
8	SGU028388	CD59	CD59 glycoprotein	SGU035083	RPL39	60S ribosomal protein L39
9	SGU025576	HBA	Hemoglobin subunit alpha	SGU023233	MT-CO1	Cytochrome c oxidase subunit 1
10	SGU035083	RPL39	60S ribosomal protein L39	SGU036307	CCL2	C-C motif chemokine 2
11	SGU021087	UBIQP_XENLA	Polyubiquitin	SGU053409	LYG_EPICO	Lysozyme g
12	SGU002473	RPS28	40S ribosomal protein S28	SGU002473	RPS28	40S ribosomal protein S28
13	SGU002418	FABP7	Fatty acid-binding protein, brain	SGU021087	UBIQP_XENLA	Polyubiquitin
14	SGU028675	RPL38	60S ribosomal protein L38	SGU037323	RPL29	60S ribosomal protein L29
15	SGU040127	RPS14	40S ribosomal protein S14	SGU003425	RPS27a	40S ribosomal protein S27a
16	SGU028033	RPLP2	60S acidic ribosomal protein P2	SGU016297	-	-
17	SGU026642	HBB2	Hemoglobin subunit beta-2	SGU005492	UBA52	Ubiquitin-60S ribosomal protein L40
18	SGU003425	RPS27a	40S ribosomal protein S27a	SGU034075	RPL21	60S ribosomal protein L21
19	SGU034075	RPL21	60S ribosomal protein L21	SGU040127	RPS14	40S ribosomal protein S14
20	SGU003965	RPL32	60S ribosomal protein L32	SGU008676	-	-

3.3. Functional Annotations

Putative annotations of these transcripts were performed using BlastX as mentioned in the method section. After gene annotation by using BlastX against the non-redundant (nr) database, the putative functions of 43,280 sequences (41.5%) of 104,348 unigenes sequences were identified.

3.4. Immune Relevant DEGs Involved in NNV Infection

A total of 3418 unigenes were differentially expressed based on DEG analysis using the TCC package. A total of 372 genes from the total of 3418 DEGs were annotated (Table S1). Immune relevant DEGs were further manipulated manually (Table 3). In immune relevant genes, a variety of cytokines were intensely up-regulated after NNV infection.

Several cytokine genes induced by NNV infection belonged to the chemokine family, including C-C motif Chemokine Ligand 2 (CCL2), CCL34, CCL19, CCL4, C-X-C motif chemokine ligand 13 (CXCL13), CXCL6, CXCL8, CXCL9, Interleukin-12 subunit alpha (IL12A) and beta (IL12B), and IL18B. CCL2 was the most critically expressed gene in the infected group showing 10.66 Log Fold Change (FC). CCL2 is involved in neuroinflammatory processes taking place in the central nervous system in various diseases [28].

Cathepsins are lysosomal cysteine enzymes with important roles in cellular homeostasis and innate immune response [29]. Among a dozen members of the Cathepsin family, subtypes L, H, K, O, S and Z were up-regulated in the brain of sevenband grouper after NNV infection. Specifically, Cathepsin L was highly expressed in the NNV-infected group showing 8.3 Log FC.

Several lectins were expressed in higher levels in the NNV-infected group compared to the mock group, including C-type lectins (CLEC4M, CLEC10A), galectins (LGALS9, LGALS3), fucolectin (FUCL4), and mannose-binding lectin (MBL). In the case of C-type lectins, its receptor (CD209) was also highly expressed in the infected group (Table 3) indicating that C-type lectin might play specific roles in the response of sevenband grouper to NNV infection.

As expected, a number of antiviral proteins also showed high levels of expression in the NNV-infected group. For example, radical S-adenosyl methionin domain-containing protein 2 (RSAD2), also known as viperin, was highly expressed in the NNV-infected group with 10.40 Log FC.

Mx gene (*MX*), one of the important downstream effectors of interferon (IFN), was also expressed more in the infected group with 8.64 Log FC. Besides Mx, a lot of IFN-induced proteins were upregulated by NNV infection, including IFN-induced protein 44 (IFI44), IFN-induced protein with tetratricopeptide repeats 5 (IFIT5), IFN-induced very large GTPase 1 (GVINP1), IFN-induced double-stranded RNA-activated protein kinase (EIF2Ak2), and IFN-induced helicase C domain-containing protein 1 (IFIH1). Interestingly, of the various heat shock proteins (HSPs), only HSP30 was significantly upregulated in the NNV-infected group with Log 8.42 FC. NK-Lysin, a known antibacterial protein, was also highly expressed in the NNV-infected group with 8.85 Log FC.

Table 3. Immune relevant differentially expressed genes (DEGs) after NNV infection.

Name	Description	Expression Level (FPKM)		Log FC	*p*-Value	FDR
		Mock	Infected			
Cytokine						
CCL2	C-C Motif Chemokine 2	51.7	83,844.2	10.66	2.3×10^{-6}.	6.0×10^{-3}
IL12B	Interleukin-12 Subunit Beta	12.5	14,158.0	10.15	5.6×10^{-6}	7.1×10^{-3}
CCL34A.4	Chemokine (C-C Motif) Ligand 34a, Duplicate 4	5.8	7038.8	10.24	6.2×10^{-6}	7.4×10^{-3}
CXCL13	C-X-C Motif Chemokine Ligand 13	10.0	7500.3	9.55	1.4×10^{-5}	1.3×10^{-2}
CCL19	C-C Motif Chemokine 19	15.0	4523.8	8.24	8.2×10^{-5}	3.0×10^{-2}
CXCL6	C-X-C Motif Chemokine Ligand 6	60.8	11,405.2	7.55	1.8×10^{-4}	5.3×10^{-2}
IL18R1	Interleukin-18 Receptor 1	9.2	1814.2	7.63	2.3×10^{-4}	6.0×10^{-2}
CXCL8	C-X-C Motif Chemokine Ligand 8 (Interleukin-8)	6.7	1307.0	7.61	2.7×10^{-4}	6.4×10^{-2}
EBI3	Interleukin-27 Subunit Beta	9.2	1506.8	7.36	3.4×10^{-4}	7.5×10^{-2}
CXCL9	C-X-C Motif Chemokine Ligand 9	12.5	1438.8	6.85	6.4×10^{-4}	1.2×10^{-1}
CCL4	C-C Motif Chemokine 4	11.7	1328.8	6.83	6.7×10^{-4}	1.2×10^{-1}
IL1R2	Interleukin-1 Receptor Type 2	6.7	458.6	6.10	2.7×10^{-3}	2.9×10^{-1}
CXCR4	C-X-C Chemokine Receptor Type 4	35.0	1016.3	4.86	8.4×10^{-3}	6.1×10^{-1}
CRLF1	Cytokine Receptor-like Factor 1	9.2	326.3	5.15	9.1×10^{-3}	6.5×10^{-1}
XCR1	Chemokine Xc Receptor 1	0.8	97.5	6.88	1.0×10^{-2}	7.0×10^{-1}
IL13RA1A	Il-13 Receptor-alpha-1-a Precursor	117.5	2426.3	4.37	1.4×10^{-2}	8.6×10^{-1}
CXCR3	C-X-C Chemokine Receptor Type 3	56.7	1140.0	4.33	1.6×10^{-2}	9.3×10^{-1}
IL12A	Interleukin-12 Subunit Alpha	3.5	128.3	5.20	1.8×10^{-2}	9.9×10^{-1}
Cathepsin						
CTSL	Cathepsin L	7.5	2370.0	8.30	9.2×10^{-5}	3.2×10^{-2}
CTSH	Cathepsin H	418.3	9396.3	4.49	1.1×10^{-2}	7.5×10^{-1}
CTSK	Cathepsin K	170.0	8802.5	5.69	2.3×10^{-3}	2.7×10^{-1}
CTSO	Cathepsin O	20.0	485.0	4.60	1.4×10^{-2}	8.6×10^{-1}
CTSS	Cathepsin S	388.3	20,089.4	5.69	2.2×10^{-3}	2.7×10^{-1}
CTSZ	Cathepsin Z	604.2	10,606.8	4.13	1.8×10^{-2}	9.8×10^{-1}
Cluster of differentiation						
CD274	Programmed Cell Death 1 Ligand 1	5.8	2255.0	8.60	6.7×10^{-5}	2.8×10^{-2}
CD4	T-cell Surface Glycoprotein CD4	3.3	256.9	6.27	3.7×10^{-3}	3.6×10^{-1}
CD209A	CD209 Antigen-like Protein A	3.3	251.6	6.24	3.9×10^{-3}	3.8×10^{-1}
CD48	CD48 Antigen	30.8	1969.6	6.00	1.7×10^{-3}	2.3×10^{-1}
CD209D	CD209 Antigen-like Protein D	68.3	3346.3	5.61	2.7×10^{-3}	3.0×10^{-1}
TSPAN6	Tetraspanin-6	9.7	387.8	5.32	7.2×10^{-3}	5.6×10^{-1}
Complement						
C4A	Complement C4-a	9.6	2062.4	7.74	2.0×10^{-4}	5.6×10^{-2}
C1QA	Complement C1q Subcomponent Subunit A	74.2	5517.8	6.22	1.1×10^{-3}	1.8×10^{-1}
C4	Complement C4	1.2	220.1	7.53	1.3×10^{-3}	1.9×10^{-1}
C1QC	Complement C1q Subcomponent Subunit C	61.7	4210.0	6.09	1.4×10^{-3}	2.0×10^{-1}
C1S	Complement C1s Subcomponent	126.4	8272.5	6.03	1.4×10^{-3}	2.0×10^{-1}
C1QB	Complement C1q Subcomponent Subunit B	116.7	7135.2	5.93	1.7×10^{-3}	2.2×10^{-1}
C7	Complement Component C7	30.0	1288.2	5.42	3.9×10^{-3}	3.8×10^{-1}
CFB	Complement Factor B	144.2	4880.0	5.08	5.3×10^{-3}	4.6×10^{-1}
Lectin						
CLEC4M	C-type Lectin Domain Family 4 Member M	10.0	5250.0	9.04	2.9×10^{-5}	1.9×10^{-2}
LGALS9	Galectin-9	89.2	8098.8	6.50	7.6×10^{-4}	1.3×10^{-1}
FUCL4_ANGJA	Fucolectin-4	32.5	1838.8	5.82	2.2×10^{-3}	2.7×10^{-1}
CLEC10A	C-type Lectin Domain Family 10 Member A	12.5	603.8	5.59	4.1×10^{-3}	3.9×10^{-1}
MBL	Mannose-binding Lectin	1.7	175.0	6.71	4.2×10^{-3}	4.0×10^{-1}
LGALS3	Galectin-3	10.8	415.0	5.26	7.1×10^{-3}	5.5×10^{-1}
LGALS3BPA	Galectin-3-binding Protein A	1083.3	27,214.6	4.65	8.9×10^{-3}	6.5×10^{-1}

Table 3. *Cont.*

Name	Description	Expression Level (FPKM)		Log FC	*p*-Value	FDR
Ubiquitination						
UBL4A	Ubiquitin-like Protein 4a	31.7	87,332.8	11.43	8.0×10^{-7}	6.0×10^{-3}
HERC5	E3 Ubiquitin-protein Ligase Herc5	3.1	4851.1	10.61	5.6×10^{-6}	7.1×10^{-3}
HERC6	E3 Ubiquitin-protein Ligase Herc6	65.0	34,045.8	9.03	2.2×10^{-5}	1.6×10^{-2}
USP18	Ubiquitin Carboxyl-terminal Hydrolase 18	14.2	4250.9	8.23	8.4×10^{-5}	3.0×10^{-2}
USP12	Ubiquitin Carboxyl-terminal Hydrolase 12	9.2	1986.2	7.76	1.9×10^{-4}	5.4×10^{-2}
UBR1	E3 Ubiquitin-protein Ligase Ubr1	18.4	2933.8	7.32	2.9×10^{-4}	6.6×10^{-2}
TRIM21	Tripartite Motif-containing Protein 21	6.7	1061.3	7.31	4.2×10^{-4}	8.7×10^{-2}
TRIM47	Tripartite Motif-containing Protein 47	9.2	1332.5	7.18	4.4×10^{-4}	8.8×10^{-2}
RNF213	E3 Ubiquitin-protein Ligase Rnf213	40.8	4928.8	6.92	4.6×10^{-4}	8.9×10^{-2}
TRIM29	Tripartite Motif-containing Protein 29	58.0	2261.1	5.29	4.3×10^{-3}	4.0×10^{-1}
TRIM39	Tripartite Motif-containing Protein 39	7.0	354.2	5.67	4.7×10^{-3}	4.2×10^{-1}
TRIM25	Tripartite Motif-containing Protein 25	2.5	182.5	6.19	5.8×10^{-3}	4.8×10^{-1}
HERC4	E3 Ubiquitin-protein Ligase Herc4	788.6	17,971.6	4.51	1.1×10^{-2}	7.3×10^{-1}
TRIM16	Tripartite Motif-containing Protein 16	17.5	476.7	4.77	1.2×10^{-2}	7.7×10^{-1}
TRIM14	Tripartite Motif-containing Protein 14	85.7	1728.9	4.33	1.5×10^{-2}	8.9×10^{-1}
Others						
RSAD2	Radical S-adenosyl Methionine Domain-containing Protein 2	30.8	41,672.6	10.40	3.4×10^{-6}	7.1×10^{-3}
IFI44	Interferon-induced Protein 44	62.8	73,980.2	10.20	4.3×10^{-6}	7.1×10^{-3}
IFIT5	Interferon-induced Protein With Tetratricopeptide Repeats 5	4.2	3119.8	9.55	2.0×10^{-5}	1.6×10^{-2}
SOCS1	Suppressor Of Cytokine Signaling 1	25.8	13,362.6	9.01	2.5×10^{-5}	1.7×10^{-2}
MX	Interferon-induced GTP-binding Protein Mx	49.2	19,664.4	8.64	3.9×10^{-5}	2.3×10^{-2}
NKL	Antimicrobial Peptide Nk-lysin	8.3	3838.8	8.85	4.0×10^{-5}	2.3×10^{-2}
FCGR1A	High Affinity Immunoglobulin Gamma Fc Receptor I	33.3	12,632.2	8.57	4.5×10^{-5}	2.5×10^{-2}
FCER1A	High Affinity Immunoglobulin Epsilon Receptor Subunit Alpha	13.4	5255.2	8.62	4.8×10^{-5}	2.5×10^{-2}
PIGR	Polymeric Immunoglobulin Receptor	5.0	2332.5	8.87	4.9×10^{-5}	2.5×10^{-2}
LYG_EPICO	Lysozyme G	352.5	120,945.0	8.42	5.0×10^{-5}	2.5×10^{-2}
SAMD9	Sterile Alpha Motif Domain-containing Protein 9	2.5	1271.2	8.99	6.5×10^{-5}	2.8×10^{-2}
IRF4	Interferon Regulatory Factor 4	10.0	2811.3	8.14	1.1×10^{-4}	3.6×10^{-2}
GVINP1	Interferon-induced Very Large GTPase 1	53.3	11,420.6	7.74	1.4×10^{-4}	4.5×10^{-2}
TMEM173	Stimulator of Interferon Genes Protein	23.3	4917.5	7.72	1.6×10^{-4}	4.8×10^{-2}
MPEG1	Macrophage-expressed Gene 1 Protein	13.3	2433.8	7.51	2.4×10^{-4}	6.0×10^{-2}
HSP30	Heat Shock Protein 30	1.7	572.5	8.42	2.4×10^{-4}	6.0×10^{-2}
GZMA	Granzyme A	9.2	1761.4	7.57	2.4×10^{-4}	6.0×10^{-2}
IRF3	Interferon Regulatory Factor 3	5.8	1155.6	7.63	2.8×10^{-4}	6.6×10^{-2}
EIF2AK2	Interferon-induced, Double-stranded RNA-activated Protein Kinase	15.6	2438.5	7.29	3.3×10^{-4}	7.3×10^{-2}
IRF1	Interferon Regulatory Factor 1	349.2	22,350.8	6.00	1.5×10^{-3}	2.0×10^{-1}
PSMB6L-B	Proteasome Subunit Beta Type-6-b Like Protein	131.7	8506.1	6.01	1.5×10^{-3}	2.0×10^{-1}
CASP3	Caspase-3	24.2	1688.8	6.13	1.5×10^{-3}	2.1×10^{-1}
PSME1	Proteasome Activator Complex Subunit 1	220.8	12,675.8	5.84	1.8×10^{-3}	2.4×10^{-1}
IFIH1	Interferon-induced Helicase C Domain-containing Protein 1	96.7	5280.5	5.77	2.1×10^{-3}	2.6×10^{-1}
PSMB8	Proteasome Subunit Beta Type-8	86.7	3997.5	5.53	3.0×10^{-3}	3.2×10^{-1}
GRN	Granulins	275.0	11,720.0	5.41	3.3×10^{-3}	3.3×10^{-1}
MR1	Major Histocompatibility Complex Class I-related Gene Protein	153.6	6162.8	5.33	3.8×10^{-3}	3.6×10^{-1}
SOCS3	Suppressor Of Cytokine Signaling 3	125.7	4607.5	5.20	4.6×10^{-3}	4.2×10^{-1}
IGSF3	Immunoglobulin Superfamily Member 3	11.7	408.0	5.13	8.3×10^{-3}	6.1×10^{-1}
IRGC	Interferon-inducible GTPase 5	11.9	346.9	4.86	1.2×10^{-2}	7.7×10^{-1}

Note, FPKM: fragments per kilobase of transcript per million mapped reads; Log FC: Log value of fold changes, FDR: false discovery rate.

3.5. GO Enrichment of Differentially Expressed Genes

GO is a widely used method to classify gene functions and their products in organisms. Therefore, the identified DEGs were subsequently used for GO enrichment analysis. According to GO terms, 2094 (61.3%) of the total of 3418 DEGs were classified into the three categories of molecular function, biological process, and cellular component. "Binding" (1258 genes, 46.3%) was the major subcategory in the molecular function. The largest subcategory found in the biological process category was "cellular process" (1488 genes, 12.3%) while "Cell" (1687 genes, 19.6%) and "cell part" (1687 genes, 19.6%) were the most abundant GO terms in the cellular component category (Figure 2). Because one gene could be categorized into several subcategories, the sum of genes in the subcategories could exceed 100%. GO analysis of the transcriptome revealed nine molecular function subcategories, 62 biological process subcategories, and 12 cellular component subcategories with p value of less than 1×10^{-5}) (Table S2).

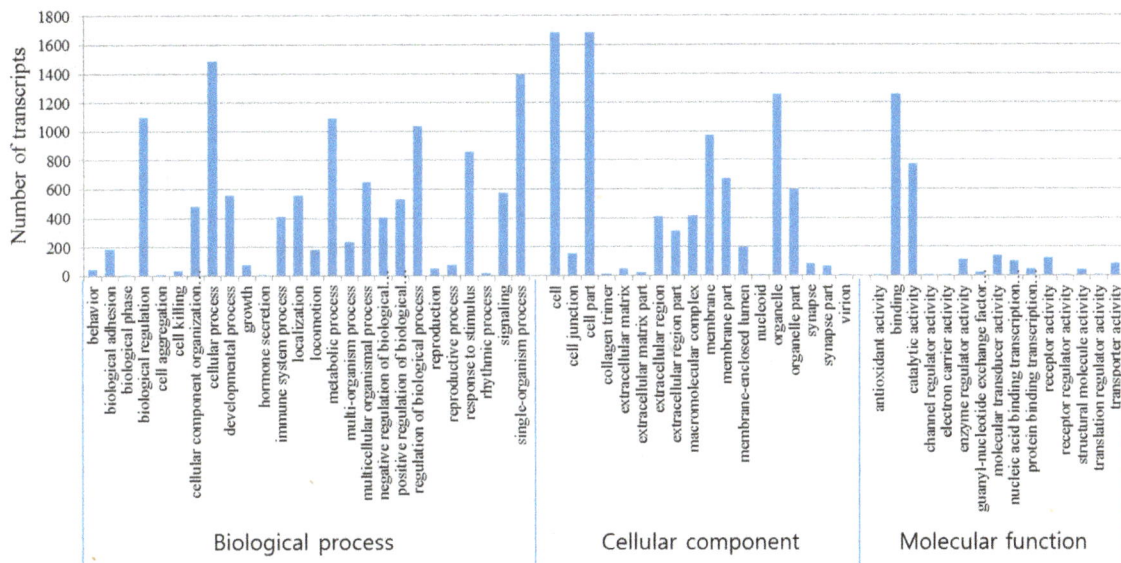

Figure 2. GO annotation of Differentially Expressed Genes (DEGs).

4. Discussion

NNV infection has caused high mortalities of sevenband groupers in aqua-farms during the summer season, especially at larval and juvenile stages. It has also caused tremendous economic losses [1]. Due to the greater damage to the sevenband grouper industry, investigation on the molecular response of NNV infection is required to understand the outbreak mechanism of disease and develop prevention methods such as vaccines. In this study, we performed a transcriptome analysis of the brain tissue of sevenband grouper infected with NNV compared to mock brain tissue using a RNA-Seq.

The total number of unigenes and the average length of the unigenes were 104,348 and 845 bp, respectively. The number and average length found in this study indicated a fairly good performance compared to other previous NGS transcriptome studies on crimson spotted rainbowfish (107,749 transcripts, 961 bp) [30] common carp (130,292 transcripts, 1401 bp) [18], blunt snout bream (253,439 transcripts, 998 bp) [31], orange-spotted grouper (116,678 transcripts, 685 bp) [32] and Asian seabass (89,026 transcripts, 1175 bp) [33].

Gene annotation by BlastX provides valuable information about the transcripts. In this study, 43,289 unigenes (41.5%) of 104,348 unigenes were annotated. This is similar to the result of orange-spooted grouper (45.8%) [32]. Liu et al. have addressed the possible reasons of such poor annotation: (1) novel genes; (2) sequencing errors; and (3) artefacts by assembly algorithm [33]. Therefore, more genetic studies are needed to understand the biological functions.

The importance of innate defense mechanisms against viral infection has been extensively reviewed [34–36]. In this study, we identified innate immune response relevant genes of sevenband grouper involved in NNV infection. Chemokines are critical components of the immune system. The roles of chemokines and their receptors in viral interactions have been reported in various studies [37]. The chemokines family comprises four subfamilies based on N-terminal cystein-motifs: C, C-C, C-X-C, and C-X3-C subfamilies [38]. In this study, we also detected significant up-regulation of *CCL2*, *CCL34*, *CCL19*, *CCL4*, *CXCL13*, *CXCL6*, *CXCL8*, and *CXCL9* in sevenband grouper brain tissue after NNV infection. Especially, *CCL2* was highly over expressed at about 10.66 Log FC. CCL2 is a pro-inflammatory chemokine that is induced during several human acute and chronic viral infections, including human immunodeficiency virus (HIV) [39], hepatitis C virus [40], Epstein–Barr virus [41], respiratory synctitial virus [42], Severe Acute Respiratory Syndrome (SARS) [43], herpes simplex virus-1 [44], and Japanese encephalitis virus [45].

Cathepsins are lysosomal cysteines that play important roles in normal metabolism for the maintenance of cellular homeostasis. Cathepsins are one of the superfamilies involved in the regulation of antigen presentation and degradation as well as immune responses, including apoptosis, inflammation, and regulation of hormone processing [46–48]. In addition, Chandran et al. have shown that Cathepsin B and Cathepsin L are involved in Ebola virus infection [49]. They are involved in the entry of reovirus [50]. Recently, Cathepsin L has also been shown to be involved in the entry mediated by the SARS coronavirus spike glycoprotein [51] as well as in the process of fusion glycoprotein of Hendra virus [52]. In this study, Cathepsin L and Cathepsin S were found to be notably expressed after NNV infection. Their functional roles in the interaction between grouper and NNV merit further studies.

Lectins are carbohydrate-binding proteins that are highly specific for sugar moieties. They mediate the attachment and binding of viruses to their targets [53]. Lectins are also known to play important roles in the immune system. Within the innate immune system, lectins can help mediate the first-line of defense against invading microorganisms. In this study, several kinds of lectins were found to be highly induced in sevenband grouper brain tissue by NNV infection, such as C-type lectins (CTLs), galectins, fucolectin, and mannose-binding lectin. CTLs are the most well studied lectins. They can promote antibacterial and antiviral immune defense [54]. Many CTLs have been identified in teleost but the exact function of CTLs remains unclear.

Hundreds of interferon stimulated genes (ISGs) were transcribed in sevenband grouper brain tissue during NNV infection. Interferon induced protein 44 (IFI44) was expressed the most. IFI44 is an interferon-alpha inducible protein associated with infection of several viruses. Power et al. have demonstrated that IFI44 can inhibit HIV-1 replication in vitro by suppressing HIV-1 LTR promoter activity [55]. Carlton-Smith and Elliott have screened ISGs related to *Bunyamwera orthbunyavirus* replication using nonstructural (NSs) protein knock out virus. One of these ISGs that have inhibitory activity is found to be *IFI44* [56]. Whether protein B2 of NNV has roles in virus replication and its relationship with ISGs such as IFI44 merit further study.

HSPs are one of the most phylogenetically conserved classes of proteins with critical roles in maintaining cellular homeostasis and protecting cells from various stresses [57]. Ironically, HSP70 has roles to suppress some virus infections, and support their replication in other viruses [58]. In this study, *HSP30* was the highly induced gene instead of *HSP70*. *HSP30* has also been reported to be the most induced gene in the NNV infected Asian seabass epithelial cell [33]. However, the function of *HSP30* in NNV infection remains unclear.

Krasnov et al. previously reported the effects of NNV on gene expression in Atlantic cod brain using a microarray [59]. Compared to our study, a number of genes show a similar up-regulation result in the study, such as Caspase, Cathepsins, IRF, Radical *S*-adenosyl methionine domain-containing protein, tripartite motif-containing protein (TRIM) and so on. However, a lot of novel genes (e.g., NK-Lysin, Ubiquitin-like protein 4, Granzyme A, etc.) were identified from our RNA-Seq result because a microarray can only evaluate the genes on a chip.

Our findings are preliminary based on the small scale of the study and the results have not yet been confirmed by an independent technique such as quantitative polymerase chain reaction (qPCR). In future studies, it will be necessary to confirm the expression level of genes and to characterize the function of genes that are highly involved in NNV infection.

5. Conclusions

In conclusion, to the best of our knowledge, this is the first study reporting the transcriptome of brain tissue of NNV-infected sevenband grouper. In this study, we obtained the transcriptome of sevenband grouper. A total of 104,348 transcripts were obtained, including 628 DEGs between NNV infected and non-infected sevenband grouper. A large number of differential expressed genes relevant to immune response were identified as well as several candidate genes (*CCL2, Cathepsins, Lectins, Hsp30,* and *Interferon-induced protein 44*) that were intensely induced by NNV. Their functions in sevenband grouper against NNV infection merit further study. The acquired data from such transcriptome analysis provide valuable information for future functional genes related to NNV infection and VNN outbreak.

Acknowledgments: This research was supported by Basic Science Research Program through the National Research Foundation of Korea (NRF) funded by the Ministry of Science, ICT & Future Planning (NRF-2015R1C1A1A01054829).

Author Contributions: Jong-Oh Kim and Myung-Joo Oh conceived and designed the experiments; Jong-Oh Kim and Jae-Ok Kim performed the experiments; Jong-Oh Kim and Wi-Sik Kim analyzed the data; Jong-Oh Kim and Myung-Joo Oh wrote the paper.

References

1. Kim, C.S.; Kim, W.S.; Nishizawa, T.; Oh, M.J. Prevalence of viral nervous necrosis (VNN) in sevenband grouper, *Epinephelus septemfasciatus* farms. *J. Fish Pathol.* **2012**, *25*, 111–116. [CrossRef]
2. Munday, B.L.; Kwang, J.; Moody, N. Betanodavirus infections of teleost fish: A review. *J. Fish Dis.* **2002**, *25*, 127–142. [CrossRef]
3. Munday, B.L.; Nakai, T. Special topic review: Nodaviruses as pathogens in larval and juvenile marine finfish. *World J. Microbiol. Biotechnol.* **1997**, *13*, 375–381. [CrossRef]
4. Muroga, K. Viral and bacterial diseases of marine fish and shellfish in Japanese hatcheries. *Aquaculture* **2001**, *202*, 23–44. [CrossRef]
5. Cha, S.J.; Do, J.W.; Lee, N.S.; An, E.J.; Kim, Y.C.; Kim, J.W.; Park, J.W. Phylogenetic analysis of betanodaviruses isolated from cultured fish in Korea. *Dis. Aquat. Org.* **2007**, *77*, 181–189. [CrossRef] [PubMed]
6. Oh, M.J.; Jung, S.J.; Kitamura, S.I. Comparison of the coat protein gene of nervous necrosis virus (NNV) detected from marine fishes in Korea. *J. World Auqac. Soc.* **2005**, *36*, 223–227. [CrossRef]
7. Sohn, S.G.; Park, M.A.; Oh, M.J.; Chun, S.K. A fish nodavirus isolated from cultured sevenband grouper, *Epinephelus septemfasciatus. J. Fish Pathol.* **1998**, *11*, 97–104.
8. Huang, Y.; Huang, X.; Yan, Y.; Cai, J.; Ouyang, Z.; Cui, H.; Wang, P.; Qin, Q. Transcriptome analysis of orange-spotted grouper (*Epinephelus coioides*) spleen in response to Singapore grouper iridovirus. *BMC Genom.* **2011**, *12*, 556. [CrossRef] [PubMed]
9. Nishizawa, T.; Mori, K.; Furuhashi, M.; Nakai, T.; Furusawa, I.; Muroga, K. Comparison of the coat protein genes of five fish nodaviruses, the causative agents of viral nervous necrosis in marine fish. *J. Gen. Virol.* **1995**, *76*, 1563–1569. [CrossRef] [PubMed]
10. Johansen, R.; Sommerset, I.; Torud, B.; Korsnes, K.; Hjortaas, M.J.; Nilsen, F.; Nerland, A.H.; Dannevig, B.H. Characterization of nodavirus and viral encephalophathy and retinopathy in farmed turbot, *Scophthalmus maximus* (L.). *J. Fish Dis.* **2004**, *27*, 591–601. [CrossRef] [PubMed]
11. Jones, S.I.; Vodkin, L.O. Using RNA-Seq to profile soybean seed development from fertilization to maturity. *PLoS ONE* **2013**, *8*, e59270. [CrossRef] [PubMed]
12. Tan, M.H.; Au, K.F.; Yablonovitch, A.L.; Wills, A.E.; Chuang, J.; Baker, J.C.; Wong, W.H.; Li, J.B. RNA sequencing reveals a diverse and dynamic repertoire of the *Xenopus tropicalis* transcriptome over development. *Genome Res.* **2013**, *23*, 201–216. [CrossRef] [PubMed]
13. Zhang, H.; Fei, C.; Wu, H.; Yang, M.; Liu, Q.; Wang, Q.; Zhang, Y. Transcriptome profiling reveals Th17-like immune responses induced in zebrafish bath-vaccinated with a live attenuated *Vibrio anguillarum. PLoS ONE* **2013**, *8*, e73871. [CrossRef] [PubMed]
14. Iwamoto, T.; Mise, K.; Takeda, A.; Okinaka, Y.; Mori, K.I.; Arimoto, M.; Okuno, T.; Nakai, T. Characterization of Striped jack nervous necrosis virus subgenomic RNA3 and biological activities of its encoded protein B2. *J. Gen. Virol.* **2005**, *86*, 2807–2816. [CrossRef] [PubMed]
15. Mu, Y.; Li, M.; Ding, F.; Ding, Y.; Ao, J.; Hu, S.; Chen, X. De novo characterization of the spleen transcriptome of the large yellow croaker (*Pseudosciaena crocea*) and analysis of the immune relevant genes and pathways involved in the antiviral response. *PLoS ONE* **2014**, *9*, e97471. [CrossRef] [PubMed]

16. Pereiro, P.; Balseiro, P.; Romero, A.; Dios, S.; Forn-Cuni, G.; Fuste, B.; Planas, J.V.; Beltran, S.; Nova, B.; Figueras, A. High- throughput sequence analysis of turbot (*Scophthalmus maximus*) transcriptome using 454-pyrosequencing for the discovery of antiviral immune genes. *PLoS ONE* **2012**, *7*, e35369. [CrossRef] [PubMed]

17. Xiang, L.X.; He, D.; Dong, W.R.; Zhang, Y.W.; Shao, J.Z. Deep sequencing- based transcriptome profiling analysis of bacteria-challenged *Lateolabrax japonicus* reveals insight into the immune-relevant genes in marine fish. *BMC Genom.* **2010**, *11*, 472. [CrossRef] [PubMed]

18. Li, G.; Zhao, Y.; Liu, Z.; Gao, C.; Yan, F.; Liu, B.; Feng, J. De novo assembly and characterization of the spleen transcriptome of common carp (*Cyprinus carpio*) using Illumina paired-end sequencing. *Fish Shellfish Immunol.* **2015**, *44*, 420–429. [CrossRef] [PubMed]

19. Kim, J.O.; Kim, J.O.; Kim, W.S.; Oh, M.J. Development and application of quantitative detection method for nervous necrosis virus (NNV) isolated from sevenband grouper *Hyporthodus septemfasciatus*. *Asian Pac. J. Trop. Med.* **2016**, *9*, 742–748. [CrossRef] [PubMed]

20. Reed, L.J.; Muench, H. A simple method of estimating fifty percent end points. *Am. J. Hyg.* **1938**, *27*, 493–497.

21. Andrews, S. FastQC a Quality Control Tool for High Throughput Sequence Data [Online]. 2010. Available online: http://www.bioinformatics.babraham.ac.uk/projects/fastqc/ (accessed on 3 March 2016).

22. Martin, J.A.; Wang, Z. Next-generation transcriptome assembly. *Nat. Rev. Genet.* **2011**, *12*, 671–682. [CrossRef] [PubMed]

23. Hass, B.J.; Papanicolaou, A.; Yassour, M.; Grabherr, M.; Blood, P.D.; Bowden, J.; Couger, M.B.; Eccles, D.; Li, B.; Lieber, M.; et al. De novo transcript sequence reconstruction from RNA-seq using the Trinity platform for reference generation and analysis. *Nat. Protoc.* **2013**, *8*, 1494–1512. [CrossRef] [PubMed]

24. Fu, L.; Niu, B.; Zhu, Z.; Wu, S.; Li, W. CD-HIT: Accelerated for clustering the next-generation sequencing data. *Bioinformatics* **2012**, *28*, 3150–3152. [CrossRef] [PubMed]

25. Li, B.; Dewey, C.N. RSEM: Accurate transcript quantification from RNA-Seq data with or without a reference genome. *BMC Bioinform.* **2011**, *4*, 323. [CrossRef] [PubMed]

26. Sun, J.; Nishiyama, T.; Shimizu, K.; Kadota, K. TCC: An R package for comparing tag count data with robust normalization strategies. *BMC Bioinform.* **2013**, *14*, 219. [CrossRef] [PubMed]

27. Kadota, K.; Nishiyama, T.; Shimizu, K. A normalization strategy for comparing tag count data. *Algorithms Mol. Biol.* **2012**, *7*, 5. [CrossRef] [PubMed]

28. Gerard, C.; Rollins, B.J. Chemokines and disease. *Nat. Immunol.* **2001**, *2*, 108–115. [CrossRef] [PubMed]

29. Choi, K.M.; Shim, S.H.; An, C.M.; Nam, B.H.; Kim, Y.O.; Kim, J.W.; Park, C.I. Cloning, characterisation, and expression analysis of the cathepsin D gene from rock bream (*Oplegnathus fasciatus*). *Fish Shellfish Immunol.* **2014**, *40*, 253–258. [CrossRef] [PubMed]

30. Smith, S.; Bernatchez, L.; Beheregaray, L.B. RNA-seq analysis reveals extensive transcriptional plasticity to temperature stress in a freshwater fish species. *BMC Genom.* **2013**, *14*, 375. [CrossRef] [PubMed]

31. Tran, N.T.; Gao, Z.X.; Zhao, H.H.; Yi, S.K.; Chen, B.X.; Zhao, Y.H.; Lin, L.; Liu, X.Q.; Wang, W.M. Transcriptome analysis and microsatellite discovery in the blunt snout bream (*Megalobrama amblycephala*) after challenge with *Aeromonas hydrophila*. *Fish Shellfish Immunol.* **2015**, *45*, 72–82. [CrossRef] [PubMed]

32. Wang, Y.D.; Huang, S.J.; Chou, H.N.; Liao, W.L.; Gong, H.Y.; Chen, J.Y. Transcriptome analysis of the effect of *Vibrio alginolyticus* infection on the innate immunity-related complement pathway in *Epinephelus coioides*. *BMC Genom.* **2014**, *15*, 1102. [CrossRef] [PubMed]

33. Liu, P.; Wang, L.; Kwang, J.; Yue, G.H.; Wong, S.M. Transcriptome analysis of genes responding to NNV infection in Asian seabass epithelial cells. *Fish Shellfish Immunol.* **2016**, *54*, 342–352. [CrossRef] [PubMed]

34. Koyama, S.; Ishii, K.J.; Coban, C.; Akira, S. Innate immune response to viral infection. *Cytokine* **2008**, *43*, 336–341. [CrossRef] [PubMed]

35. Takeuchi, O.; Akira, S. Innate immunity to virus infection. *Immunol. Rev.* **2009**, *227*, 75–86. [CrossRef] [PubMed]

36. Thompson, M.R.; Kaminski, J.J.; Kurt-Jones, E.A.; Fitzgerald, K.A. Pattern Recognition Receptors and the Innate Immune Response to Viral Infection. *Viruses* **2011**, *3*, 920–940. [CrossRef] [PubMed]

37. Abdul, W.A.; Heiken, H.; Meyer-Olson, D.; Schmidt, R.E. CCL2: A potential prognostic marker and target of anti-inflammatory strategy in HIV/AIDS pathogenesis. *Eur. J. Immunol.* **2011**, *41*, 3412–3418.

38. Rollins, B.J. Chemokines. *Blood* **1997**, *90*, 909–928. [PubMed]

39. Weiss, J.M.; Nath, A.; Major, E.O.; Berman, J.W. HIV-1 Tat induces monocyte chemoattractant protein-1-mediated monocyte transmigration across a model of the human blood-brain barrier and up-regulates CCR5 expression on human monocytes. *J. Immunol.* **1999**, *163*, 2953–2959. [PubMed]

40. Fisher, N.C.; Neil, D.A.; Williams, A.; Adams, D.H. Serum concentrations and peripheral secretion of the beta chemokines monocyte chemoattractant protein 1 and macrophage inflammatory protein 1α in alcoholic liver disease. *Gut* **1999**, *45*, 416–420. [CrossRef] [PubMed]

41. Gaudreault, E.; Fiola, S.; Olivier, M.; Gosselin, J. Epstein-Barr virus induces MCP-1 secretion by human monocytes via TLR2. *J. Virol.* **2007**, *81*, 8016–8024. [CrossRef] [PubMed]

42. Culley, F.J.; Pennycook, A.M.; Tregoning, J.S.; Hussell, T.; Openshaw, P.J. Differential chemokine expression following respiratory virus infection reflects Th1- or Th2-biased immunopathology. *J. Virol.* **2006**, *80*, 4521–4527. [CrossRef] [PubMed]

43. Yen, Y.T.; Liao, F.; Hsiao, C.H.; Kao, C.L.; Chen, Y.C.; Wu-Hsieh, B.A. Modeling the early events of severe acute respiratory syndrome coronavirus infection in vitro. *J. Virol.* **2006**, *80*, 2684–2693. [CrossRef] [PubMed]

44. Kurt-Jones, E.A.; Chan, M.; Zhou, S.; Wang, J.; Reed, G.; Bronson, R.; Arnold, M.; Knipe, D.M.; Finberg, R.W. Herpes simplex virus 1 interaction with Toll-like receptor 2 contributes to lethal encephalitis. *Proc. Natl. Acad. Sci. USA* **2004**, *101*, 1315–1320. [CrossRef] [PubMed]

45. Mishra, M.K.; Kumawat, K.L.; Basu, A. Japanese encephalitis virus differentially modulates the induction of multiple pro-inflammatory mediators in human astrocytoma and astroglioma cell-lines. *Cell Biol. Int.* **2008**, *32*, 1506–1513. [CrossRef] [PubMed]

46. Dixit, A.K.; Dixit, P.; Sharma, R.L. Sharma Immunodiagnostic/protective role of Cathepsin L cysteine proteinases secreted by *Fasciolaspecies*. *Vet. Parasitol.* **2008**, *154*, 177–184. [CrossRef] [PubMed]

47. Hsing, L.C.; Rudensky, A.Y. The lysosomal cysteine proteases in MHC class II antigen presentation. *Immunol. Rev.* **2005**, *207*, 229–241. [CrossRef] [PubMed]

48. Yasothornsrikul, S.; Greenbaum, D.; Medzihradszky, K.F.; Toneff, T.; Bundey, R.; Miller, R.; Schilling, B.; Petermann, I.; Dehnert, J.; Logvinova, A.; et al. Cathepsin L in secretory vesicles functions as a prohormone-processing enzyme for production of the enkephalin peptide neurotransmitter. *Proc. Natl. Acad. Sci. USA* **2003**, *100*, 9590–9595. [CrossRef] [PubMed]

49. Chandran, K.; Sullivan, N.J.; Felbor, U.; Whelan, S.P.; Cunningham, J.M. Endosomal proteolysis of the Ebola virus glycoprotein is necessary for infection. *Science* **2005**, *308*, 1643–1645. [CrossRef] [PubMed]

50. Ebert, D.H.; Deussing, J.; Peters, C.; Dermody, T.S. Cathepsin L and cathepsin B mediate reovirus disassembly in murine fibroblast cells. *J. Biol. Chem.* **2002**, *277*, 24609–24617. [CrossRef] [PubMed]

51. Simmons, G.; Gosalia, D.N.; Rennekamp, A.J.; Reeves, J.D.; Diamond, S.L.; Bates, P. Inhibitors of cathepsin L prevent severe acute respiratory syndrome coronavirus entry. *Proc. Natl. Acad. Sci. USA* **2005**, *102*, 11876–11881. [CrossRef] [PubMed]

52. Pager, C.T.; Dutch, R.E. Cathepsin L is involved in proteolytic processing of the Hendra virus fusion protein. *J. Virol.* **2005**, *79*, 12714–12720. [CrossRef] [PubMed]

53. Bartenschlager, R.; Sparacio, S. Hepatitis C virus molecular clones and their replication capacity in vivo and in cell culture. *Virus Res.* **2007**, *127*, 195–207. [CrossRef] [PubMed]

54. Zhou, Z.J.; Sun, L. CsCTL1, a teleost C-type lectin that promotes antibacterial and antiviral immune defense in a manner that depends on the conserved EPN motif. *Dev. Comp. Immunol.* **2015**, *50*, 69–77. [CrossRef] [PubMed]

55. Power, D.; Santoso, N.; Dieringer, M.; Yu, J.; Huang, H.; Simpson, S.; Seth, I.; Miao, H.; Zhu, J. IFI44 suppresses HIV-1 LTR promoter activity and facilitates its latency. *Virology* **2015**, *481*, 142–150. [CrossRef] [PubMed]

56. Carlton-Smith, C.; Elliott, R.M. Viperin, MTAP44, and protein kinase R contribute to the interferon-induced inhibition of *Bunyamwera Orthobunyavirus* replication. *J. Virol.* **2012**, *86*, 11548–11557. [CrossRef] [PubMed]

57. Hartl, F.U. Molecular chaperones in cellular protein folding. *Nature* **1996**, *381*, 571–579. [CrossRef] [PubMed]

58.   Kim, M.Y.; Oglesbee, M. Virus-heat shock protein interaction and a novel axis for innate antiviral immunity. *Cells* **2012**, *1*, 646–666. [CrossRef] [PubMed]

59.   Krasnov, A.; Kileng, O.; Skugor, S.; Jorgensena, S.M.; Afanasyev, S.; Timmerhaus, G.; Sommer, A.I.; Jensen, I. Genomic analysis of the host response to nervous necrosis virus in Atlantic cod (*Gadus morhua*) brain. *Mol. Immunol.* **2013**, *54*, 443–452. [CrossRef] [PubMed]

5

Potential Inhibitory Influence of miRNA 210 on Regulatory T Cells during Epicutaneous Chemical Sensitization

Carrie Mae Long [1,2], Ewa Lukomska [2], Nikki B. Marshall [2,†], Ajay Nayak [2] and Stacey E. Anderson [2,*]

[1] Immunology and Microbial Pathogenesis Graduate Program, West Virginia University, Morgantown, WV 26505, USA; clong14@mix.wvu.edu

[2] Centers for Disease Control and Prevention, National Institute for Occupational Safety and Health, Allergy and Clinical Immunology Branch, Morgantown, WV 26505, USA; uvm3@cdc.gov (E.L.); nikki.marshall@inovio.com (N.B.M.); ajay.nayak@jefferson.edu (A.N.)

* Correspondence: dbx7@cdc.gov

† Current address: Inovio Pharmaceuticals, Inc., 660 West Germantown Pike, Plymouth Meeting, PA 19462, USA.

Academic Editor: Paolo Cinelli

Abstract: Toluene diisocyanate (TDI) is a potent low molecular weight chemical sensitizer and a leading cause of chemical-induced occupational asthma. The regulatory potential of microRNAs (miRNAs) has been recognized in a variety of disease states, including allergic disease; however, the roles of miRNAs in chemical sensitization are largely unknown. In a previous work, increased expression of multiple miRNAs during TDI sensitization was observed and several putative mRNA targets identified for these miRNAs were directly related to regulatory T-cell (T_{reg}) differentiation and function including Foxp3 and Runx3. In this work, we show that miR-210 expression is increased in the mouse draining lymph node (dLN) and T_{reg} subsets following dermal TDI sensitization. Alterations in dLN mRNA and protein expression of T_{reg} related genes/putative miR-210 targets (foxp3, runx3, ctla4, and cd25) were observed at multiple time points following TDI exposure and in ex vivo systems. A T_{reg} suppression assay, including a miR-210 mimic, was utilized to investigate the suppressive ability of T_{regs}. Cells derived from TDI sensitized mice treated with miR-210 mimic had less expression of miR-210 compared to the acetone control suggesting other factors, such as additional miRNAs, might be involved in the regulation of the functional capabilities of these cells. These novel findings indicate that miR-210 may have an inhibitory role in T_{reg} function during TDI sensitization. Because the functional roles of miRNAs have not been previously elucidated in a model of chemical sensitization, these data contribute to the understanding of the potential immunologic mechanisms of chemical induced allergic disease.

Keywords: microRNA; regulatory T cell; immunotoxicology; toluene diisocyanate; isocyanate; miR-210

1. Introduction

Occupational allergic disease is a significant health burden. A variety of diseases can be caused by workplace chemical exposures, including asthma, conjunctivitis, dermatitis, rhinitis, and urticaria [1]. Diisocyanates are a group of highly reactive chemicals characterized by the presence of double isocyanate functional groups; many of these chemicals are potent sensitizers and major causative agents of occupational allergic disease [2–4]. Toluene diisocyanate (TDI) is frequently utilized in the automobile industry and in the manufacture of polyurethane foams, paints, and coatings [5–8]. TDI is widely used throughout the U.S. and around the globe; the U.S. Environmental Protection Agency reports that U.S. production and importation of 2–4 and 2–6 TDI isomers rose above one billion pounds in 2006 [6]. While TDI has generally been classified as a T-helper type 2 (Th2) sensitizer, the immune response following exposure is more accurately characterized as a mixed Th1/Th2 response in both rodents and humans [1,2,9]. The principal routes of human exposure to TDI are inhalation and dermal contact [6] and sensitization leading to allergic disease has been documented for both routes [7,10]. Once the induction threshold has been reached, skin sensitization is thought to be sufficient for subsequent allergic disease of the respiratory tract [7], illustrating the systemic nature of TDI sensitization. Although TDI-induced allergic disease is an extremely relevant occupational health concern the pathogenic mechanisms of diisocyanate-induced allergic disease are not fully understood [5]. The identification of novel mediators of allergic disease may be necessary to obtain a full and complete understanding of the disease. Because of the occupational significance of and lack of validated identification strategies for chemical respiratory sensitizers like TDI [4,11], it is necessary to investigate and identify functional pathways and mechanisms that are involved in TDI sensitization. Specific understanding of disease mechanism may have direct implications in risk assessment, hazard communication and guidance used in the selection of safe products, work place interventions and training programs to warn workers about potential risks, the identification of safer alternatives, and the selection of proper personal protective equipment (PPE).

An emerging class of epigenetic regulatory elements that have been the subject of recent scientific focus are microRNAs (miRNAs). These molecules are single stranded, noncoding RNA molecules that are approximately 19–23 nucleotides long [12]. miRNAs exhibit functional significance though posttranscriptional gene regulation due to their ability to bind to the target mRNA and destabilize and inhibit protein translation when in complex with Argonaute proteins and the RNA-induced silencing complex (RISC). The seed sequence of a mature miRNA interacts with the 3′ untranslated region of target mRNA and leads to the translational repression or degradation of the target mRNA, influencing gene expression. Recently, it has been shown that miRNAs play a major role in a variety of immune responses [13–16]. These regulatory factors have not been functionally investigated in the context of chemical sensitization, with the exception of recent work by our group that profiled the general expression kinetics of miRNAs involved in TDI sensitization [1] and research profiling miRNA expression in both human and murine diphenylcyclopropenone-induced dermatitis [17].

The identification of increased expression of several miRNAs in the draining lymph nodes (dLN) of mice during epicutaneous TDI sensitization prompted further investigation into the potential functional roles of these molecules in the allergic response. Target analysis for upregulated miRNAs (miR-210, -31, and -155) revealed several putative and confirmed regulatory T cell (T_{reg})-related targets including *foxp3* and *runx3* in murine and human genomes [1]. These transcription factors and the signaling molecules CD25 and CTLA4 are integral to T_{reg} differentiation and function. The expression of these molecules allows T_{reg} to differentiate in response to allergens and exert immunoregulatory functions, dampening inappropriate inflammatory and adaptive immune responses. In addition, miR-31 and -155 have been implicated as regulators of T_{reg} in a variety of contexts [18,19]. A role for T_{regs} has been suggested in models of chemical-induced contact hypersensitivity [20,21] and, in a recent manuscript, Long et al. demonstrated the increased expression and functional capability of T_{regs} during TDI sensitization [21]. While the collection of data regarding roles for T_{regs} and miRNAs in chemical allergy is growing, it is still limited. Recently published data suggests an important role for T_{regs} in dermal TDI sensitization, yet the interaction between these cells and selected miRNAs has not been investigated. While miR-210 is well characterized in the hypoxia response, its specific role in allergic disease has not yet been defined. In the present study, we utilized a murine model of epicutaneous TDI sensitization in order to elucidate the expression kinetics and role of miR-210 and its putative mRNA targets in a murine model of epicutaneous TDI sensitization, specifically in relation to the T_{reg} subset.

2. Materials and Methods

2.1. Mice

Female BALB/c mice (6–8 weeks of age) were obtained from Taconic (Germantown, NY, USA), acclimated for 5 days, and then randomly assigned to treatment group; homogenous weight distribution was ensured across treatment groups. Mice were housed in ventilated plastic shoebox cages with hardwood chip bedding at a maximum of five animals per cage. A NIH-31 modified 6% irradiated rodent diet (Harlan Teklad, Frederick, MD, USA) and tap water were administered ad libitum. Housing facilities were maintained at 68–72°F and 36%–57% relative humidity, and a 12 h light–dark cycle was maintained. All animal experiments were performed in the Association for Assessment and Accreditation of Laboratory Animal Care (AAALAC) accredited National Institute for Occupational Safety and Health (NIOSH) animal facility in accordance with an Institutional Animal Care and Use Committee-approved protocol (protocol number 15-SA-M-004, date of approval 1 August 2015).

2.2. TDI Sensitization Model

Toluene 2,4-diisocyanate (TDI, CAS# 584-84-9) was obtained from Sigma-Aldrich (Milwaukee, WI, USA). Animals were exposed to a single dose of 0%, 0.5%, and 4% TDI (v/v) on the dorsal surface of each ear (25 μL per ear). The chosen TDI concentrations (0.5% and 4% v/v) and dosing regimen was previously shown to induce sensitization [1,21] and 4% TDI (1000 μg TDI/cm^2) was previously reported as the minimum single dose concentration of TDI that could induce maximum sensitization in the absence of systemic toxicity [1]. Acetone was selected as the vehicle control and has been historically utilized in our laboratory to evaluate chemical sensitization [21–23]. It is important to note that due to hydrolysis, slight variations in the concentration of TDI dosing solutions may have occurred between preparation and application. However, animals were exposed within 30 min of TDI preparation and no visualization of hydrolysis was observed subsequent to exposures.

2.3. Euthanasia, Tissue Collection, and Processing

Animals were weighed, euthanized via CO_2 asphyxiation at time points ranging from 1 to 11 days post chemical exposure, and examined for gross pathology. Left and right auricular draining lymph nodes (dLNs; drain the site of chemical application) were collected in 4 mL sterile phosphate-buffered saline (PBS, pH 7.4) and manually dissociated using the frosted ends of two microscope slides. Cells were counted using a Cellometer (Nexcelom Bioscience, Lawrence, MA, USA) and size exclusion parameters (3.5 to 36 μm) with a combined acridine orange/propidium iodide solution to identify viable cells. For isolation of specific cellular subsets Stemcell magnetic isolation kits (Vancouver, BC, Canada) were utilized, including the T_{reg} isolation kit (CD4 negative and CD25 positive selection).

2.4. Ex Vivo miRNA Transfection Assays

Following dLN processing and counting, miR-31, -155, and -210 mimics (25 pmol in Lipofectamine transfection reagent, ThermoFisher, Waltham, MA, USA) were reverse transfected in flat-bottom 24-well plates for 24 h. Then, 1.5×10^6 naïve or acetone-treated dLN cells/mL (in RPMI-1640 media) were added to plates followed by general T cell stimulation (2 μg/mL α-CD3, 1 μg/mL α-CD28 and incubated in fresh media at 37 °C and 5% CO_2. Following a 72 h incubation, cells were washed with RPMI-1640 and RNA was isolated as described in the following section. RNA was assayed for intracellular miRNA expression to conform the intracellular uptake of miRNA mimics as well as T_{reg}-related gene expression via RT-PCR.

2.5. RNA Isolation, Reverse Transcription, and RT-PCR

Total RNA was isolated from the dLN using the miRNeasy kit (Qiagen, Hilden, Germany) according to the manufacturer's directions. A QiaCube (Qiagen) automated RNA isolation machine was utilized in conjunction with the specified RNA isolation kit. The concentration and purity of the RNA was determined using a ND-1000 spectrophotometer (Thermo Scientific Nanodrop, Wilmington, DE, USA). For gene and primary miRNA expression analysis, first strand cDNA synthesis was performed using a High-Capacity cDNA Synthesis Kit (Applied Biosystems, Carlsbad, CA, USA) according to the manufacturer's recommendations. For mature miRNA, reverse transcription TaqMan MicroRNA Assays (looped-primer RT-PCR; Applied Biosystems) were utilized according to manufacturer's recommendations (both multiplex and singleplex protocols were utilized).

For analysis of mRNA and primary miRNA expression, TaqMan Universal Fast master mix (Life Technologies, Carlsbad, CA, USA), cDNA, and mouse-specific mRNA primers (TaqMan Custom PCR Arrays, Carlsbad, CA, USA) were combined and PCR was performed according to the manufacturer's protocol (TaqMan Gene Expression Analysis). For analysis of miRNA expression, TaqMan Universal $2\times$ master mix, No AmpErase UNG (Life Technologies), cDNA, and mouse-specific miRNA primers (TaqMan Custom PCR Arrays) were combined and PCR was performed according to manufacturer protocol (TaqMan miRNA Assays both Single- and Multi-Plex). Primers used include: *β-actin, cd25 (il2rα), ctla4, foxp3, mature miR-210, -31, -155, primary miR-210, runx3,* and *sno234*. MicroAmp Fast Optical 96-well reaction plates were analyzed on an Applied Biosystems 7500 Fast Real Time PCR system using cycling conditions as specified by the manufacturer. *β-actin* (mRNA and primary miRNA) and snoRNA234 (miRNA) were used as the endogenous reference control gene as expression was determined to be stable following chemical exposure (data not shown). RT-PCR data were collected and represented as relative fold change over vehicle control, calculated by the following formula: $2^{-\Delta\Delta Ct} = \Delta Ct_{Sample} - \Delta Ct_{Control}$. $\Delta Ct = Ct_{Target} - Ct_{\beta\text{-ACTIN}}$, where Ct = cycle threshold as defined by manufacturer.

2.6. Flow Cytometric Analysis and T_{reg} Phenotyping

Single cell suspensions were prepared from tissues and a minimum of 150,000 dLN cells were aliquoted into 96-well U-bottom plates and washed in fluorescence-activated cell sorting (FACS) staining buffer (PBS + 1% bovine serum albumin + 0.1% sodium azide). Cells were resuspended in staining buffer containing anti-mouse CD16/32 antibody (clone 2.4G2; BD Biosciences, San Jose, CA, USA) for blocking of F_c receptors to minimize nonspecific binding. Cells were resuspended in staining buffer containing a cocktail of fluorochrome-conjugated antibodies specific for cell surface antigens including: CD3 (500A2, V500, BD Biosciences, Franklin Lanes, NJ, USA), CD4 (RM4-5, AF700, BD), CD8a (53-6.7, AF488, BioLegend, San Diego, CA, USA), CD25 (PC61, APC Cy7, BioLegend), CD45 (30-F11, PE, BD). Following surface staining, cells were washed in staining buffer and fixed using the Foxp3 fixation buffer set (eBioscience, San Diego, CA, USA). After overnight incubation in staining buffer, cells were permeablilized using the Foxp3 fixation buffer set (eBioscience) and re-suspended in permeabilization buffer containing a cocktail of fluorochrome-conjugated antibodies specific for intracellular antigens including: RUNX3 (R3-5G4, PE, BD), Foxp3 (FLK-16s, eF450 and APC, eBioscience). Following staining, cells were re-suspended in staining buffer and analyzed on an LSR II flow cytometer using FacsDiva software (BD Biosciences). Data analysis was performed with FlowJo 10.0 software (TreeStar Inc., Ashland, OR, USA). A minimum of 10,000 events were captured for each sample. Leukocytes were first identified by their expression of CD45. The T_{reg} subset was further identified as $CD3^+CD4^+CD8^-CD25^+Foxp3^+$. Numerical population values were calculated by applying subset frequencies to the initial cell count obtained following lymph node homogenization. Compensation controls were performed using single stained cellular suspensions and OneComp beads (eBioscience, San Diego, CA, USA) and fluorescence minus one (FMO) staining controls were included to help set gating boundaries.

2.7. T_{reg} Suppression Assay

The suppressive ability of T_{regs} was analyzed using an ex vivo T_{reg} suppression assay as described by Long et al. [21] with modifications. This assay evaluates the ability of naïve, conventional dLN-derived T cells (T_{cons}) to proliferate in the presence of varying numbers of T_{regs} isolated from acetone- or TDI-exposed mice. Mice were exposed to acetone ($n = 7$–11) or TDI (4%) ($n = 4$–5) as previously described and following sacrifice at 7 days (peak of the expansion) post TDI exposure the dLN and spleens were removed. T_{regs} ($CD4^+CD25^+$) and T_{cons} ($CD4^+CD25^-$) were isolated from the lymph nodes and $CD4^-$ accessory cells were isolated from naïve spleens using CD4 negative and CD25 positive selection-based magnetic separation kits (Stemcell, Vancouver, BC, USA). Average T_{reg} purity is as follows for 7 days. Acetone: 96.5% $\pm$ 0.8% of $CD3^+CD4^+$ cells. 4% TDI: 97.35% $\pm$ 0.25% of $CD3^+CD4^+$ cells. Following isolation from naïve mouse dLNs, T_{cons} were labeled with 2 μM carboxyfluorescein succinimidyl ester (CFSE). miR-210 mirVana mimic (2 pmol; ThermoFisher) in Lipofectamine RNAiMAX (ThermoFisher) or Lipofectamine only control (LO) were added to a 96-well U-bottom plate in order to reverse transfect cells. T_{cons} and T_{regs} were cultured in a 96-well U-bottom plate with anti-CD3 (0.2 μg/mL; BD Biosciences) and accessory cells (naïve $CD4^-$ splenocytes treated with mitomycin C) at a variety of T_{con}:T_{reg} ratios (1:1, 2:1, 4:1, and 8:1). Additional controls included stimulated T_{cons} only to assess baseline proliferation, T_{regs} only, accessory cells only, and T_{cons} only with no stimulation nor accessory cells. Cells from each treatment group were pooled and added to triplicate wells of the culture plate. Seventy-two hours following plating, cells were stained with anti-CD4 and Live/Dead Violet (Life Technologies). T_{cons} were defined as $CD4^+CFSE^+$ cells and suppression was measured based on changes in the frequency of dividing $CFSE^+$ cells based on the dilution of CFSE. T_{regs} were analyzed for purity based on their expression of CD3, CD4, and Foxp3 as determined by flow cytometric analysis as previously described.

2.8. Statistical Analysis

Statistical analyses were generated using GraphPad Prism version 5.0 (San Diego, CA, USA). Data were analyzed by a Student t-test comparing groups as indicated in the figure legends. Figure 4C was analyzed by analysis of variance using PROC MIXED. In some cases, data were transformed using the natural log to meet the assumptions of the analysis. Significant interactions were explored utilizing the "slice" option in PROC MIXED and pairwise differences were assessed using a Fisher's Least Significant Difference Test. All differences were considered significant at $p < 0.05$; representative significance symbols varied by figure, as indicated in the legend.

3. Results

3.1. Examination of miR-210 Expression during TDI Sensitization

The kinetics of mature miR-210 expression were investigated via RT-PCR in the dLN following 0.5% and 4% TDI exposure. As previously reported [1], dLN miR-210 increased at various time points during 0.5% and 4% TDI sensitization, including four (4%), seven, and nine (0.5% and 4%) days post single exposure, with expression appearing to peak at four days post 4% TDI exposure (Figure 1A). Primary miRNA (pri-miRNA) transcripts represent the early precursor stage of mature miRNA structure prior to modification and cleavage by Drosha and Dicer. Pri-miR-210 levels were analyzed in the dLN at four days post TDI exposure, which represents the relative peak of dLN miR-210 expression. Although expression of pri-miR-210 was detected in the dLN, no significant alterations were observed. Alternatively, pri-miR-210 expression significantly decreased in the $CD4^+$ subset during 4% TDI sensitization (Figure 1B).

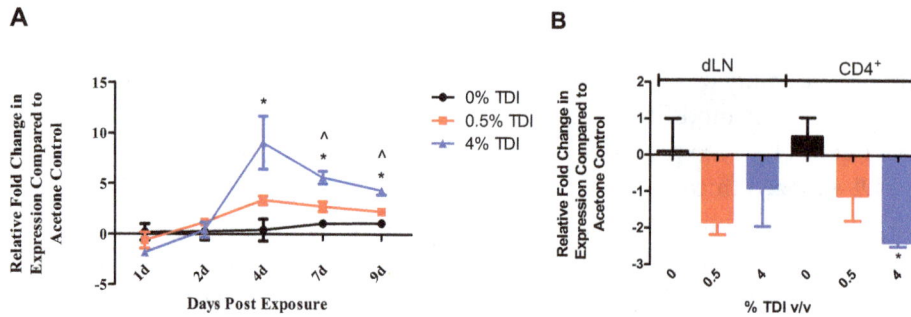

Figure 1. Mature and primary miR-210 expression increases in the draining lymph node (dLN) and CD4+ subsets following toluene diisocyanate (TDI) sensitization: RT-PCR analysis of mature miR-210 expression in the dLN at various time points post TDI exposure (**A**); and primary miR-210 expression in the whole dLN and dLN CD4+ subset four days post TDI exposure (**B**). Cellular purity was assessed via flow cytometric staining for CD4+ Cells (% CD4+ of all cells; Acetone: 89.6% ± 2.2%, 0.5% TDI: 91.5% ± 0.3%, and 4% TDI: 90.4% ± 0.3%). Bars represent mean relative fold change (± standard error, SE) of 3–5 mice per group. Statistical significance is represented by ^ (0.5% TDI) and * (4% TDI) ($p < 0.05$) compared to vehicle control.

3.2. Ex Vivo miR-210 Mimic Transfection Reveals Potential Inhibitory Effect of miR-210 on T_{reg}-Related Genes

Due to the putative link between miR-210 and T_{reg}-related targets in both human and mouse-based studies and target algorithms [1], an ex vivo transfection and stimulation assay was set up to directly examine miR-210 and T_{reg}-related gene expression. Analysis of T_{reg}-related gene expression induced by the addition of excess levels of miR-210 was examined. Although not statistically significant, Figure 2A reveals increased miR-210 expression in cells treated with miR-210 mimic (101 fold) as compared to lipofectamine only (LO) control (1.2 fold). Analysis of T_{reg}-related genes revealed decreased trends of expression of T_{reg}-related genes following miR-210 mimic transfection compared to the LO control, including *foxp3* (~−2.0 fold compared to 0.6 fold) and *cd25* ((~−3.2 fold compared to 0.2 fold) (Figure 2B)). No apparent changes in *runx3* or *ctla4* were observed (data not shown).

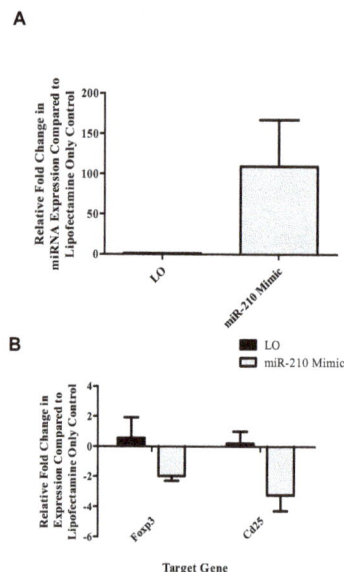

Figure 2. Ex vivo miR-210 mimic augmentation reveals potential inhibitory effect of miR-210 on regulatory T cell (T_{reg})-related genes in dLN cells. (**A**) Intracellular miR-210 was quantified via RT-PCR in ex vivo stimulated samples 72 h following transfection with lipofectamine only control (LO) or miR-210 mimic. (**B**) *foxp3* and *cd25* (*il2rα*) were investigated as potential miRNA targets via RT-PCR. Bars represent mean relative fold change (± SE) of 3 replicates per group.

3.3. In Vivo dLN Target Expression Reveals Decreased Expression of Several Key T_{reg}-Related Genes

The expression of potential miR-210 targets and key T_{reg} genes potentially affected by miR-210 was investigated in the whole dLN at various time points following 4% TDI exposure. Similar to the findings of the ex vivo study and previously reported findings, dLN target/key player mRNA expression was significantly decreased at four (*runx3*), seven (*foxp3*), and nine (*foxp3*) days post 4% TDI exposure (Figure 3A,B). *Foxp3* expression was decreased by approximately threefold in comparison to the acetone control at seven and nine days post TDI exposure (Figure 3A). Early statistically significant increases in *foxp3* (one day) and *runx3* (two day) were also observed but only for a single time point and not determined to be of biological significance. dLN *ctla4* and *cd25* expression significantly increased with peak expression at four and two days post 4% TDI exposure, respectively. However, these increases in expression were not maintained at later time points (Figure 3C,D).

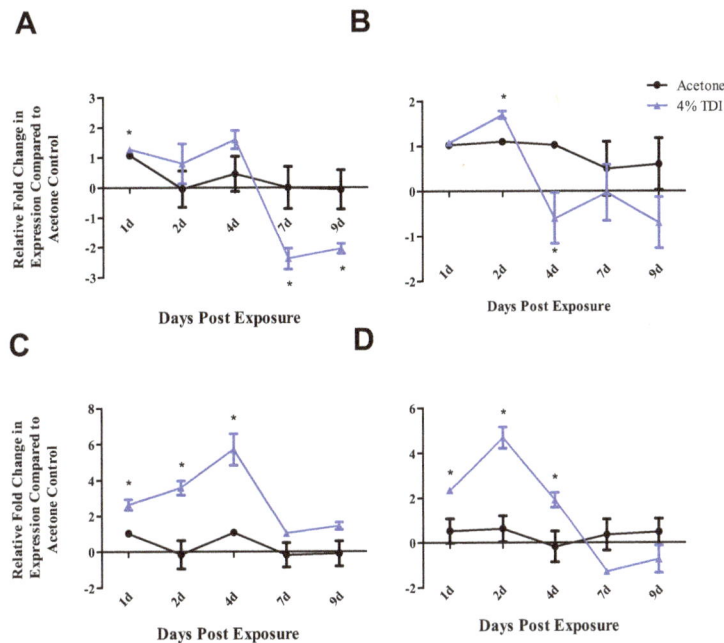

Figure 3. In vivo dLN mRNA expression of potential miR-210 targets reveals altered expression during TDI sensitization. RT-PCR analysis of: *foxp3* (**A**); *runx3* (**B**); *ctla4* (**C**); and *cd25* (**D**) expression in the whole dLN at various time points post 0% and 4% TDI exposure. Bars represent mean relative fold change ($\pm$ SE) of five mice per group. Statistical significance is represented by * ($p < 0.05$) compared to the vehicle control.

3.4. T_{reg}-Specific miRNA and Gene Expression

Due to the suspected influence of miR-210 on T_{regs}, the levels of this miRNA were examined in T_{regs} (CD4$^+$CD25$^+$) from the dLN of mice treated with 0%, 0.5%, and 4% TDI at two and seven days post exposure. These time points were selected to reflect early and peak miR-210 and T_{reg} responses in the dLN (Figure 1A and [1]). While no statistically significant changes were observed following two days of exposure, miR-210 levels were significantly increased in T_{regs} during 0.5% and 4% TDI sensitization in a dose responsive fashion at seven days post exposure (Figure 4A). Expression of T_{reg}-related genes was also examined in isolated T_{regs} following TDI exposure. Significant decreases in *foxp3*, *ctla4*, *runx3*, and *cd25* were observed at both concentrations (except for *ctla4* at 4%) at two days post exposure (Figure 4B). Similarly, significant decreases in *foxp3*, *ctla4*, and *cd25* mRNA were observed in T_{regs} from 4% TDI-exposed mice at seven days post 4% TDI exposure (Figure 4C), further indicating that these genes may be influenced by miR-210 (Figure 1A). T_{reg}-specific expression of *runx3* was assayed in dLN T_{regs} during TDI sensitization. In contrast to the mRNA levels, increases in frequency and numbers of Runx3$^+$ T$_{regs}$ at all time points were observed following 0.5% and 4% TDI exposure (Figure 4D,E).

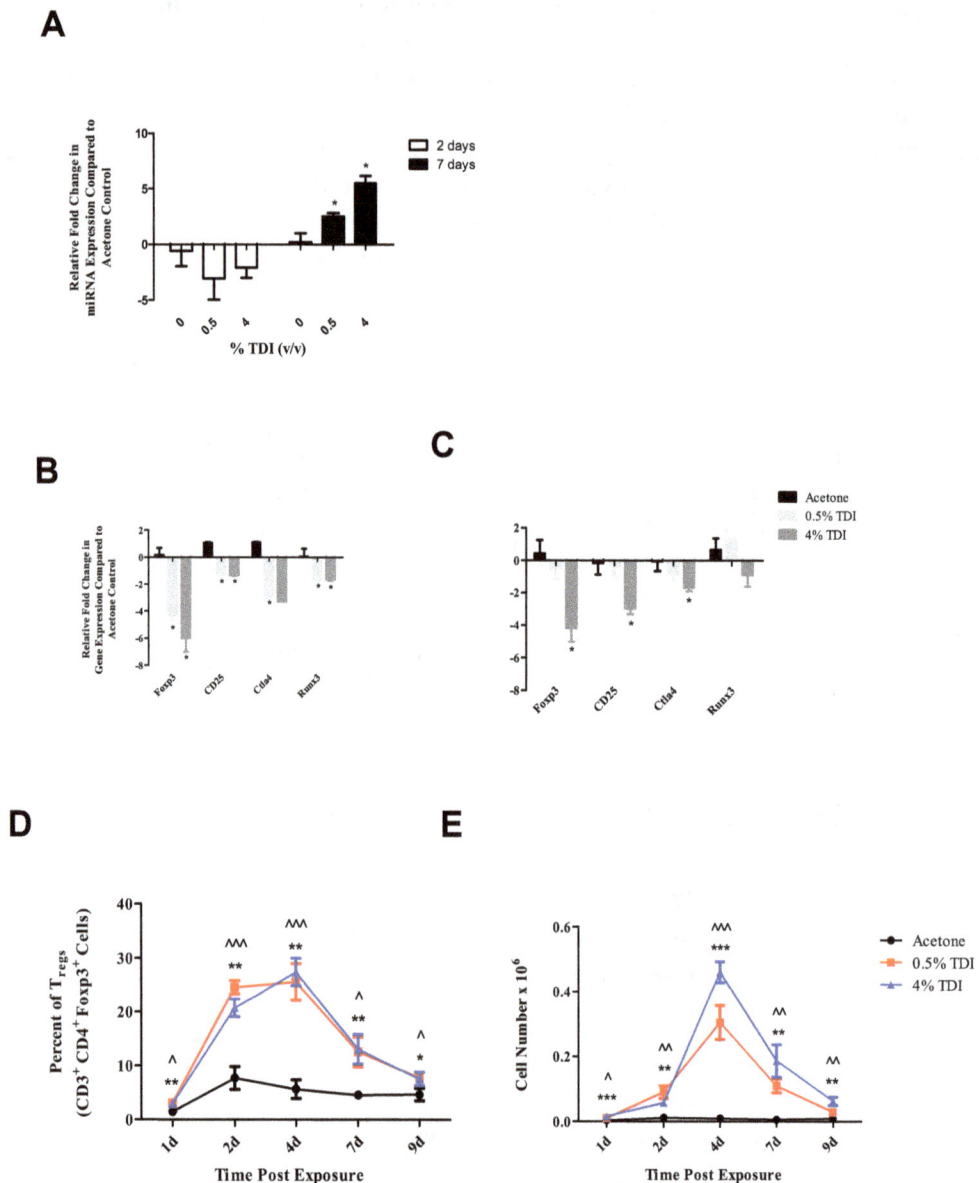

Figure 4. Mature miRNA and Runx3 expression increases in T$_{regs}$ during TDI sensitization. RT-PCR analysis of miR-210 (**A**) and T$_{reg}$-related gene expression in the dLN T$_{reg}$ subset at two (**B**) and seven days (**C**) post TDI exposure. T$_{regs}$ were isolated as CD4$^+$CD25$^+$ cells and cellular purity was assessed via flow cytometric staining for CD3$^+$CD4$^+$Foxp3$^+$ events (2 days: % T$_{regs}$ of all CD4$^+$ events, mean, n = 4–5 per group: Acetone: 83.8% ± 1.6%, 0.5% TDI: 67.1% ± 4.2%, and 4% TDI: 72.3% ± 1.9%; 7 days: % T$_{regs}$ of all CD4$^+$ events: Acetone: 91%, 0.5% TDI: 88.4%, and 4% TDI: 86.7%). Flow cytometric analysis of dLN Runx3$^+$ T$_{reg}$ frequency (**D**) and number (**E**) following TDI sensitization. Bars represent mean relative fold change (±SE) of 4–5 mice per group. Significance is indicated by * $p \leq 0.05$ for (**A–C**) and $p \leq 0.05$ (*), $p \leq 0.01$ (**), $p \leq 0.001$ (***), and $p \leq 0.0001$ (****) for 4% TDI or $p \leq 0.05$ (^), $p \leq 0.01$ (^^), $p \leq 0.001$ (^^^), and $p \leq 0.0001$ (^^^^) for 0.5% TDI compared to vehicle control (**D,E**).

3.5. The Ex Vivo Suppressive Capability of T$_{regs}$ Is Influenced by miR-210 Levels

In order to examine the suspected inhibitory role of miR-210 on T$_{regs}$ during TDI sensitization, the functional capabilities of T$_{regs}$ were tested in an ex vivo suppression assay. miR-210 mimic was transfected in selected wells utilizing a lipid-based transfection strategy and intracellular miR-210 mimic uptake was confirmed in these wells at the time of cellular harvest (Figure 5A). Interestingly, following addition of the miR-210 mimic, miRNA-210 levels were almost 10-fold higher in wells containing acetone T_{regs} compared to those with TDI T_{regs} (Figure 5A). No differences were noted between miR-210 levels among the acetone and TDI groups treated with LO (Figure 5A). Significant baseline T_{reg} suppressive ability was observed in wells with Lipofectamine only and acetone-derived T_{regs} as well as 4% TDI-derived T_{regs} (data not shown). This was not altered significantly upon the addition of miR-210 mimic with the exception of the 1:1 ratio in wells with 4% TDI-derived T_{regs} (data not shown). As previously evidenced [21], TDI-derived T_{regs} exhibited greater suppressive capability than acetone-derived T_{regs} in mimic-treated wells for the 1:1, 2:1, 4:1, and 8:1 ratios (Figure 5B). This increased function was observed in the presence of lower miR-210 levels suggesting that TDI may induce additional factors that might influence the functional capabilities of the T_{regs}.

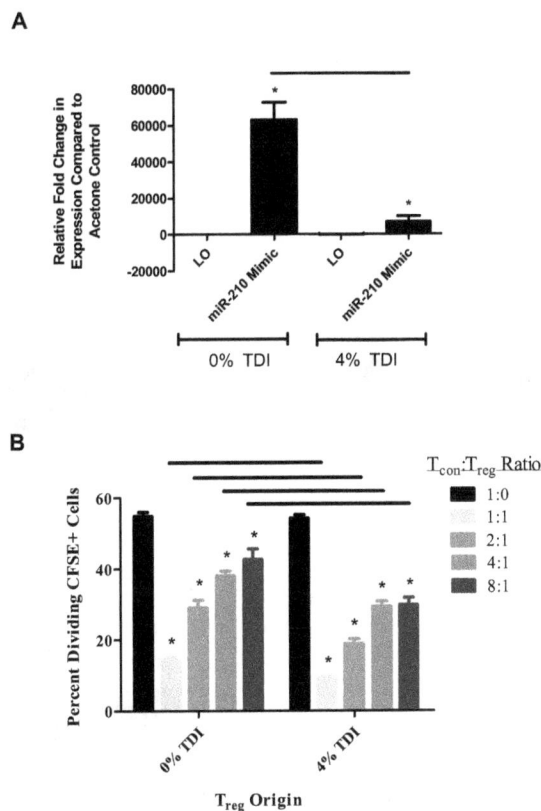

Figure 5. Increased T_{reg} suppression is associated with decreased miR-210 levels. A CFSE-based T_{reg} suppression assay was performed with T_{regs} from mice treated with acetone or TDI (seven days post exposure) and the addition of miR-210 mimic. RT-PCR analysis of intracellular miR-210 expression for acetone or TDI treated mice following addition of miR-210 mimic or Lipofectamine only (LO) control (**A**). Statistical significance is represented by horizontal lines comparing indicated groups and asterisks (compared to LO control) ($p < 0.05$). Functional capacity of T_{regs} based on percent dividing CFSE$^+$ naive conventional T cells (T_{con}) at indicated ratios following addition of miR-210 mimic (**B**). p values are represented by * ($p < 0.05$; comparison of each treatment group to 1:0 ratio from the same chemical treatment group) or horizontal bars (comparison of identical ratios between different mimic treatment groups). Bars represent mean relative fold change ($\pm$ SE) of three replicates per group.

3.6. Further Investigation of miR-31 and -155 Suggest Additional Regulation of the TDI Sensitization Response, Potentially Impacting miR-210 and T_{regs}

Evaluation of additional miRNAs was performed in an attempt to identify other potential factors that might influence miR-210 and T_{reg} function. miR-31 and -155 were selected since they have been identified to significantly increase in dLN expression earlier in TDI sensitization compared to miRNA-210 (Figure 6A,B). Ex vivo target analysis was performed utilizing additional miRNA mimics including miR-31 and -155 in order to further pursue the potential regulation of the T_{reg} subset by these miRNAs in addition to miR-210. Significantly increased intracellular levels of miR-31 (Figure 6C) and miR-155 (Figure 6D) were confirmed 72 h following transfection and stimulation. Although not statistically significant, apparent decreases in *foxp3* expression were observed following miR-31 (Figure 6E) and miR-155 (Figure 6F) mimic transfection. No changes in expression were observed for *runx3*, *ctla4*, or *cd25* following mimic addition (data not shown).

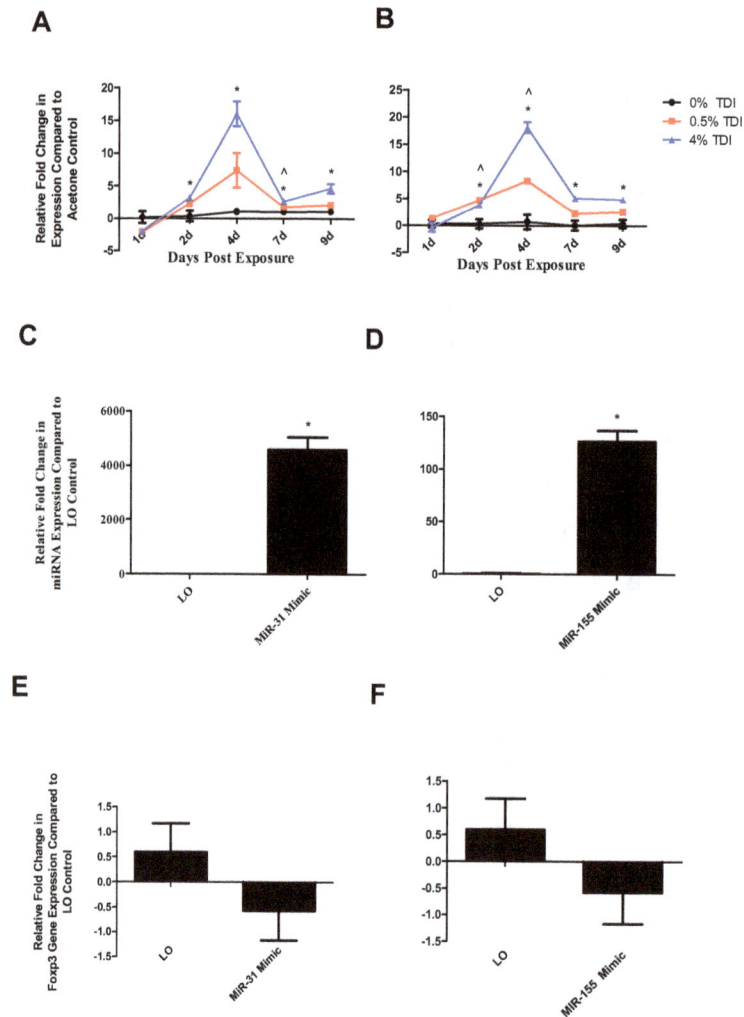

Figure 6. In vivo dLN miR-31 and -155 levels increase during TDI sensitization. Whole dLN miR-31 (**A**) and -155 (**B**) levels were quantified via RT-PCR at various time points post TDI sensitization. Statistical significance is represented by ^ (0.5% TDI) and * (4% TDI) compared to vehicle control ($p < 0.05$). Intracellular miR-31 (**C**), -155 (**D**) was quantified via RT-PCR in ex vivo stimulated samples (with miR-31 or -155 mimic addition, respectively) after 72 h. (**D–F**) *foxp3* expression was investigated in cultures following miR-31 (**D**) and -155 (**E**) mimic treatment. Statistical significance is represented by * when compared to lipofectamine only (LO) control ($p < 0.05$). ($n = 3$–5/group).

T_{reg}-specific expression of miR-31 and miR-155 were also analyzed, revealing increased miR-31 levels two and seven days post 0.5% TDI exposure (Figure 7A,B) and increased miR-155 levels seven days post 4% TDI exposure (Figure 7D). Since miR-31 was identified to increase in expression earlier than miR-210 in the dLN and T_{regs}, miR-210 expression was evaluated in dLN cells treated with miR-31 and -155 mimics. However, no changes in expression were observed. Similarly, expression of miR-31 and -155 was analyzed in cells treated with miR-210 mimic but no significant changes in expression were observed (data not shown).

2 Days   **7 Days**

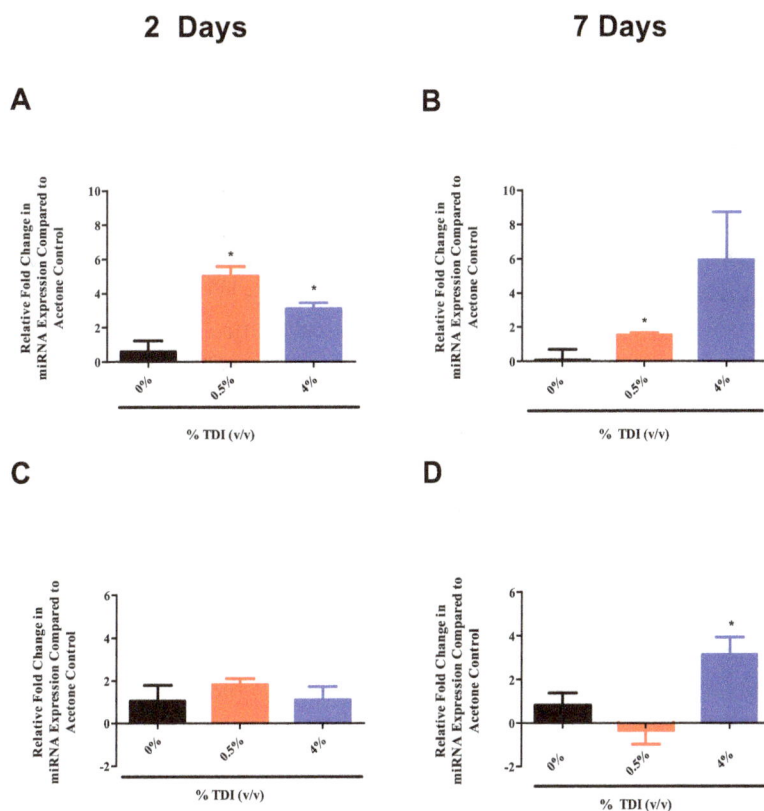

Figure 7. Mature miR-31 and -155 expression increases in T_{regs} during TDI sensitization. RT-PCR analysis of miR-31 (**A,B**) and -155 (**C,D**) expression in the dLN T_{reg} subset at two and seven days post TDI exposure. T_{regs} were isolated as $CD4^+CD25^+$ cells and cellular purity was assessed via flow cytometric staining for $CD3^+CD4^+Foxp3^+$ events (2 days: % T_{regs} of all $CD4^+$ events, mean, $n = 4$–5 per group: Acetone: 83.8% ± 1.6%, 0.5% TDI: 67.1% ± 4.2%, and 4% TDI: 72.3% ± 1.9%; 7 days: % T_{regs} of all $CD4^+$ events, Acetone: 91%, 0.5% TDI: 88.4%, and 4% TDI: 86.7%). Bars represent mean relative fold change (± SE) of 4–5 mice per group. Statistical significance is represented by * ($p < 0.05$).

4. Discussion

The occupational use of sensitizing chemicals such as TDI remains a significant public health concern. There are no validated hazard identification strategies for respiratory sensitizers like TDI and the complete immunologic mechanisms of sensitization have not been elucidated for these agents, hindering development of appropriate preventative assays. This justifies research pertaining to the identification of novel cellular subsets and epigenetic regulatory mechanisms such as miRNAs that may be involved in the respiratory chemical sensitization process. Following the identification of several upregulated dLN miRNAs during TDI sensitization, these molecules were investigated, specifically in relation to T_{reg} development and functionality. To our knowledge, this is the first work that functionally investigates miRNAs in a model of TDI-induced chemical sensitization.

miRNAs are powerful regulatory molecules which have been implicated in a number of immunologic states and conditions, including allergic disease [24–26]. Specifically, miR-155 has demonstrated a critical role in the development of antibody responses and germinal center function [27], miR-326 has been shown to regulate Th17 differentiation, exhibiting critical involvement in multiple sclerosis pathogenesis [28], and in vivo miR-126 inhibition reduces a house dust mite-induced asthmatic phenotype, demonstrating the importance of this miRNA in the regulation of Th2 responses and allergic asthma [24]. Vennegaard et al. described upregulation of several miRNAs, including miR-21, in skin biopsies from patients with allergic responses to diphenylcyclopropenone and in a murine model of dinitrofluorobenzene (DNFB) allergic contact dermatitis [17]. Additionally, previous work from our groups identified upregulation of several miRNAs, including miR-31, -155, and -210, in a murine model of epicutaneous TDI sensitization [1]. These ubiquitous signaling molecules are well-established mediators of many of signaling pathways in a number of cell types; however, their role in chemical sensitization is not well understood. For the work described in this manuscript, miR-210 was selected for additional investigation since it has been predicted and demonstrated to target T_{reg}-related genes. In addition, its role in chemical sensitization and T_{reg} regulation has not yet been described.

The expression of miR-210 was quantified in a variety of tissues and cellular subsets in a murine model of TDI sensitization. Consistent with previous findings [1], increased expression of miR-210 in the dLN was also demonstrated in the present study during TDI sensitization (Figure 1A). In addition, increased expression of miR-210 was also identified in T_{regs} during TDI sensitization (Figure 4A). Since miRNA can be transported to cells via mechanisms such as exosomal transport, experiments to determine if cells are actively producing miR-210 following TDI exposure were conducted. The expression of pri-miRNA indicates gene level expression, presumably within the cell type tested. Pri-miR-210 was detected in the dLN and CD4$^+$ subsets four days post TDI exposure (Figure 1B), indicating that miR-210 is being expressed in this tissue by CD4$^+$ T cells. Interestingly, although they were detectable, pri-miR-210 levels significantly decreased in CD4$^+$ T cells four days post 4% TDI exposure compared to equivalent cells in acetone-exposed mice, potentially indicating that the majority of mature miR-210 in the dLN is being produced by another cell type, is being transported from another tissue, or is being transcriptionally downregulated at this point, the peak of mature miR-210 levels in the dLN. While pri-miR-210 levels in T_{regs} were not investigated in the current study, previous studies have demonstrated miR-210 expression following the polarization of naïve T cells into T_{regs} [29].

Since T_{regs} have been implicated as regulators of TDI sensitization [21] and miR-210 expression has been shown to increase in T_{regs} following TDI sensitization (Figure 4A), an ex vivo target analysis system was designed in order to directly examine the effects of miR-210 on T_{reg}-related genes. Selected genes identified as target of miR-210 included *foxp3*, the master transcription factor of the T_{reg} subset; and *runx3*, a transcription factor that signals upstream of *foxp3* by binding to this gene's promoter [30]. Additionally, *cd25* and *ctla4* were investigated as important T_{reg}-related genes as they are involved in IL-2 signaling and proliferation along with direct suppressive functions, respectively. This setup revealed a potential role for miR-210 in the downregulation of T_{reg}-associated genes (*foxp3* and *cd25*; Figure 2B) following the addition of miR-210 mimic. In addition *foxp3* and *runx3* dLN expression decreased in the dLN at various time points following TDI exposure (Figure 3A,B) which is consistent with our previously reported findings [21]. The earlier decrease in *runx3* expression observed at four days post 4% TDI exposure may be reflective of the upstream signaling activity of this transcription factor in relation to *foxp3* (Figure 3B). While the findings for the ex vivo assay did not reach statistical significance, further support for T_{reg}-associated genes as miRNA targets was provided by in vivo data. Interestingly, whole dLN expression of *ctla4* and *cd25* increased at one, two and four days post 4% TDI exposure (Figure 3C,D) which is likely a reflection of the activation of both T_{regs} and conventional T cells as elevated protein expression of these molecules is observed in T_{regs} at these time points [21] and would likely be increased in conventional T cells involved in TDI sensitization as well. In T_{regs}, decreases in *foxp3*, *cd25* and *ctla4* were observed at two and seven days with *runx3* only being decreased at the earlier time point post TDI exposure (Figure 4B). These early changes in T_{reg} factors provide further support that additional factors might be involved in T_{reg} regulation since peak increases in miR-210 occur later than two days; accordingly, miR-31 was shown be increased at this time point. In contrast to the transcript, the expression kinetics of Runx3 increased following TDI exposure (Figure 4D). This expression pattern was similar to other T_{reg} proteins such as CD25 and Foxp3 (which are represented by the general T_{reg} population) and T_{reg}-specific CTLA4 expression, which have previously been investigated during TDI sensitization [21]. The kinetics of T_{regs} bearing these molecules tended to peak at four days post TDI exposure with a relative decrease in both cellular frequency and number at seven days post TDI exposure [21]. In relation to miR-210 expression kinetics, this data may suggest that miR-210 has a regulatory role on the T_{reg} subset, as its expression wanes in concert with the general T_{reg} population as well as CTLA4$^+$ and Runx3$^+$ T_{regs}. Additionally, miR-210 is a putative *runx3* target, suggesting a potential direct effect on this gene [1]. Collectively, this data is suggestive of T_{reg} regulation with visible effects on the expression of proteins beginning at Day 7 post TDI exposure. miR-210 expression remains elevated in the dLN throughout nine days post 0.5% and 4% TDI exposure (Figure 1A) and in T_{regs} at seven days post 0.5% and 4% TDI exposure (Figure 4A).

Due to the potential link between miR-210 and T_{reg}-related gene expression, the functional capabilities of T_{regs} (acetone and TDI-derived) were examined in the presence and absence of miR-210 mimic. Interestingly, it appeared that miR-210 levels were lower in wells containing T_{regs} from TDI-treated mice and miR-210 mimic compared to wells with acetone-derived T_{regs} and miR-210 mimic (Figure 5A). The increased suppressive capability of TDI T_{regs} with miR-210 mimic (Figure 5B) may be a reflection of reduced miR-210 levels, as we hypothesize that miR-210 is inhibiting T_{reg} differentiation and/or function. This finding suggested that other regulatory factors including other miRNAs might be involved in the regulation of T_{reg} function. Complex interactions and interplay have often been reported for other miRNAs [31], therefore this concept was evaluated in the current study. Typically, direct miRNA–miRNA interactions are mediated by reverse complementary binding, resulting in the formation of duplexes [31]. Additionally, indirect miRNA–miRNA interaction may occur via target gene interaction; e.g., if a miRNA targets a gene that induces a different miRNA, this miRNA is being regulated by its own species.

miRNA-31 and -155 were further investigated for the potential to regulate the expression of miR-210 as they were identified to increase at early time points in TDI sensitization in the dLN (Figure 6A (miR-31) and B (miR-155)) and T_{regs} (Figure 7A (miR-31)). Similarly to miR-210, miR-31 and -155 were shown to potentially downregulate *foxp3* expression in this assay (Figure 6E,F). Although limitations in the assay sensitivities did not reflect significant changes, this may be reflective of a direct effect on *foxp3* or an indirect effect on this gene via other signaling pathways such as miR-210. These alterations are in accordance with recent findings pertaining to miRNA–mRNA interactions. miR-31 may indirectly target *foxp3*, leading to suppressed iT_{reg} development [19], accounting for the potential decreases in this gene evidenced following miR-31 mimic transfection. In addition, miR-31 increases earlier when more persistent decreases in *foxp3* were observed. Additionally, miR-155 expression appears to be controlled by *foxp3* in T_{regs} via binding to the intron within the DNA sequence encoding Bic, the precursor transcript of miR-155; accordingly, T_{reg} miR-155 levels have been shown to be highly responsive to *foxp3* levels [32]. This regulation may be interrupted by abnormally high levels of miR-155 in the mimic transfection system, resulting in decreased *foxp3* expression in these conditions via signaling feedback. This data suggests that miR-31 and -155 may be influencing the expression of miR-210 and/or T_{regs}, possibly acting as early signaling mediators in the TDI sensitization response.

The lack of significance associated with the ex vivo experiments conducted in this work could be a reflection of experimental variability associated with similar assays and temporal discrepancies associated with signaling events. The ex vivo system displays several limitations, explaining the utilization of the in vivo TDI sensitization model in the target investigation as well. The T_{reg}-related gene alterations that were observed as a consequence of miRNA mimic transfections in this system may indicate direct and/or indirect targeting by the miRNA. We propose that for the majority of miRNA–mRNA interactions investigated in our model, regulation is indirect, as few, if any, putative binding sites were identified for many of the potential targets and the corresponding miRNA. For example, miR-210 is predicted to target the 3' UTR of *runx3* [1] and although we did not observe significant alterations in *runx3* expression following miR-210 mimic transfection in our ex vivo system, the in vivo expression kinetics of *runx3* suggest potential regulation. The ubiquitous nature of miRNAs and their involvement in various signaling processes accounts for their functional significance but can also cloud investigations into their mechanistic functions.

It is important to note that the increases in miRNA expression were not dependent on the irritant response, as dLN miR-210 levels significantly increased (Figure 4A) following the non-irritating [21] 0.5% TDI exposure. As 4% TDI exposure causes significant dermal irritation [21], ear miR-210 expression was analyzed at both non-irritant (0.5%) and irritant (1%, 2%, and 4%) TDI concentrations, revealing significant increases in dLN at both non-irritant and irritant doses (Figure 1A). In addition, other miRNAs including miR-22, -31, and -301a were also shown to increase significantly in expression regardless of the irritant status of the TDI dose (data not shown). This data prompts insight into the concept of the "two-signal" sensitization hypothesis which states that antigen delivery alone is insufficient for effective immunological priming but rather a second, innate signal is necessary to ensure the development of sensitization [33,34]. As noted in previous studies, the irritant response appears to be a prerequisite for strong sensitization responses in the case of dermal TDI sensitization [21]. Regardless, the expression of multiple miRNAs in the dLN appears to be due to the sensitization response alone and not significantly influenced by the irritant component of this response, which may be revealing as to their supposed functional roles in the sensitization response and may suggest potential utility as biomarkers of sensitization.

These studies reveal a potential role for miR-210 in a murine model of dermal TDI sensitization (Figure 8). Additionally, miR-31 and miR-155 were investigated for their regulatory potential in this response. The investigation of novel mediators of chemical-induced allergic disease is important for the overall understanding of the mechanisms involved in these responses. Therefore, this data may result in enhanced understanding of the mechanisms involved in chemical sensitization and could potentially aid in the development of hazard identification strategies for respiratory chemical sensitizers. In conclusion, we have demonstrated that miR-210 may negatively influence the differentiation and/or function of T_{regs} via direct targeting of *runx3* and/or indirect actions on other T_{reg}-related genes. These findings suggest that these miRNAs may work in concert to affect the differentiation and function of T_{regs} as well as the expression and function of miR-210.

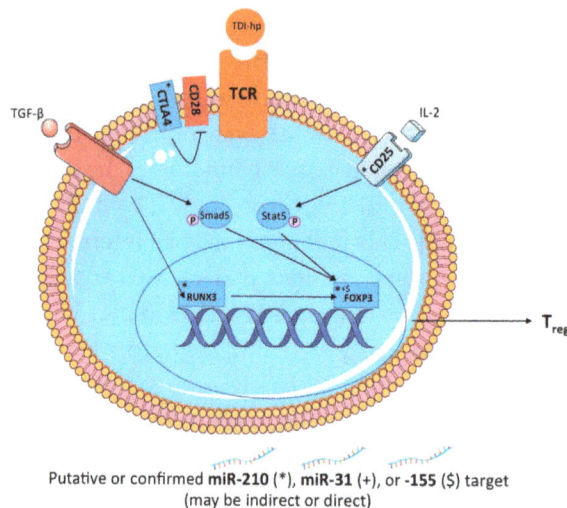

Putative or confirmed **miR-210** (*), **miR-31** (+), or **-155** ($) target (may be indirect or direct)

Figure 8. Proposed model of selected miRNA action on the regulatory T cell pathway. Integral components of the T_{reg} activation pathway are highlighted alongside putative direct targets or key players indirectly impacted by miRNA-210 (indicated by *), -31 (indicated by +), and -155 (indicated by $). Legend: CD28, Cluster of differentiation 28; CD25, IL-2 receptor alpha (in complex with IL-2Rβ and γc); CTLA4, Cytotoxic T-lymphocyte associated protein 4; Foxp3, Forkhead box P3; P, Phosphate group; Runx3, Runt related transcription factor 3; Smad5, Mothers against decapentaplegic homolog 5; Stat5, Signal transducer and activator of transcription 5, TCR, T cell receptor; TDI-hp, Toluene diisocyanate haptenated complex; TGF-β1, Transforming growth factor beta 1.

Acknowledgments: This work was supported by National Institute for Occupational Safety and Health (NIOSH) intramural funds. The findings and conclusions of this article do not necessarily represent the views of the National Institute of Occupational Safety and Health.

Author Contributions: C.M.L., N.B.M., and S.E.A. conceived and designed the experiments; C.M.L., E.L., N.B.M., and S.E.A. performed the experiments; C.M.L. and S.E.A. analyzed the data; A.N. contributed reagents and analysis; and C.M.L. wrote the paper.

References

1. Anderson, S.E.; Beezhold, K.; Lukomska, E.; Richardson, J.; Long, C.; Anderson, K.; Franko, J.; Meade, B.J.; Beezhold, D.H. Expression kinetics of miRNA involved in dermal toluene 2,4-diisocyanate sensitization. *J. Immunotoxicol.* **2013**, *11*, 250–259. [CrossRef] [PubMed]

2. Johnson, V.J.; Yucesoy, B.; Reynolds, J.S.; Fluharty, K.; Wang, W.; Richardson, D.; Luster, M.I. Inhalation of toluene diisocyanate vapor induces allergic rhinitis in mice. *J. Immunol.* **2007**, *179*, 1864–1871. [CrossRef] [PubMed]

3. U.S. Environmental Protection Agency. *Toluene Diisocyanate (TDI) and Related Compounds Action Plan*; RIN 2070-za14; U.S. Environmental Protection Agency: Washington, DC, USA, 2011.

4. Mapp, C.E. Agents, old and new, causing occupational asthma. *Occup. Environ. Med.* **2001**, *58*, 354. [CrossRef] [PubMed]

5. Kim, S.H.; Choi, G.S.; Ye, Y.M.; Jou, I.; Park, H.S.; Park, S.M. Toluene diisocyanate (TDI) regulates haem oxygenase-1/ferritin expression: Implications for toluene diisocyanate-induced asthma. *Clin. Exp. Immunol.* **2010**, *160*, 489–497. [CrossRef] [PubMed]

6. Network Time Protocol (NTP). *Report on Carcinogens*, 12th ed.; National Toxicology Program; U.S. Department of Health and Human Services: Research Triangle Park, NC, USA, 2011; p. 499.

7. Bello, D.; Herrick, C.A.; Smith, T.J.; Woskie, S.R.; Streicher, R.P.; Cullen, M.R.; Liu, Y.; Redlich, C.A. Skin exposure to isocyanates: Reasons for concern. *Environ. Health Perspect.* **2007**, *115*, 328–335. [CrossRef] [PubMed]

8. Anderson, S.E.; Meade, B.J. Potential health effects associated with dermal exposure to occupational chemicals. *Environ. Health Insights* **2014**, *8*, 51–62. [CrossRef] [PubMed]

9. Vandebriel, R.J.; De Jong, W.H.; Spiekstra, S.W.; Van Dijk, M.; Fluitman, A.; Garssen, J.; Van Loveren, H. Assessment of preferential T-helper 1 or T-helper 2 induction by low molecular weight compounds using the local lymph node assay in conjunction with RT-PCR and ELISA for interferon-γ and interleukin-4. *Toxicol. Appl. Pharmacol.* **2000**, *162*, 77–85. [CrossRef] [PubMed]

10. Karol, M.H.; Hauth, B.A.; Riley, E.J.; Magreni, C.M. Dermal contact with toluene diisocyanate (TDI) produces respiratory tract hypersensitivity in guinea pigs. *Toxicol. Appl. Pharmacol.* **1981**, *58*, 221–230. [CrossRef]

11. Anderson, S.E.; Siegel, P.D.; Meade, B.J. The LLNA: A brief review of recent advances and limitations. *J. Allergy* **2011**, *2011*, 424203. [CrossRef] [PubMed]

12. Kopriva, S.E.; Chiasson, V.L.; Mitchell, B.M.; Chatterjee, P. TLR3-induced placental miR-210 down-regulates the STAT6/interleukin-4 pathway. *PLoS ONE* **2013**, *8*, e67760. [CrossRef] [PubMed]

13. Qi, J.; Qiao, Y.; Wang, P.; Li, S.; Zhao, W.; Gao, C. MicroRNA-210 negatively regulates LPS-induced production of proinflammatory cytokines by targeting NF-κB1 in murine macrophages. *FEBS Lett.* **2012**, *586*, 1201–1207. [CrossRef] [PubMed]

14. Taganov, K.D.; Boldin, M.P.; Baltimore, D. MicroRNAs and immunity: Tiny players in a big field. *Immunity* **2007**, *26*, 133–137. [CrossRef] [PubMed]

15. Baltimore, D.; Boldin, M.P.; O'Connell, R.M.; Rao, D.S.; Taganov, K.D. MicroRNAs: New regulators of immune cell development and function. *Nat. Immunol.* **2008**, *9*, 839–845. [CrossRef] [PubMed]

16. Xiao, C.; Rajewsky, K. MicroRNA control in the immune system: Basic principles. *Cell* **2009**, *136*, 26–36. [CrossRef] [PubMed]

17. Vennegaard, M.T.; Bonefeld, C.M.; Hagedorn, P.H.; Bangsgaard, N.; Lovendorf, M.B.; Odum, N.; Woetmann, A.; Geisler, C.; Skov, L. Allergic contact dermatitis induces upregulation of identical microRNAs in humans and mice. *Contact Dermat.* **2012**, *67*, 298–305. [CrossRef] [PubMed]

18. Lodish, H.F.; Zhou, B.; Liu, G.; Chen, C.Z. Micromanagement of the immune system by microRNAs. *Nat. Rev. Immunol.* **2008**, *8*, 120–130. [CrossRef] [PubMed]

19. Zhang, L.; Ke, F.; Liu, Z.; Bai, J.; Liu, J.; Yan, S.; Xu, Z.; Lou, F.; Wang, H.; Zhu, H.; et al. MicroRNA-31 negatively regulates peripherally derived regulatory T-cell generation by repressing retinoic acid-inducible protein 3. *Nat. Commun.* **2015**, *6*, 7639. [CrossRef] [PubMed]

20. Christensen, A.D.; Skov, S.; Kvist, P.H.; Haase, C. Depletion of regulatory T cells in a hapten-induced inflammation model results in prolonged and increased inflammation driven by T cells. *Clin. Exp. Immunol.* **2015**, *179*, 485–499. [CrossRef] [PubMed]

21. Long, C.M.; Marshall, N.B.; Lukomska, E.; Kashon, M.L.; Meade, B.J.; Shane, H.; Anderson, S.E. A role for regulatory T cells in a murine model of epicutaneous toluene diisocyanate sensitization. *Toxicol. Sci. Off. J. Soc. Toxicol.* **2016**, *152*, 85–98. [CrossRef] [PubMed]

22. Anderson, S.E.; Umbright, C.; Sellamuthu, R.; Fluharty, K.; Kashon, M.; Franko, J.; Jackson, L.G.; Johnson, V.J.; Joseph, P. Irritancy and allergic responses induced by topical application of ortho-phthalaldehyde. *Toxicol. Sci. Off. J. Soc. Toxicol.* **2010**, *115*, 435–443. [CrossRef] [PubMed]

23. Franko, J.; Jackson, L.G.; Hubbs, A.; Kashon, M.; Meade, B.J.; Anderson, S.E. Evaluation of furfuryl alcohol sensitization potential following dermal and pulmonary exposure: Enhancement of airway responsiveness. *Toxicol. Sci. Off. J. Soc. Toxicol.* **2012**, *125*, 105–115. [CrossRef] [PubMed]

24. Mattes, J.; Collison, A.; Plank, M.; Phipps, S.; Foster, P.S. Antagonism of microRNA-126 suppresses the effector function of Th2 cells and the development of allergic airways disease. *Proc. Natl. Acad. Sci. USA* **2009**, *106*, 18704–18709. [CrossRef] [PubMed]

25. Lu, T.X.; Rothenberg, M.E. Diagnostic, functional, and therapeutic roles of microRNA in allergic diseases. *J. Allergy Clin. Immunol.* **2013**, *132*, 3–13. [CrossRef] [PubMed]

26. Zech, A.; Ayata, C.K.; Pankratz, F.; Meyer, A.; Baudiss, K.; Cicko, S.; Yegutkin, G.G.; Grundmann, S.; Idzko, M. MicroRNA-155 modulates P2R signaling and Th2 priming of dendritic cells during allergic airway inflammation in mice. *Allergy* **2015**, *70*, 1121–1129. [CrossRef] [PubMed]

27. Thai, T.H. Regulation of the germinal center response by microRNA-155. *Science* **2007**, *316*, 604–608. [CrossRef] [PubMed]

28. Du, C.; Liu, C.; Kang, J.; Zhao, G.; Ye, Z.; Huang, S.; Li, Z.; Wu, Z.; Pei, G. MicroRNA miR-326 regulates Th-17 differentiation and is associated with the pathogenesis of multiple sclerosis. *Nat. Immunol.* **2009**, *10*, 1252–1259. [CrossRef] [PubMed]

29. Wang, H.; Flach, H.; Onizawa, M.; Wei, L.; McManus, M.T.; Weiss, A. Negative regulation of *HIF1A* expression and Th17 differentiation by the hypoxia-regulated microRNA miR-210. *Nat. Immunol.* **2014**, *15*, 393–401. [CrossRef] [PubMed]

30. Bruno, L.; Mazzarella, L.; Hoogenkamp, M.; Hertweck, A.; Cobb, B.S.; Sauer, S.; Hadjur, S.; Leleu, M.; Naoe, Y.; Telfer, J.C.; et al. Runx proteins regulate Foxp3 expression. *J. Exp. Med.* **2009**, *206*, 2329–2337. [CrossRef] [PubMed]

31. Guo, L.; Zhao, Y.; Yang, S.; Zhang, H.; Chen, F. Integrative analysis of miRNA-mRNA and miRNA-miRNA interactions. *BioMed Res. Int.* **2014**, *2014*, 907420. [CrossRef] [PubMed]

32. Lu, L.-F.; Thai, T.-H.; Calado, D.P.; Chaudhry, A.; Kubo, M.; Tanaka, K.; Loeb, G.B.; Lee, H.; Yoshimura, A.; Rajewsky, K.; et al. Foxp3-dependent microRNA155 confers competitive fitness to regulatory T cells by targeting SOCS1 protein. *Immunity* **2009**, *30*, 80–91. [CrossRef] [PubMed]

33. Kimber, I.; Basketter, D.A.; McFadden, J.P.; Dearman, R.J. Characterization of skin sensitizing chemicals: A lesson learnt from nickel allergy. *J. Immunotoxicol.* **2011**, *8*, 1–2. [CrossRef] [PubMed]

34. McFadden, J.P.; Basketter, D.A. Contact allergy, irritancy and 'danger'. *Contact Dermat.* **2000**, *42*, 123–127. [CrossRef]

6

Integrative miRNA-Gene Expression Analysis Enables Refinement of Associated Biology and Prediction of Response to Cetuximab in Head and Neck Squamous Cell Cancer

Loris De Cecco [1,*], **Marco Giannoccaro** [1], **Edoardo Marchesi** [1], **Paolo Bossi** [2], **Federica Favales** [2], **Laura D. Locati** [2], **Lisa Licitra** [2], **Silvana Pilotti** [3] **and Silvana Canevari** [1,*]

[1] Functional Genomics and Bioinformatics, Department of Experimental Oncology and Molecular Medicine, Fondazione IRCCS Istituto Nazionale dei Tumori, Milan 20133, Italy; marco.giannoccaro@istitutotumori.mi.it (M.G.); edoardo.marchesi@istitutotumori.mi.it (E.M.)

[2] Head and Neck Medical Oncology Unit, Fondazione IRCCS Istituto Nazionale dei Tumori, Milan 20133, Italy; paolo.bossi@istitutotumori.mi.it (P.B.); federica.favales@istitutotumori.mi.it (F.F.); laura.locati@istitutotumori.mi.it (L.D.L.); lisa.licitra@istitutotumori.mi.it (L.L.)

[3] Laboratory of Experimental Molecular Pathology, Department of Diagnostic Pathology and Laboratory, Fondazione IRCCS Istituto Nazionale dei Tumori, Milan 20133, Italy; silvana.pilotti@istitutotumori.mi.it

[*] Correspondence: loris.dececco@istitutotumori.mi.it (L.D.C.); silvana.canevari@istitutotumori.mi.it (S.C.)

Academic Editor: Roel Ophoff

Abstract: This paper documents the process by which we, through gene and miRNA expression profiling of the same samples of head and neck squamous cell carcinomas (HNSCC) and an integrative miRNA-mRNA expression analysis, were able to identify candidate biomarkers of progression-free survival (PFS) in patients treated with cetuximab-based approaches. Through sparse partial least square–discriminant analysis (sPLS-DA) and supervised analysis, 36 miRNAs were identified in two components that clearly separated long- and short-PFS patients. Gene set enrichment analysis identified a significant correlation between the miRNA first-component and EGFR signaling, keratinocyte differentiation, and p53. Another significant correlation was identified between the second component and RAS, NOTCH, immune/inflammatory response, epithelial–mesenchymal transition (EMT), and angiogenesis pathways. Regularized canonical correlation analysis of sPLS-DA miRNA and gene data combined with the MAGIA2 web-tool highlighted 16 miRNAs and 84 genes that were interconnected in a total of 245 interactions. After feature selection by a smoothed t-statistic support vector machine, we identified three miRNAs and five genes in the miRNA-gene network whose expression result was the most relevant in predicting PFS (Area Under the Curve, AUC = 0.992). Overall, using a well-defined clinical setting and up-to-date bioinformatics tools, we are able to give the proof of principle that an integrative miRNA-mRNA expression could greatly contribute to the refinement of the biology behind a predictive model.

Keywords: miRNA; microarray; head and neck squamous cell carcinomas (HNSCC); cetuximab; drug sensitivity

1. Introduction

Head and neck cancers develop in the mucosal linings of the upper aerodigestive tract and over 90% are squamous cell carcinomas (HNSCC). The disease is diagnosed in advanced stages (stage III and IV) in the majority of patients and their treatment usually requires surgery, radiation therapy, or chemotherapy in different combinations [1]. However, 27%–50% of cases relapse within two years after treatment, and platinum-based chemotherapy (CT) plus a targeted therapy with an anti-EGFR (epidermal growth factor receptor) monoclonal antibody (mAb), cetuximab, is usually offered to recurrent or metastatic (RM) patients.

This systemic treatment still represents that achieving major improvement in RM-HNSCC patients in the last decade [2]. Since only about 40% of patients benefit from a response to this drug combination, and only a small portion of this group experience durable responses, clinicians are faced with the problem of choosing the best curative treatment without the help of reliable predictive biomarkers.

In the last 15 years we have observed continuous improvements in genomics, enabling the molecular profiling of over 20,000 genes and 2000 microRNAs (miRNAs). Similarly, there have been improvements in bioinformatics through the continuous development of new algorithms to integrate these two types of highly dimensional "omics" data [3].

Recent work has shown that by analyzing gene expression profiling of a selected retrospective series of HNSCC patients, and by associating it to the response/sensitivity to cetuximab-CT, we were able to propose that long-progression-free survival (PFS) cases behave in a manner consistent with a defined molecular subgroup. This subgroup has a poor prognosis, which can be rescued by cetuximab/CT treatment, while short-PFS cases are characterized by an over-activation of RAS signaling [4].

Taking advantage of the same HNSCC clinical materials, and on the basis of our successful analyses of other cancer types for biological characterization [5] or for prognostication [6], we profiled miRNAs and we performed an integrated miRNA-mRNA expression analysis to further gain insight into the biology and prediction of the cetuximab-CT response.

2. Materials and Methods

2.1. Patients and Study Design

Forty tumor specimens from RM-HNSCC patients were divided according to PFS following cetuximab-CT treatment in 14 long-PFS patients and 26 short-PFS patients [4]. The study design includes two groups balanced for known prognostic factors [7] (primary tumor site, performance status, weight loss, prior radiotherapy, tumor grade, residual disease at primary tumor site, age, and gender). Long-PFS had a median PFS of 19 months (range 12–36), while short-PFS had a median PFS of three months (range 1–5.5). The study was conducted in accordance with the Declaration of Helsinki and was approved by the Independent Ethics Committee of Fondazione IRCCS Istituto Nazionale dei Tumori (Approval Number 55/12).

2.2. miRNA Profiling

Total RNA was extracted from formalin fixed paraffin embedded (FFPE) tissues using the miRNeasy FFPE kit (Qiagen, Valencia, CA, USA), and concentration was assessed with the NanoDrop ND-100 Spectrophotometer (NanoDrop Technologies, Wilmington, DE, USA). RNA was processed for miRNA profiling according to the manufacturer's recommendations, and miRNA expression analysis was performed using SurePrint G3 Human miRNA 8×60K microarrays from Agilent Technologies (Santa Clara, CA, USA). RNA was dephosphorylated in the amount of 100 ng with calf intestinal alkaline phosphatase and denatured by DMSO treatment. Samples were labeled with cyanine 3-pCp using T4 RNA ligase and hybridized on miRNA array. Arrays were washed in Agilent's Wash Buffers and scanned at a resolution of 2 mm using an Agilent DNA microarray scanner (Agilent Technologies). Primary data were collected using Agilent's Feature Extraction software v10.7 (Agilent Technologies). Raw miRNA expression data were processed using an optimized version of the Robust Multi-array Average (RMA) algorithm implemented in AgiMicroRna package [8]. Based on gIsGeneDetected information provided by the Agilent's Feature Extraction software, miRNAs detected in at least 10% of samples were selected, yielding a data matrix containing 614 miRNAs. Microarray data were deposited and are available on NCBI Gene Expression Omnibus (GEO) database [9] with the accession number GSE92595.

2.3. Statistical and Bioinformatics Analyses

2.3.1. miRNA Expression Analysis

To identify expression patterns related to patient's PFS, we applied two different approaches: (i) discriminant analysis aimed at identifying features whose expression differences help stratifying patients; (ii) differential expression analysis aimed at obtaining a list of features whose expression differs among classes. In general, a discriminant feature is differentially expressed, but the reverse is not always true. Among discriminant methods, we used the sparse partial least square–discriminant analysis (sPLS-DA) [10] applying mixOmics R package [11] from Bioconductor [12], since in comparative studies it resulted in efficient generalization ability [13]. sPLS-DA is a pattern recognition technique that enables disclosing a set of features in a supervised classification context that are summarized in appropriate linear combinations performing variable selection and classification in a single step procedure. This approach has already been successfully applied to miRNA analysis in a setting where the number of samples is limited compared to the number of tested miRNAs [14]. The discriminative miRNAs were selected on the miRNA data matrix named X ($n \times p$), where n is the number of samples and p is the total number of variables (i.e., miRNAs). In our analysis, the size is $n = 40$ samples and $p = 614$ miRNAs. The selection was based on the response dummy matrix partitioned in K groups ($K = 2$), where K is the number of classes. Two parameters should be tuned in sPLS-DA: the number of discriminant vectors and the number of variables to select on each dimension (PLS component). According to further literature [10], it is conventional to choose the number of sPLS-DA dimensions $H \leq \min(p, K)$. We tested the performance in terms of the classification error rate with a maximum distances prediction method for the first 10 sPLS-DA dimensions, which included a number of miRNAs ranging from 1 to 50 in each dimension. A 10-fold cross-validation setting was imposed and the "optimal" number of variables was finally determined when the lowest error rate is obtained. Differential expression analysis, imposing the same criteria of Bossi et al. [4], was performed on miRNA expression data, and through a random variance t-test that improved estimates without assuming that all miRNA have the same variance. A False Discovery Rate (FDR, [15]) correction was applied on raw p-values and a threshold of 0.15 was set. A global test to ascertain the differences in expression patterns between classes was performed through random permutations of the class labels. For each random permutation, all t-tests are re-computed for each miRNA and the proportion of the random permutations that gave as many miRNA significant at FDR < 0.15 as were found in comparing the true class labels was assessed providing a p-value. BRB-ArrayTools (version 4.3.1) developed by Dr. Richard Simon and the BRB-ArrayTools Development Team, available at [16] was used for differential expression analysis.

2.3.2. Inference of miRNA Components on Gene Expression Data by GSEA

Pathway enrichment analysis was carried out by gene set enrichment analysis (GSEA) using the dataset GSE65021 [4] and the GSEA 2.2.2 software [17]. The first and second components of miRNA sPLS-DA defined the two continuous traits to which gene-expression data were ranked through GSEA, determining gene-sets having either positive or negative correlations. Pearson correlation was used as the metrics for ranking genes, and phenotype permutation was performed to assess the significance of the enrichment scores. For multiple testing adjustments, Benjamini and Hochberg's false discovery rate was applied. FDRs were calculated on the normalized enrichment scores (NES) and only gene-sets enriched with an FDR < 0.05 were retained. GSEA was performed on Molecular Signatures Database (MSigDB; v5.1 updated January 2016, http://software.broadinstitute.org/gsea/msigdb), including C5 collection associated with GO terms, C6 oncogenic signatures corresponding to cellular pathways dis-regulated in cancer by specific oncogenes, and Hallmark, a collection of gene-sets aggregating many

MSigDB redundant terms to represent well-defined biological processes. Graphical representation of the most significant gene-sets was provided by GOBubble function available in GOplot R package [18], displaying information about the significance of the enrichment ($-\log 10$ p-value) and the z-score of each gene set.

2.3.3. Differential Gene Expression Analysis by sPLS-DA

The procedure described in Section 2.3.1. for miRNA expression data was also applied on gene-expression data corresponding to a data matrix retrieved from GEO repository under the ID GSE65021 [4] denominated Z ($n \times q$) whose size is 40 (samples) $\times$ 17,378 (detected features = genes).

2.3.4. miRNA and Gene-Expression Integrative Analysis

We focused our attention on analytic methods capable of establishing the potential relationships between sets of measurements from the same samples, reducing the dimensionality of the data. Canonical correlation analysis (CCA) is a classical technique developed by Hotelling [19] that extracts linear correlations among sets of variables on the same set of subjects, including all variables from both data sets. However, limited sample size and high dimensional data, which is a common situation for genomic studies, results in inaccurate estimates and data overfitting issues [20]. Regularized CCA (rCCA) is a modified version of CCA that implements a regularization procedure to reduce the dimensionality of the data. This method has been already used [21] and resulted useful to disclose linear relationships between two sets of variables. Taking into account two data matrices X ($n \times p$) and Z ($n \times q$), representing miRNA and gene-expression data, respectively, where $n << p$ and $n << q$, regularization of covariance's matrices of X and Z consists in adding a multiple of the identity matrix (Id):

$$\text{Cov}(X) + \lambda_1\text{Id and Cov}(Z) + \lambda_2\text{Id}$$

The regularization parameters λ_1 and λ_2 were selected by 10-fold cross-validation procedure on a regular grid of size 100×100 defined on the region $0.001 < \lambda_1 < 0.05$, and $0.0001 < \lambda_2 < 0.05$ using the *tune.rcc* function available in mixOmics and resulting in $\lambda_1 = 0.0292$ and $\lambda_2 = 0.00199$ (Supplementary Materials Figure S1A). The choice of dimensions d ($1 \leq d \leq p$) to include in further analysis was done according to Gonzalez et al. [22] who suggests an empirical approach based on the inspection of the plot of canonical correlations versus dimensions and on selection of the appropriate number of dimensions before a clear gap among canonical correlations. On the basis of the obtained λ_1 and λ_2 parameters, we observed a clear gap between the 23rd and the 24th canonical correlations (Supplementary Materials Figure S1B).

The relationships between miRNAs and genes were displayed by correlation circle plots that allow recognition of the correlation structure between the two sets of variables X and Z. In this type of graphical representation, the variables X and Z are projected as vectors onto a plane and the relationship (correlation) between variables is approximated by the inner product between vectors (i.e., the product of the two vector lengths and their cosine angles). Since the variables X and Z correspond to unit of variance, their projections are inside a circle of radius 1 centered at the origin of the correlation plot. In this way, the correlation circle plots allow to reveal inherent features embedded into the correlation structure of the variables. In detail, the association among variables is disclosed by: (i) the distance from the origin (the greater the distance is, the stronger the association is); (ii) the direction of the projection of variables from the origin (sharp angle in the same direction meaning a positive correlation; obtuse angle in an opposite direction meaning a negative correlation; a right angle meaning null correlation). To improve interpretability, two circles are drawn: one external circle, radius = 1, and one internal circle. The area between the two circles reveals the most important variables and the variables close to the center are assumed to be not relevant. By imposing in our analysis an inner circle with radius = 0.3, the area from 0.3 to 1 contains the relevant miRNA-gene expression relationships. rCCA results were also displayed by clustered image maps (CIM) as a heatmap that represents the correlation between two matched datasets. The similarity matrix obtained by miRNA and gene-expression was organized in a bi-cluster hierarchical structure illustrating the interplay among sets of variables [23]. The Euclidian distance and the average agglomeration method were used for the hierarchical clustering.

2.3.5. Target Prediction

We integrated target predictions with correlation-based miRNA and gene expression profiles based on the hypothesis that the expression profile of a given miRNA is expected to be inversely correlated with its mRNA target if the miRNA acts on mRNA stability. Data from miRNA and gene expression identified by rCCA were analyzed by MAGIA2 web-tool [24] to build mixed miRNA-gene expression networks. MAGIA2 was run on the expression data for the top 75% of genes with greatest variation in expression among samples. Three target prediction databases, DIANA-microT [25], TargetScan [26], and microrna.org [27], were selected for miRNA target prediction using default settings. The predictions shared by at least two of the three databases are retained and the final network is visualized by Cytoscape [28]. To investigate the experimentally validated miRNA/gene interactions, we retrieved the data from miRTarBase, a comprehensive repository of evidences based on literature [29].

2.3.6. miRNA-Gene Integrative Predictive Signature

To test the feasibility in defining a miRNA-gene integrative predictive signature, a method is required satisfying the criteria to be able to: (i) select pivotal features avoiding unnecessary redundancy; (ii) integrate data from different sources (i.e., miRNA and gene expression). Smoothed t-statistic support vector machine (stSVM) is a feature selection method that proved its efficiency in merging network data by smoothing feature-wise t-statistics using a random walk kernel and the integration of data from miRNA and gene expression profiles [30]. The method was designed to select features by a permutation test with subsequent SVM training. Prediction performance was assessed by 10-time repeated 10-fold cross-validation. All calculations were performed through *netClass* R-package [31].

2.3.7. Comparison Analysis with Publically Available Data

Level 3 files of HNSCC miRNA and gene expression data present in The Cancer Genome Atlas (TCGA) [32] were downloaded along with the clinical annotations from the TCGA website [33] and used for the analysis. Subtype molecular classification was performed as described in De Cecco et al. [34] identifying Cl2-mesenchymalmesenchymal and Cl3-hypoxia subtypes.

3. Results

3.1. miRNA Expression Patterns in HNSCC Patients Treated with Cetuximab-CT

To disclose miRNA patterns associated with PFS, a supervised analysis was applied to our data matrix of 614 detected miRNAs using sPLS-DA and imposing the "a priori" knowledge of a patient's PFS category (long and short PFS) [4]. sPLS-DA allows selection for variables that best separate the two PFS categories, reducing the dimensionality of those patterns in a few discriminant components. We assessed the classification error rate using the maximum distance method to determine the optimal number of dimensions and features to retain. As such, the best performance (lowest misclassification rate) was obtained, including the first two components which explained 68% of the variance in miRNA expression; the first component (24 miRNAs), which retained a large part of the variance, explained 49% of the variance, compared to the second component (12 miRNAs), which explained 19% of the variance (miRNA list in Supplementary Materials Table S1, panel A). The results of this analysis showed that miRNA profiles were able to clearly separate long- and short-PFS patients (Figure 1A) and each component significantly divided the two categories of patients (Figure 1B).

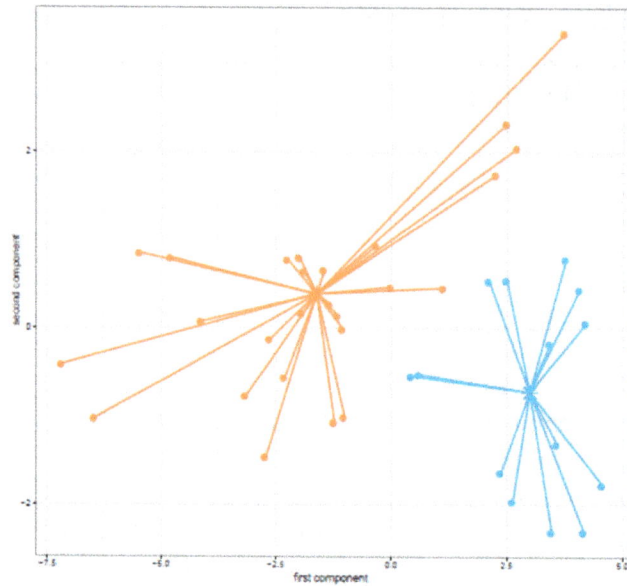

A

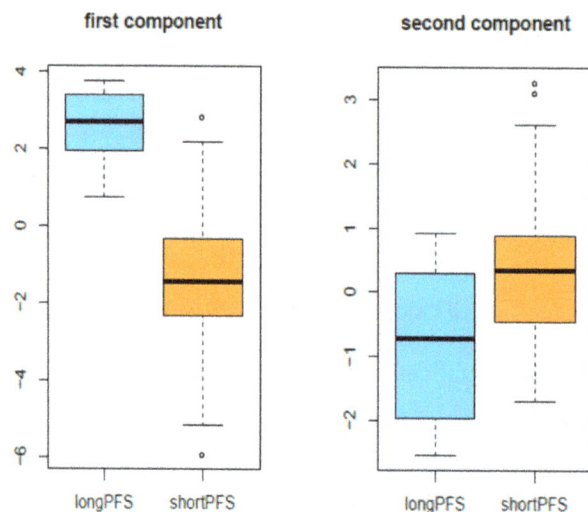

B

Figure 1. miRNA expression patterns in HNSCC tumors associated with PFS after cetuximab-CT treatment: sPLS-DA identified 24 and 12 miRNAs in first and second components, respectively (see Supplementary Materials Table S1 for list). (**A**) The score plot of sPLS-DA miRNA profiles of the two components is shown and each individual patient was plotted. The well-defined clusters, corresponding to the 14 long-PFS patients (blue dots) and the 26 short-PFS patients (orange dots), can be identified. The lines indicate the distance from the respective centroid for samples of each class; (**B**) The box-plot analysis shows the loading vector values for the first component and second component with samples divided on the basis of class (long PFS = blue; short PFS = orange). A significant difference was observed in each component (first component, p = 2.97E-07; second component, p = 0.00529).

When the same data matrix was analyzed by supervised class comparison, we identified 166 differentially expressed miRNAs (the probability of finding 166 miRNAs significant by chance was 0.013, as determined by the global test). After filtering based on $|\log_2(\text{fold change})| > 1$, 39 miRNAs, four and 35 upregulated in long- and in short-PFS samples, respectively, were retained (volcano plot in

Supplementary Materials Figure S2; fold change and *p*-values in Supplementary Materials Table S1, panel B). The comparison of the two methods (see Supplementary Materials Table S1) indicated that 18 of 39 differentially expressed miRNAs were included in the first component of the sPLS-DA while none from the second component were differentially expressed according to the selection criteria.

3.2. Biological Relevance of sPLS-DA miRNAs Inferred by GSEA

In order to investigate the biological information embedded in the miRNA sPLS-DA components, these components were correlated to gene expression data [4] and pathway recognition was performed through GSEA. miRNAs of the first component correlated with EGFR signaling, keratinocyte differentiation and ectoderm development (GO terms), and p53 and Myc (oncogenic signatures); miRNAs of the second component were correlated with the oncogenic signatures RAS and NOTCH and with numerous GO terms, including immune/inflammatory response, EMT, and angiogenesis pathways (Figure 2).

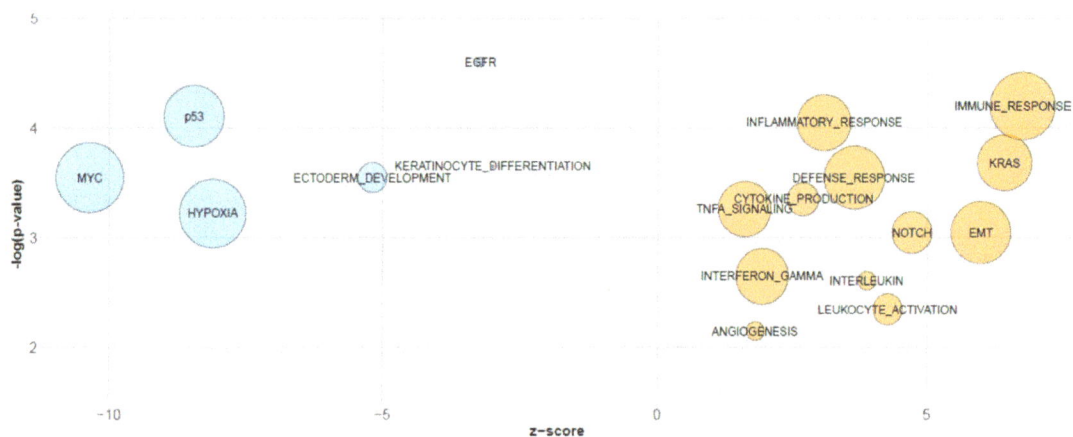

Figure 2. Bubble plot of gene sets associated with miRNAs' first and second components identified by sPLS-DA: An overview of GSEA-enriched networks was depicted. The x-axis indicates the z-score for each term, while the y-axis is the negative logarithm of the adjusted *p*-value. The area of the displayed circles is proportional to the number of genes assigned to each term. The colors display the positive association to the first (blue) and second (orange) components.

3.3. Integrated miRNA and mRNA Networks

To proceed in an integrative analysis, we first applied sPLS-DA on whole-gene expression data obtained from the same samples (see Bossi et al. for details; [4]) and following the criteria of the miRNA profile. A lower misclassification rate was found including the first two components containing 428 unique genes; Supplementary Materials Figure S3 shows the performance in stratifying patients based on long and short PFS. Through rCCA we tested the linear combinations between X and Z data matrices, representing miRNA and gene expression patterns, respectively; rCCA allowed us to highlight the most relevant correlation structures embedded into the expression data. Since the first two variates explained most of the data variability (45% and 14% variates 1 and 2, respectively) and retained the ability to stratify patient outcome, we restricted the analysis to these dimensions to identify relevant miRNA–gene expression associations. The results of the integrated analysis are displayed through a correlation circle plot (Figure 3A) in which a total of 27 miRNA and 250 genes were shown as significantly correlated (see list in Supplementary Materials Table S2). The results showed a separation of miRNAs and correlated genes that can be summarized in four clusters: (i) a core of 22 miRNAs with a long distance from the center (range: from −0.945 to −0.731) in variate 2; (ii) a cluster of 168 genes negatively correlated to miRNA variate 1 and constituting gene variate 1; (iii) a small cluster of five miRNAs in variate 2 (range: from −032 to −0.201); (iv) a cluster of 82 genes negatively correlated to miRNA variate 2 and constituting gene variate 2 (see Supplementary Materials

Table S2 for details about miRNAs and genes belonging to the four clusters). To improve our understanding of the connections between miRNAs and genes, a pair-wise similarity matrix was computed and displayed by CIM (Figure 3B).

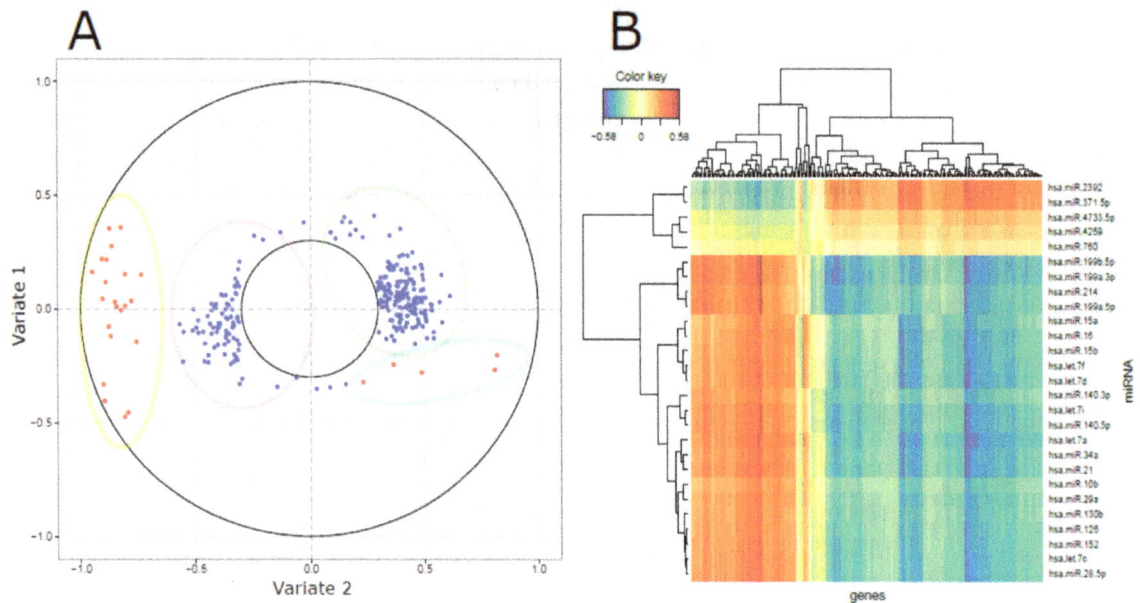

Figure 3. Identification of integrated miRNA and mRNA networks: (**A**) The correlation circle plot displays the features (miRNAs = red; genes = blue) and the position derived by the combination of the first and second variates in our rCCA. The features in the area outside the inner ring (radius < 0.3) were retained as significant and shown in the plot. The ellipses depicted four miRNA/gene clusters and the list can be found in Supplementary Materials Table S2; (**B**) The CIM plot displays the correlation structure between the features. Each colored block represents an association between miRNAs and genes spanning a range of colors from blue (negative correlation), to yellow (weak correlation), to red (positive correlation). The dendrograms on the top and left side indicate how genes and miRNAs, respectively, are connected.

3.4. Computational Integration of miRNAs and Genes by MAGIA 2

The miRNA genes identified by rCCA were integrated for network analysis using three prediction target algorithms (DIANA-microT, TargetScan, microrna.org). By imposing a q-value < 0.1 and Pearson correlation as an association measure, 16 miRNAs and 84 genes were interconnected (Figure 4 and Supplementary Materials Table S3). Analysis of the 245 identified interactions highlighted that: (i) four miRNAs (hsa-miR-130b-3p, hsa-miR-199a-3p, hsa-miR-214-3p, and hsa-miR-28-5p) accounted for 41% of the interactions; (ii) four genes (CDK5R1, KLK10, LYNX1, and TMEM79) were co-targeted by eight miRNAs; (iii) several genes were targeted by several miRNAs; (iv) a survey using the miRTarBase repository proved that CD24/miR-34a-5p, ITGA6/miR-29a-3p, and L1CAM/miR-34a-5p were experimentally validated interactions.

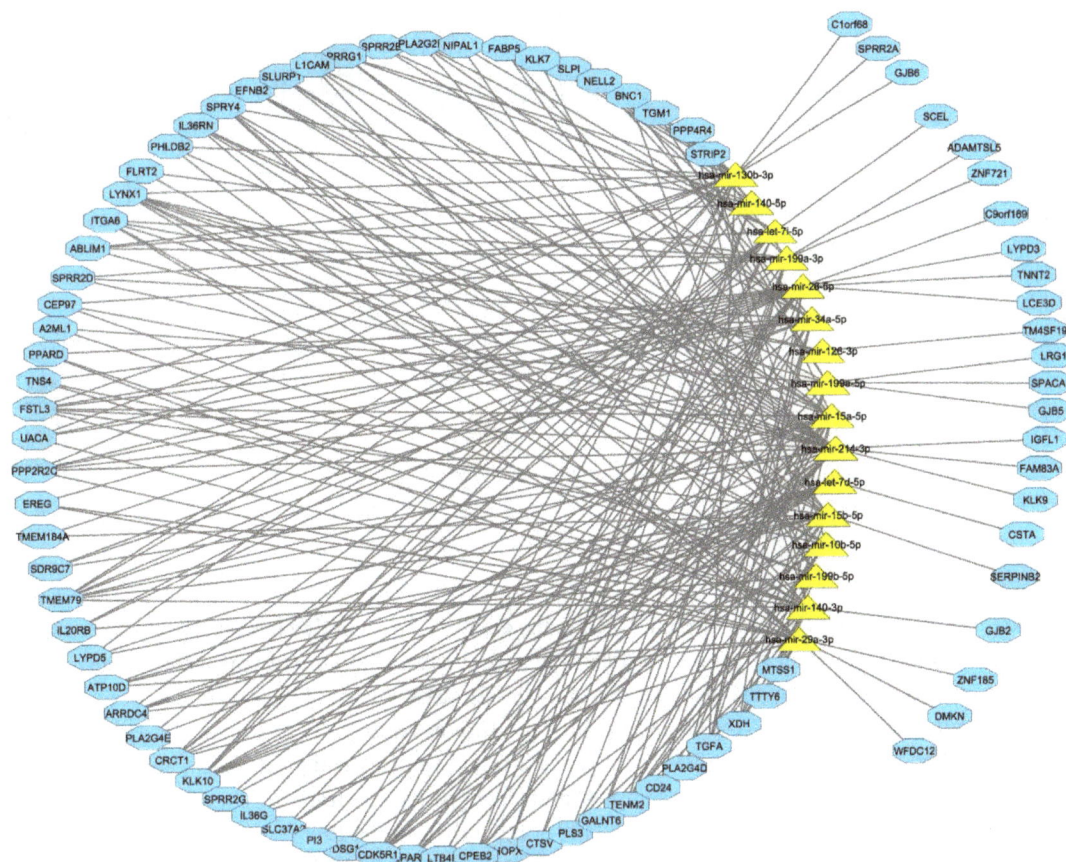

Figure 4. Computational integration of genes and miRNAs by MAGIA2 bioinformatics tool: The network of anticorrelated miRNA genes was computed by rCCA and integrated by MAGIA2 tools with the information of target prediction derived from DIANA-microT, TargetScan, and microrna.org. Two hundred and forty-five interactions were found among 16 miRNAs and 84 genes. The Cytoscape tool was applied to display the interconnections between features. Yellow triangles indicate the miRNAs, and blue boxes indicate the genes. If it is predicted that a gene is to be targeted by at least two out of 16 miRNAs, this gene is located in the central ring, comprising the miRNAs. Otherwise, it is placed outside.

3.5. Development of An Integrated miRNA-Gene Expression Predictive Model

We investigated to what extent the miRNA-gene integrated network could predict the outcome in our cohort of patients. A feature selection was imposed by means of stSVM to the miRNA-gene networks identified by rCCA, along with the adjacency matrix representing the biological interactions found by at least two of the three target prediction algorithms applied; we expected to select the most relevant features for predicting the outcome. As such, the results highlighted eight relevant features (three miRNAs and five genes): hsa-miR-199a-3p, hsa-miR-199a-5p, hsa-miR-199b-5p, ARRDC4, CRCT1, IL36G, KLK10, and PLA2G4E. The performance of the identified signature, as assessed through Receiver Operator Characteristic (ROC) analysis, reached an AUC of 0.992 (Figure 5A).

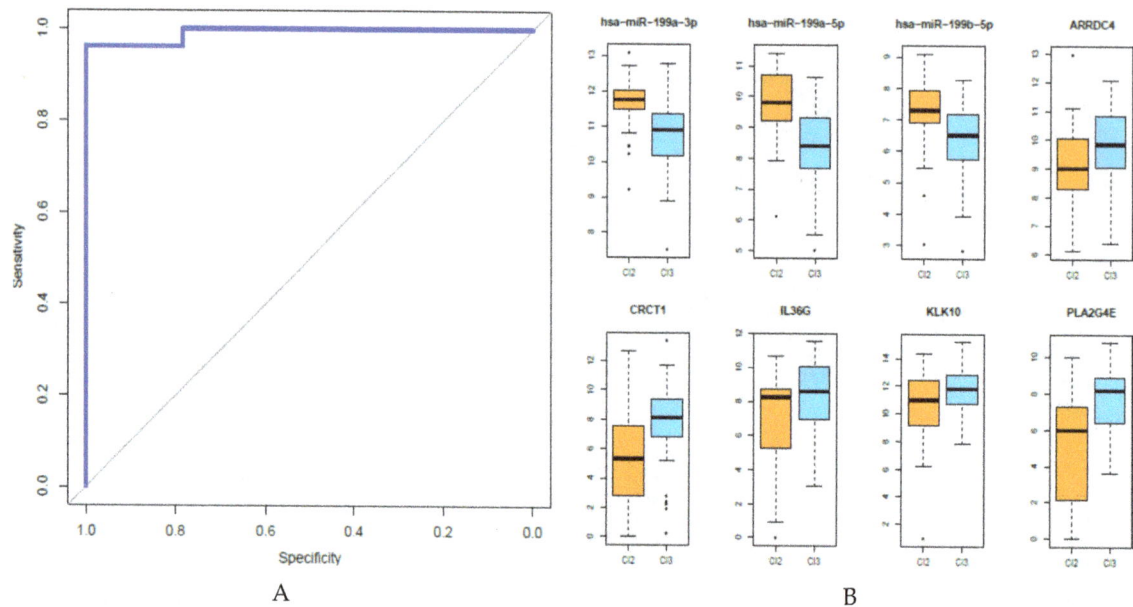

Figure 5. (A) Performance of the eight miRNA–gene expression predictive model: The performance of our signature (blue line) in predicting PFS after cetuximab-CT treatment was tested in terms of sensitivity and specificity by ROC analysis (AUC = 0.992); **(B)** Analysis of the expression of the eight features entered in the miRNA-gene integrated signature in HNSCC TCGA data set: The tumors were stratified based on our six HNSCC subtypes [34] and miRNA and gene expression data were retrieved for the Cl2-mesenchymal (30 samples) and Cl3-hypoxia (59 samples) subtypes. A significant upregulation of hsa-miR-199a-3p, hsa-miR-199a-5p, and hsa-miR-199b-5p in Cl2 ($p = 7.25E-05$, $p = 3.8E-06$, $p = 0.0019$, respectively) and of ARRDC4, CRCT1, IL36G, KLK10, and PLA2G4E in Cl3 ($p = 0.021$, $p = 4.66E-05$, $p = 0.00199$, $p = 0.0135$, $p = 2.42E-06$, respectively) was observed.

3.6. Analysis of the Eight miRNA-Gene Integrated Signatures in TCGA Data

Our previous analysis of gene expression of the same patients [4] demonstrated that long- and short-PFS patients are characterized by prevalently belonging to HNSCC tumor subtypes Cl2-mesenchymal and Cl3-hypoxia, associated with long- and short-PFS, respectively. The utilized subtype stratification was previously identified using a wide meta-analysis of publically available gene expression HSNCC datasets and was validated in other datasets including TCGA [34]. Since TCGA provides both miRNA and gene expression profiles, we investigated whether the three miRNAs and the five genes entered in the integrated signature could be associated with the Cl2 and Cl3 subtypes. The TCGA gene expression profile was used for subtype stratification, resulting in 30 and 59 samples classified as Cl2 and Cl3, respectively. These two sets of TCGA samples were used to analyze the expression of the eight features of our predictive signatures; in agreement with our hypothesis, a significant upregulation of the three miRNAs (hsa-miR-199a-5p, hsa-miR-199a-3p, and hsa-miR-199b-5p) was observed in the Cl2-mesenchymal subset of samples, while a significant upregulation of the five genes (ARRDC4, CRCT1, IL36G, KLK10, and PLA2G4E) was found in the Cl3-hypoxia subset of samples (Figure 5B).

4. Discussion

Combining data from multiple sources has the potential to draw a more comprehensive view of the biological systems, eventually enabling an improved prediction performance [35]. To our knowledge, the integrative analysis of miRNA and mRNA expression presented here is the first attempt on a cohort of HNSCC patients treated with EGFR inhibitors. Overall, using a well-defined clinical setting of pretreatment recurrent/metastatic HNSCC FFPE specimens and up-to-date bioinformatics tools, we demonstrated the proof of principle that an integrative miRNA-mRNA expression analysis could greatly contribute to a refinement in the identification of the biology behind the response to palliative cetuximab-based treatment and to the development of a potential predictive model.

High-throughput technologies have proved to be of great potential for gaining valuable insights into tumor biology, and even to build predictive models in HNSCC [36]. However, these data involve a number of variables (i.e., miRNA, genes), by far over-exceeding the number of samples, resulting in a high degree of multicollinearity and ill-conditioned issues. Many bioinformatics pipelines have been developed to solve these issues, mainly reducing the dimensionality of the data [37]. Taking into account a supervised classification setting, the approach involves the introduction of a new artificial variable able to summarize most of the information present in highly correlated miRNAs or genes. In this context, sparse partial–least squares discriminant analysis (sPLS-DA) is a well-known, supervised, pattern-recognition method able to unravel the information contained in a data matrix in relation to a qualitative response variable indicating classes of samples [10].

Long- and short-PFS cases were compared on the basis of their miRNA profiles using sPLS-DA. Our results clearly indicate that miRNA profiles separated the two groups of patients well, and the lowest misclassification rate was obtained considering a total of 36 miRNAs in two components. When this signature was compared to a supervised class comparison, most of the miRNAs differentially expressed by a random variance t-test–based class comparison were included in the first component and none in the second.

In general, miRNA expression profiles provide limited information about their biological role. In fact, miRNAs being non-coding regulators of gene expression, their function is linked not only to their expression level, but also to those of their target genes. In this context, an integrative analysis could help uncover the biological regulatory interactions between miRNAs and gene expression.

First of all, we investigated the association of miRNA sPLS-DA with gene expression data. A significant correlation between first-component miRNAs and pathways enriched in long-PFS patients (EGFR signaling, keratinocyte differentiation, ectoderm development pathways and p53 and Myc oncogenic signatures) recapitulates our previous observations obtained by directly analyzing gene expression [4], but also identified the association with hypoxia. An activated EGFR pathway leads to a response to cetuximab. In this context, CD24, a mucin-like membrane glycoprotein, regulates the expression of EGFR by preventing its internalization and RhoA-dependent degradation [38]. Our data clearly demonstrated an interaction between hsa-miR-34a-5p and its experimentally validated target CD24, as hsa-miR-34a-5p was downregulated while CD24 was upregulated in long-PFS patients. The combination of platinum-based chemotherapy and targeted therapy with an anti-EGFR inhibitor represents a strategy in treating RM-HNSCC. However, little is known about the mechanism leading to chemo-resistance [39]. It has been demonstrated that inhibition of L1 cell adhesion molecule (L1CAM) and EGFR sensitizes cells to cisplatin [40] and our findings support a molecular loop among EGFR signaling activation, the downregulation of hsa-miR-34a-5p and the upregulation of its L1CAM target in long-PFS patients.

Through a meta-analysis approach of gene expression profiling data, we identified subtypes with distinct molecular features including subtypes over-expressing hypoxia and mesenchymal pathways [34]. By association with drug sensitivity, the hypoxia subtype shows the greatest sensitivity to EGFR inhibitors [34]. In Bossi et al. [4], in a series of cetuximab-treated patients, we confirmed a molecular link between long-PFS patients and the Cl3-hypoxia subtype, and between short-PFS patients and the Cl2-mesenchymal subtype. Similar results were obtained following Keck's subtype stratification [41]. In addition, our integrated analysis disclosed the interplay between ITGA6, a hypoxia-regulated gene [42], and its experimentally validated target hsa-miR-29a-3p. Remarkably, the miRNA sPLS-DA second component, associated with short-PFS, confirmed the role of the RAS pathway [4], and the results mainly correlated with an immune/inflammatory response hub that comprises eight functionally related networks (immune response, inflammatory response, defense response, cytokine production, TNFA signaling, interferon gamma, interleukin, and leukocyte activation). In addition, inferring the gene networks associated with the miRNA sPLS-DA second component, we noticed a significant activation and enrichment of the epithelial–mesenchymal transition, angiogenesis, and NOTCH pathways, in agreement with their involvement in the Cl2-mesenchymal subtype [34]. Overall, our data strongly suggests that the miRNA sPLS-DA not only supports the biology previously identified as associated with long and short PFS [4], but provides new evidence. In particular, the sPLS-DA second component provides a completely new layer of information, not captured by differential expression analysis which strongly contributed to improving our knowledge about the biology behind the response to cetuximab-CT, and could open the way to new therapeutic suggestions. Anti–PD-1 therapies were reported to evade the checkpoint blockade, resulting in the reactivation of the immune system, enabling the eradication of the host tumor and improving, eventually, patient survival [43,44]. The significant enrichment of the immune/inflammatory pathway might suggest a potential benefit to immune therapy in short-PFS patients.

In systems biology, a major challenge in the integration effort of multi-omics and highly dimensional data is the extraction of meaningful information. Since it is expected that similar expression patterns across a set of samples potentially have a functional relationship, we applied a regularized version of canonical correlation analysis (rCCA) to explore correlation structures between miRNA and gene quantitative features assessed on the same cohort of patients [20]. Exploratory approaches with correlation circle plots and clustered image maps highlighted the association between 27 miRNAs and 250 genes, proving the existence of distinct anti-correlated expression patterns. These associations were confirmed by target prediction algorithms and in more detail. As such, all four members of the miR-199/214 cluster, and five out of the 12 members of the let-7 family, participate in the miRNA-mRNA network; a regulatory hub defined by four miRNAs (hsa-miR-130b-3p, hsa-miR-199a-3p, hsa-miR-214-3p, and hsa-miR-28-5p) and their mRNA targets, involving 41% of identified interactions, was observed.

Focusing our attention on the miRNA-gene integrated features able to stratify patients according to their response to cetuximab-based treatment, a signature based on three miRNAs and five genes was identified. The height miRNA-gene integrated signature showed an excellent accuracy in predicting treatment response (AUC = 0.992), while the single features reached an AUC ranging from 0.85 to 0.90. It was noteworthy that all three miRNAs belonged to the miR-199/214 cluster, and all five genes were regulated by two members of the let7 family (let7d and let7i). These data, together with the observation that when their expression was tested in HNSCC TCGA, the three miRNAs were upregulated in the Cl2-mesenchymal subtype, while the five genes were upregulated in the Cl3-hypoxia subtype, strongly support the hypothesis that these miRNAs and their regulated genes have a pivotal role in HSNCC progression and sensitivity to therapy.

We are aware that miRNAs, by binding to mRNA molecules, can inhibit their translation or induce degradation. However, most of the available tools and methods for miRNA target prediction, such as those here applied, have been developed following the assumption of an inverse correlation between miRNA/mRNA and they tend to miss the targets where expression repression is caused by translation repression. Since in the future we will expect a strong improvement in the knowledge on miRNA biology, the public availability of our miRNA and gene expression data for the scientific community at the GEO website will enable any future analysis of the interplay between miRNAs and genes.

5. Conclusions

Overall, through gene and miRNA expression profiling of the same samples of HNSCC and an integrative miRNA-mRNA expression analysis, we identified candidate biomarkers of PFS in patients treated with cetuximab-based approaches. The use of a well-defined clinical settings and up-to-date bioinformatics tools enabled to give the proof of principle that an integrative miRNA–mRNA expression could greatly contribute to further gain insight into the biology and prediction, and could have potential direct implications on clinical care. Since we selected patients with marked opposite outcomes after treatment, to transpose our predictor of response into a useful clinical grade assay, additional work is needed.

Acknowledgments: This work was supported by the Associazione Italiana Ricerca Cancro (AIRC IG 14750 to SC and IG 18519 to LDC). The authors thank Marco Siano (Cantonal Hospital St Gallen, Department of Internal Medicine, Clinic for Medical Oncology, St Gallen, Switzerland), Wolfram Jochum (Institute of Pathology, Kantonsspital St Gallen, St Gallen, Switzerland), Maria Cossu Rocca (Division of Medical Oncology, European Institute of Oncology, Milan, Italy), Andrea P. Sponghini (SC of Oncology, AOU Maggiore della Carità, Novara, Italy), and Federica Perrone (Department of Diagnostic Pathology and Laboratory, Fondazione IRCCS Istituto Nazionale dei Tumori, Milan, Italy) for help in case material collection and pathological review.

Author Contributions: Conception and design: S. Canevari, L. De Cecco; development of methodology: M. Giannoccaro, E. Marchesi, L. De Cecco; acquisition of data (acquired and managed patients, constructing databases): P. Bossi, F. Favales, L.D. Locati, L. Licitra, S. Pilotti; analysis and interpretation of data (e.g., statistical analysis, biostatistics, computational analysis): P. Bossi, S. Canevari, L. De Cecco; writing, review and/or revision of the manuscript: S. Canevari, L. De Cecco, P. Bossi, F. Favales, M. Giannoccaro, L. D. Locati, L. Licitra, E. Marchesi, S. Pilotti; study supervision: S. Canevari, L. De Cecco.

References

1. Pignon, J.P.; le Maître, A.; Maillard, E.; Bourhis, J.; MACH-NC Collaborative Group. Meta-analysis of chemotherapy in head and neck cancer (MACH-NC): An update on 93 randomised trials and 17,346 patients. *Radiother. Oncol.* **2009**, *92*, 4–14. [CrossRef] [PubMed]

2. Fung, C.; Grandis, J.R. Emerging drugs to treat squamous cell carcinomas of the head and neck. *Exp. Opin. Emerg. Drugs* **2010**, *15*, 355–373. [CrossRef] [PubMed]

3. Ritchie, M.D.; Holzinger, E.R.; Li, R.; Pendergrass, S.A.; Kim, D. Methods of integrating data to uncover genotype-phenotype interactions. *Nat. Rev. Genet.* **2015**, *16*, 85–97. [CrossRef] [PubMed]

4. Bossi, P.; Bergamini, C.; Siano, M.; Rocca, M.C.; Sponghini, A.P.; Favales, F.; Giannoccaro, M.; Marchesi, E.; Cortelazzi, B.; Perrone, F.; et al. Functional genomics uncover the biology behind the responsiveness of head and neck squamous cell cancer patients to cetuximab. *Clin. Cancer Res.* **2016**, *22*, 3961–3970. [CrossRef] [PubMed]

5. De Cecco, L.; Capaia, M.; Zupo, S.; Cutrona, G.; Matis, S.; Brizzolara, A.; Orengo, A.M.; Croce, M.; Marchesi, E.; Ferrarini, M.; et al. Interleukin 21 controls mRNA and microRNA expression in CD40-activated chronic lymphocytic leukemia cells. *PLoS ONE* **2015**, *10*, e0134706. [CrossRef] [PubMed]

6. Bagnoli, M.; Canevari, S.; Califano, D.; Losito, S.; Maio, M.D.; Raspagliesi, F.; Carcangiu, M.L.; Toffoli, G.; Cecchin, E.; Multicentre Italian Trials in ovarian cancer (MITO) translational group; et al. Development and validation of a microRNA-based signature (MiROvaR) to predict early relapse or progression of epithelial ovarian cancer: A cohort study. *Lancet Oncol.* **2016**, *17*, 1137–1146. [CrossRef]

7. Argiris, A.; Li, Y.; Forastiere, A. Prognostic factors and long-term survivorship in patients with recurrent or metastatic carcinoma of the head and neck. *Cancer* **2004**, *101*, 2222–2229. [CrossRef] [PubMed]

8. López-Romero, P. Pre-processing and differential expression analysis of agilent microRNA arrays using the AgiMicroRna Bioconductor library. *BMC Genomics* **2011**. [CrossRef] [PubMed]

9. Gene Expression Omnibus. Available online: www.ncbi.nlm.nih.gov/geo/ (accessed on 20 July 2016).

10. Lê Cao, K.A.; Boitard, S.; Besse, P. Sparse PLS discriminant analysis: Biologically relevant feature selection and graphical displays for multiclass problems. *BMC Bioinform.* **2011**. [CrossRef]

11. Le Cao, K.A.; Gonzalez, I.; Dejean, S. integrOmics: An R package to unravel relationships between two omics datasets. *Bioinformatics* **2009**, *25*, 2855–2856. [CrossRef] [PubMed]

12. Bioconductor. Available online: www.bioconductor.org (accessed on 20 July 2016).

13. Lê Cao, K.-A.; LeGall, C. Integration and variable selection of "omics" data sets with PLS: A survey. *J. de la Société Francaise de Statistique* **2011**, *152*, 77–96.

14. Murria Estal, R.; Palanca Suela, S.; de Juan Jiménez, I.; Egoavil Rojas, C.; García-Casado, Z.; Juan Fita, M.J.; Sánchez Heras, A.B.; Segura Huerta, A.; Chirivella González, I.; Sánchez-Izquierdo, D.; et al. MicroRNA signatures in hereditary breast cancer. *Breast Cancer Res. Treat.* **2013**, *142*, 19–30. [CrossRef] [PubMed]

15. Benjamini, Y.; Hochberg, Y. Controlling the False Discovery Rate—A practical and powerful approach to multiple testing. *J. R. Stat. Soc.* **1995**, *57*, 289–300.

16. Biometric Research Program (BRP). Available online: http://linus.nci.nih.gov/ (accessed on 10 September 2015).

17. Subramanian, A.; Tamayo, P.; Mootha, V.K.; Mukherjee, S.; Ebert, B.L.; Gillette, M.A.; Paulovich, A.; Pomeroy, S.L.; Golub, T.R.; Lander, E.S.; et al. Gene set enrichment analysis: A knowledge-based approach for interpreting genome-wide expression profiles. *Proc. Natl. Acad. Sci. USA* **2005**, *102*, 15545–15550. [CrossRef] [PubMed]

18. Wencke, W.; Sánchez-Cabo, F.; Ricote, M. GOplot: An R package for visually combining expression data with functional analysis. *Bioinformatics* **2015**, *31*, 2912–2914.

19. Hotelling, H. Relations between two sets of variates. *Biometrika* **1936**, *28*, 321–377. [CrossRef]

20. Parkhomenko, E.; Tritchler, D.; Beyene, J. Sparse canonical correlation analysis with application to genomic data integration. *Stat. Appl. Genet. Mol. Biol.* **2009**, *8*, 1–34. [CrossRef] [PubMed]

21. González, I.; Le Cao, K.A.; Davis, M.J.; Déjean, S. Visualising associations between paired "omics" data sets. *BioData Min.* **2012**. [CrossRef] [PubMed]

22. Gonzalez, I.; Dejean, S.; Martin, P.; Baccini, A. CCA: An R package to extend canonical correlation analysis. *J. Stat. Software* **2008**. [CrossRef]

23. Insightful graphical outputs to explore relationships between two "omics" data sets. Available online: http://mixomics.org/wp-content/uploads/2012/03/Visualising_JDataMining.pdf (accessed on 20 July 2016).

24. Bisognin, A.; Sales, G.; Coppe, A.; Bortoluzzi, S.; Romualdi, C. MAGIA^2: From miRNA and genes expression data integrative analysis to microRNA-transcription factor mixed regulatory circuits (2012 update). *Nucleic Acids Res.* **2012**. [CrossRef] [PubMed]

25. Maragkakis, M.; Alexiou, P.; Papadopoulos, G.L.; Reczko, M.; Dalamagas, T.; Giannopoulos, G.; Goumas, G.; Koukis, E.; Kourtis, K.; Simossis, V.A.; et al. Accurate microRNA target prediction correlates with protein repression levels. *BMC Bioinform.* **2009**. [CrossRef] [PubMed]

26. Grimson, A.; Farh, K.K.-H.; Johnston, W.K.; Garrett-Engele, P.; Lim, L.P.; Bartel, D.P. MicroRNA targeting specificity in mammals: Determinants beyond seed pairing. *Mol. Cell* **2007**, *27*, 91–105. [CrossRef] [PubMed]

27. Betel, D.; Wilson, M.; Gabow, A.; Marks, D.S.; Sander, C. The microRNA.org resource: Targets and expression. *Nucleic Acids Res.* **2007**, *36*, D149–D153. [CrossRef] [PubMed]

28. Shannon, P.; Markiel, A.; Ozier, O.; Baliga, N.S.; Wang, J.T.; Ramage, D.; Amin, N.; Schwikowski, B.; Ideker, T. Cytoscape: A software environment for integrated models of biomolecular interaction networks. *Genome Res.* **2003**, *13*, 2498–2504. [CrossRef] [PubMed]

29. Chou, C.-H.; Chang, N.-W.; Shrestha, S.; Hsu, S.D.; Lin, Y.L.; Lee, W.H.; Yang, C.D.; Hong, H.C.; Wei, T.Y.; Tu, S.J.; et al. miRTarBase 2016: Updates to the experimentally validated miRNA-target interactions database. *Nucleic Acids Res.* **2016**, *44*, D239–D247. [CrossRef] [PubMed]

30. Cun, Y.; Frohlich, H. Network and data integration for biomarker signature discovery via network smoothed t-statistics. *PLoS ONE* **2013**. [CrossRef] [PubMed]

31. Cun, Y.; Frohlich, H. netClass: An R-package for network based, integrative biomarker signature discovery. *Bioinformatics* **2014**, *30*, 1325–1326. [CrossRef] [PubMed]

32. Cancer Genome Atlas Network. Comprehensive genomic characterization of head and neck squamous cell carcinomas. *Nature* **2015**, *517*, 576–582.

33. The Cancer Genome Atlas—Cancer Genome. Available online: http://cancergenome.nih.gov/ (accessed on 20 June 2013).

34. De Cecco, L.; Nicolau, M.; Giannoccaro, M.; Daidone, M.G.; Bossi, P.; Locati, L.; Canevari, S. Head and neck cancer subtypes with biological and clinical relevance: Meta-analysis of gene-expression data. *Oncotarget* **2015**, *6*, 9627–9642. [CrossRef] [PubMed]

35. Gomez-Cabrero, D.; Abugessaisa, I.; Maier, D.; Teschendorff, A.; Merkenschlager, M.; Gisel, A.; Ballestar, E.; Bongcam-Rudloff, E.; Conesa, A.; Tegnér, J. Data integration in the era of omics: Current and future challenges. *BMC Syst. Biol.* **2014**. [CrossRef] [PubMed]

36. De Cecco, L.; Bossi, P.; Locati, L.; Canevari, S.; Licitra, L. Comprehensive gene expression meta-analysis of head and neck squamous cell carcinoma microarray data defines a robust survival predictor. *Ann. Oncol.* **2014**, *25*, 1628–1635. [CrossRef] [PubMed]

37. Bersanelli, M.; Mosca, E.; Remondini, D.; Giampieri, E.; Sala, C.; Castellani, G.; Milanesi, L. Methods for the integration of multi-omics data: mathematical aspects. *BMC Bioinform.* **2016**, *17*. [CrossRef] [PubMed]

38. Deng, W.; Gu, L.; Li, X.; Zheng, J.; Zhang, Y.; Duan, B.; Cui, J.; Dong, J.; Du, J. CD24 associates with EGFR and supports EGF/EGFR signaling via RhoA in gastric cancer cells. *J. Transl. Med.* **2016**. [CrossRef] [PubMed]

39. Brown, D.; Wang, R.; Russell, P. Antiepidermal growth factor receptor antibodies augment cytotoxicity of chemotherapeutic agents on squamous cell carcinoma cell lines. *Otolaryngol. Head Neck Surg.* **2000**, *122*, 75–83. [CrossRef]

40. Yoon, H.; Min, J.K.; Lee, D.G.; Kim, D.G.; Koh, S.S.; Hong, H.J. L1 cell adhesion molecule and epidermal growth factor receptor activation confer cisplatin resistance in intrahepatic cholangiocarcinoma cells. *Cancer Lett.* **2012**, *316*, 70–76. [CrossRef] [PubMed]

41. Keck, M.K.; Zuo, Z.; Khattri, A.; Stricker, T.P.; Brown, C.D.; Imanguli, M.; Rieke, D.; Endhardt, K.; Fang, P.; Brägelmann, J.; et al. Integrative analysis of head and neck cancer identifies two biologically distinct HPV and three non-HPV subtypes. *Clin. Cancer Res.* **2015**, *21*, 870–881. [CrossRef] [PubMed]

42. Brooks, D.L.; Schwab, L.P.; Krutilina, R.; Parke, D.N.; Sethuraman, A.; Hoogewijs, D.; Schörg, A.; Gotwald, L.; Fan, M.; Wenger, R.H.; et al. ITGA6 is directly regulated by hypoxia-inducible factors and enriches for cancer stem cell activity and invasion in metastatic breast cancer models. *Mol. Cancer* **2016**. [CrossRef] [PubMed]

43. Chow, L.Q.; Haddad, R.; Gupta, S.; Mahipal, A.; Mehra, R.; Tahara, M.; Berger, R.; Eder, J.P.; Burtness, B.; Lee, S.H.; et al. Antitumor activity of pembrolizumab in biomarker-unselected patients with recurrent and/or metastatic head and neck squamous cell carcinoma: results from the phase Ib KEYNOTE-012 expansion Cohort. *J. Clin. Oncol.* **2016**, *34*, 3838–3845. [CrossRef] [PubMed]

44. Ferris, R.L.; Blumenschein, G.; Fayette, J.; Guigay, J.; Colevas, A.D.; Licitra, L.; Harrington, K.; Kasper, S.; Vokes, E.E.; Even, C.; et al. Nivolumab for recurrent squamous-cell carcinoma of the head and neck. *N. Engl. J. Med.* **2016**, *375*, 1856–1867. [CrossRef] [PubMed]

Tissue Non-Specific Genes and Pathways Associated with Diabetes

Hao Mei [1,2,*,†], Lianna Li [3,†], Shijian Liu [2,*], Fan Jiang [2], Michael Griswold [1] and Thomas Mosley [4]

[1] Department of Data Science, School of Population Health, University of Mississippi Medical Center, Jackson, MS 39216, USA; mgriswold@umc.edu
[2] Shanghai Children's Medical Center, School of Public Health and School of Medicine, Shanghai Jiaotong University, Shanghai 200127, China; jiangfan@scmc.com.cn
[3] Department of Biology, Tougaloo College, Jackson, MI 39216, USA; lli@tougaloo.edu
[4] Department of Neurology, University of Mississippi Medical Center, Jackson, MS 39216, USA; tmosley@umc.edu
* Correspondence: hmei@umc.edu (H.M.); arrow64@163.com (S.L.)
† These authors contributed equally to this work.

Academic Editor: Roel Ophoff

Abstract: We performed expression studies to identify tissue non-specific genes and pathways of diabetes by meta-analysis. We searched curated datasets of the Gene Expression Omnibus (GEO) database and identified 13 and five expression studies of diabetes and insulin responses at various tissues, respectively. We tested differential gene expression by empirical Bayes-based linear method and investigated gene set expression association by knowledge-based enrichment analysis. Meta-analysis by different methods was applied to identify tissue non-specific genes and gene sets. We also proposed pathway mapping analysis to infer functions of the identified gene sets, and correlation and independent analysis to evaluate expression association profile of genes and gene sets between studies and tissues. Our analysis showed that *PGRMC1* and *HADH* genes were significant over diabetes studies, while *IRS1* and *MPST* genes were significant over insulin response studies, and joint analysis showed that HADH and MPST genes were significant over all combined data sets. The pathway analysis identified six significant gene sets over all studies. The KEGG pathway mapping indicated that the significant gene sets are related to diabetes pathogenesis. The results also presented that 12.8% and 59.0% pairwise studies had significantly correlated expression association for genes and gene sets, respectively; moreover, 12.8% pairwise studies had independent expression association for genes, but no studies were observed significantly different for expression association of gene sets. Our analysis indicated that there are both tissue specific and non-specific genes and pathways associated with diabetes pathogenesis. Compared to the gene expression, pathway association tends to be tissue non-specific, and a common pathway influencing diabetes development is activated through different genes at different tissues.

Keywords: gene expression; pathway; diabetes

1. Introduction

Diabetes is a chronic metabolic disease of hyperglycemia resulting from defects in insulin secretion, action, or both: the type I diabetes (T1D) is mainly caused by beta-cell destruction and the type II diabetes (T2D) is characterized by defects in insulin action and/or secretion. In the T1D, the cell destruction will eventually eliminate insulin production and lead to absolute insulin deficiency [1]. In contrast, people with the T2D are often resistant to the insulin action. A good understanding of genetics underlying diabetes pathogenesis will play an important role in developing effective prevention, diagnosis and therapy strategy to manage diabetes and relieve its public heath burden.

Diabetes progress and impaired insulin action are accompanied with pathochanges at multiple tissues, including the pancreas, skeletal muscles, the liver and adipose. Of these tissues, the pancreas plays a central role in the diabetes development, and either destruction of its beta cells or reduction of insulin production will lead to the impaired glucose homeostasis. Insulin resistance is a major predictor of T2D and plays an important role in diabetes pathogenesis [2]. More than 80% of insulin-stimulated glucose uptake occurs in skeletal muscle and ~5%–10% of the uptake happens in the adipose tissue [3]. Impaired insulin responses at these tissues will cause abnormal glucose metabolism and the followed hyperglycemia.

In the skeletal muscle, diabetic myotubes is often accompanied with mitochondrial dysfunction, presenting decreased rates of mitochondrial ATP production and substrate oxidation [4]. Hyperglycemia in diabetic patients increases the production of superoxide, resulting in the endothelial dysfunction and decreased numbers of endothelial progenitor cells (EPCs), and diabetes and the impaired progenitor cells are considered to have common pathogenesis [5]. Diabetes progression is associated with arterial pathology, including extracellular matrix changes and increased stiffness in the nonatherosclerotic arterial tissue [6,7]. The liver is also a major tissue taking important roles in glucose homeostasis. Hepatic lipid accumulation is associated with insulin resistance, and the liver can produce various secretory proteins, termed hepatokines, associated with insulin resistance and clinical manifestations of diabetes [8,9]. Pathogenetic study of these tissues contributes to understanding the etiology of impaired insulin action and diabetes development. Although the T1D and T2D have different etiology with pathochanges at multiple tissues, they present some common clinical manifestations and the gene expression study showed that both types of diabetes share pathogenic mechanisms [10].

Advances of high-throughput technology have led to an explosion of gene expression data collected from different tissues in the past decade. The Gene Expression Omnibus (GEO) database has served as a public repository to archive these expression measures, which are generated mostly by microarray technology, and to facilitate retrieval and mining of published expression data [11]. The continuous increase of archived GEO data offers the opportunity to pool gene expression from different studies and tissues, which will help to improve identification of gene signatures associated with a disease that may lack sufficient evidence in a single study before. For this study, we hypothesize that those tissues involved in diabetes pathogenesis share common genetic regulations, and aim to identify tissue non-specific genes and pathways based on expression datasets from the GEO by meta-analysis.

2. Materials and Methods

2.1. Gene Expression Datasets

We searched the GEO database for gene expression datasets that are related to diabetes and insulin response. Our analysis is focused only on those manually curated GEO datasets (GDSs) that were directly downloaded and parsed through R package of GEO query [12]. The expression level of every gene, measured as M value (i.e., log 2-expression level), was extracted for follow-up analysis. GDSs will be merged to a single study if the datasets are expression measures from the same samples but different microarray platforms. Gene expression of every study is measured as the largest M value if the gene was assayed on multiple platforms [13]. To make gene expression comparable across samples, all probe M values were normalized by quantile normalization from the R package, preprocessCore [14,15].

2.2. Gene Expression Association Test and Meta-Analysis

The empirical Bayes-based linear regression method [16] was applied to test differential gene expression based on the null hypothesis that expression of M value is equal across all K phenotypes:

$M_1 = M_2 = \ldots = M_i \ldots = M_k$ ($K \geq 2$), where M_i can be the case status of T1D and T2D, healthy control, insulin resistant or insulin sensitive. A significant test will suggest gene association with diabetes and insulin response. The analysis was performed by the R package, limma [16], and the standard errors of tests were moderated across genes by empirical Bayes model to calculate F statistic and p-value for every gene. The U-score [17] of the i-th gene (U_i) is calculated as $U_i = (\sum_j I(p_j < p_i) + 0.5 \cdot \sum_j I(p_j = p_i))/N$, where p_i is p-value of the i-th gene and N is the total number of measured genes. The U-score approximately follows uniform distribution, estimating the percentage of genes with stronger expression association than the tested one. We hypothesize that 5% of genes are associated with diabetes, and a gene with U-score ≤ 0.05 is defined as significant for following pathway test and meta-analysis.

Meta-analysis was conducted by the binomial test to evaluate differential gene expressions over studies and tissues. The binomial test counted the number of significant genes with U-score ≤ 0.05 over studies as a random variable X, which follows a binomial distribution with probability of 0.05, and the meta-analysis p-value was calculated as $Bin\_P = Pr(X \geq \sum_{i=1}^{M} I(U_i \leq 0.05))$. The significance of Bin_P is based on the Bonferroni adjustment for the total number of genes.

2.3. Pathway Expression Association Test and Meta-Analysis

Pathway expression association was examined by testing enrichment of knowledge-based gene sets for significant genes. The test was based on the MSigDB knowledge base [18] that contains curated information of over 10,000 gene sets extracted from different public pathway databases, e.g., the Kyoto Encyclopedia of Genes and Genomes (KEGG) [19]. The enrichment analysis was conducted by the hypergeometric test of significant genes with U-score ≤ 0.05 using the R package snpGeneSets [17,20]. The pathway effect was estimated as the proportion of significant genes in the gene set minus 5%. The pathway p-value ($path\_p$) was calculated based on hypergeometric distribution, and the adjusted p-value ($path\_p_a$) was obtained by a permutation test to adjust for multiple testing.

Meta-analysis was conducted by the fixed-effect model and the binomial test to measure pathway expression associations across studies. The fixed-effect model with inverse of variance as study-specific weight was applied to estimate pathway enrichment effect over all studies, and meta-analysis p-value ($Fixed\_p$) was calculated to test the null hypothesis of effect = 0. The analysis was performed by the R package of metaphor [21]. The binomial test calculates meta-analysis p-value based on unadjusted ($path\_p$) and adjusted pathway p-value ($path\_p_a$), respectively, over M studies as $Bin\_p0 = Pr(X \geq \sum_{i=1}^{M} I((path\_p)_i \leq 0.05))$ and $Bin\_p1 = Pr(X \geq \sum_{i=1}^{M} I((path\_p_a)_i \leq 0.05))$, where X is a random variable following binomial distribution with size M and probability of 0.05. The significance of $Fixed\_p$ and Bin_p0 is based on the Bonferroni adjustment for the number of tested gene sets, while Bin_p1 is significant if the value is ≤ 0.05.

2.4. KEGG Pathway Mapping Analysis

The KEGG [22] pathway database describes manually curated molecular interaction and reaction networks, and provides pathway maps for common human diseases. The mapping analysis, similar to the enrichment analysis above, applied hypergeometric test by the R package snpGeneSets to examine if the MSigDB gene set significantly overlaps a KEGG pathway [20]. The mapping effect estimates the higher probability for a gene of the MSigDB gene set than a random gene, while they also belong to the KEGG pathway [20]. The mapping p-value is based on the hypergeometric distribution and the adjusted p-value is obtained by 10,000 permutation tests. A significant test with adjust p-value ≤ 0.05 suggests that the KEGG pathway is correlated with the MSigDB gene set, and they potentially share common functions.

2.5. Correlation and Independent Analysis of Expression Association Profile between Studies

To investigate the profile of expression association with diabetes and insulin response between studies, we conducted correlation and independent analysis for both genes and MSigDB gene sets.

The correlation analysis was based on Spearman's rank correlation test by the *R* function of *cor.test*, which aimed to examine similar expression association profiles between studies. The independent analysis was based on the *U*-score of gene set association and tested by the McNemar's method with the *R* function of *mcnemar.test*. The pathway *U*-score was calculated the same as the gene *U*-score above, and the value ≤ 0.05 indicated that its association strength ranked at the top 5%. The independent analysis counted the inconsistent number of genes and gene sets with *U*-score ≤ 0.05 at one study but *U*-score > 0.05 at the compared study, which aimed to examine different expression association between studies. For both types of analyses, the Bonferroni method was applied to adjust for multiple testing.

3. Results

3.1. Characteristics of the Gene Expression Datasets and Studies

Our search against the GEO database formed 13 gene expression studies based on 14 GDSs of diabetes states, including T1D and T2D, and the tissues include skeletal muscles, myotube, pancreas, liver, blood cells, endothelial progenitor cells (EPCs), arteries and adipose. The search also generated five expression studies based on 11 GDSs of insulin actions, and the tissues include skeletal muscles and adipose. The characteristics of all expression studies were summarized at the Table 1, showing the study ID, the GDS ID, the microarray platform, the PUBMED ID, the number of genes measured, the sample size, the contrast test for differential gene expression and the tissue.

Table 1. Characteristics of the gene expression studies.

Study	GDS_ID	GPL_ID	Pub_ID	N_genes	Size	Contrast	Tissue
						Diabetes State	
1	GDS3665	GPL2986		16,075	10	T2D vs. control	adipose
2	GDS3980	GPL571	21926180 [6]; 22340758 [7]	12,778	21	T2D vs. control	artery
3	GDS3874 GDS3875	GPL96 GPL97	17595242 [10]	18,552	117	T1D vs. healthy and T2D vs. healthy	blood
4	GDS3963	GPL6883	21829658 [23]	17,476	24	T2D vs. impaired fasting glucose vs. control	blood
5	GDS3656	GPL2700	19706161 [5]	16,778	32	T1D vs. Healthy	EPC
6	GDS3876	GPL96	19549744 [9]	12,779	18	obese T2D vs. obese no T2D	liver
7	GDS3883	GPL570	21035759 [8]	20,539	17	T2D vs. normal glucose tolerance	liver
8	GDS3681	GPL8300	18719883 [4]	8861	20	T2D vs. control	myotube
9	GDS3782	GPL1352	20644627 [24]	20,185	20	T2D vs. control	pancreas
10	GDS3882	GPL96	21127054 [25]	12,779	13	T2D vs. non-diabetes	pancreas
11	GDS4337	GPL6244	22768844 [26]	17,323	63	T2D vs. non-diabetes	pancreas
12	GDS3880	GPL570	22802091 [27]	20,539	42	T2D vs. pre-diabetes vs. normoglycemic control	skeletal muscle
13	GDS3884	GPL570	21393865 [28]	20,539	50	T2D vs. Normoglycemia with FH+ vs. Normoglycemia with FH−	skeletal muscle

Table 1. *Cont.*

Study	GDS_ID	GPL_ID	Pub_ID	N_genes	Size	Contrast	Tissue
				Insulin Action			
1	GDS157 GDS158 GDS160 GDS161 GDS162	GPL80 GPL98 GPL99 GPL100 GPL101	12436343 [29]	13,742	10	insulin resistant vs. insulin sensitive	skeletal muscle
2	GDS2790 GDS2791	GPL80 GPL96	17472435 [30]	12,885	12	Before vs. after Hyperinsulinemic-euglycemic clamp for nondiabetes	skeletal muscle
3	GDS3181	GPL96	18334611 [31]	12,779	36	-60 vs. 30 vs. 240 min of Hyperinsulinemic-euglycemic clamp for nondiabetes	skeletal muscle
4	GDS3715	GPL91	17709892 [32]; 21109598 [33]	8768	110	Diabetes vs. insulin sensitive vs. insulin resistant before and after Hyperinsulinemic-euglycemic clamp	skeletal muscle
5	GDS3781 GDS3962	GPL570	20678967 [34]	20,539	39; 19	insulin sensitive vs. insulin resistant	adipose

GDS_ID: the GDS ID of the expression dataset; GPL_ID: ID of the platform for generating the expression dataset; PUB_ID: the publication ID at the PubMed database; N_genes: the number of genes measured for expression level; Size: the number of samples at the study; Contrast: it presented the test of differential gene expression between two or more phenotypes by the regression method; EPC: endothelial progenitor cells; FH+: family history of diabetes; FH−: no family history of diabetes; T1D: type I diabetes; T2D: type II diabetes.

3.2. Tissue Non-Specific Gene Expression Association

We performed differential expression tests for 6889 and 7332 genes that were measured at all studies of diabetes and insulin actions, respectively, and the corresponding adjusted significance levels by Bonferroni correction were 7.26E−06 and 6.82E−6. The meta-analysis showed that the genes of progesterone receptor membrane component 1 (*PGRMC1*) and hydroxyacyl-CoA dehydrogenase (*HADH*) were significant over 13 studies of diabetes with *p*-values (*Bin_P*) of 1.03E−6 and 1.03E−6 respectively, and the genes of insulin receptor substrate 1 (*IRS1*) and mercaptopyruvate sulfurtransferase (*MPST*) were significant across five studies of insulin action with *Bin_P* of 3.13E−7. *U*-scores and *p*-values of the four genes at every study were summarized in Table 2 and the gene descriptions were shown in Supplementary Table S1.

The meta-analysis results (Table 2) showed that the *PGRMC1* were significant across diabetes studies, presenting six out of 13 studies with the *U*-score ≤ 0.05, and the *IRS1* gene was significant in four studies of insulin response (i.e., *U*-score ≤ 0.05). The joint analysis of all 17 studies showed that the *HADH* and the *MPST* were significant with *p*-values of 6.31E−7 and 6.28E−8, respectively. The *HADH* was significant in six diabetes studies (studies 1, 3, 5, 8, 11 and 13) of adipose, blood, EPC, myotube, pancreas and skeletal muscles, and the gene had the smallest *U*-score of 0.44% at the study 11 of pancreas that compared T2D with non-diabetes. Of the five studies of insulin response, the *HADH* had the *U*-score of 3.71% in study 5 of adipose. The *MPST* was significant in four insulin response studies of skeletal muscles with the smallest *U*-score of 0.86% (study 4), and the significant differential expression was also observed in four diabetes studies of adipose, arteries, blood and the liver (study 1, 2, 4 and 7) with the smallest *U*-score of 0.20% (study 4).

Table 2. Differential gene expression and meta-analysis.

Study	GDS_ID	PGRMC1	HADH	IRS1	MPST
		Gene U-Score (%) of Diabetes State			
1	GDS3665	*4.79*	*0.48*	9.17	*2.68*
2	GDS3980	31.31	14.46	20.64	*3.64*
3	GDS3874/GDS3875	8.55	*3.48*	24.46	97.48
4	GDS3963	45.03	48.6	*3.01E−03*	*0.2*
5	GDS3656	*0.52*	*3.35*	68.95	89.07
6	GDS3876	*4.16*	82.07	89.31	35.61
7	GDS3883	97.9	69.24	15.34	*3.85*
8	GDS3681	88.77	*2.73*	81.4	27.87
9	GDS3782	*0.34*	7.43	81	79.18
10	GDS3882	2.33	66.82	95.75	47.81
11	GDS4337	*4.66*	*0.44*	7.54	37.53
12	GDS3880	95.17	50.5	41.35	6.55
13	GDS3884	96.25	*2.97*	48.08	54.24
	Bin_P	1.03E−6	1.03E−6	0.14	2.87E−4
Study	**GDS_ID**	*Gene U-Score (%) of Insulin Action*			
1	GDS157/GDS158/GDS160/GDS161/GDS162	7.48	NA	*3.82*	2.2
2	GDS2790/GDS2791	9.59	10.14	*2.5*	*3.92*
3	GDS3181	55.03	16.78	*3.46*	*4.56*
4	GDS3715	30.9	82.89	25.76	*0.86*
5	GDS3781/GDS3962	48.65	*3.71*	*4.52*	23.59
	Bin_P	0.23	0.01	3.13E−7	3.13E−7
Joint Analysis of Combined Diabetes State and Insulin Action					
	Bin_P		6.31E−7		6.28E−8

The bold italic font indicates significantly differential gene expression (i.e., U-score ≤ 5%).

3.3. Tissue Non-Specific Pathway Expression Association

We performed meta-analysis of the pathway expression test for MSigDB gene sets over 13 diabetes studies, five insulin action studies and their combined data sets. The meta-analysis p-values of *Fixed_p* and *Bin_p0* used significant levels of 5.0×10^{-6} based on the Bonferroni correction for about 10,000 gene sets, while the p-value of *Bin_p1* directly took the significant level of 0.05, due to fact that its calculation was based on adjusted pathway p-values of individual expression studies. The analysis identified six significant gene sets at the diabetes studies and the combined datasets, including "UV response", "chronic myelogenous leukemia", "KLF1 targets", "SMARCA2 targets", "Alzheimer's disease" and "stromal stem cells". p-values of the six gene sets by different methods were shown in Table 3, and a description of these gene sets can be found at the Supplementary Table S2. The detailed results for every study were summarized at the Supplementary Tables S3–S8 with the forest plots shown at the Supplementary Figures S1–S6.

Meta-analysis of diabetes studies showed that the *Fixed_p* and *Bin_p0* of the six gene sets ranged at 1.45×10^{-38}–1.88×10^{-15} and 3.47×10^{-13}–4.01×10^{-08}, respectively, while the *Bin_p1* ranged at 3.10×10^{-3}~1.97×10^{-5} (Table 3). The gene sets were also consistently confirmed at the meta-analysis of all combined data sets: the "chronic myelogenous leukemia" had the smallest p-values of *Fixed_p* (3.91×10^{-44}) and *Bin_p1* (1.52×10^{-5}) and the second smallest p-value of *Bin_p0* (3.41×10^{-12}); and the "Alzheimer's disease" had the least significant p-values of *Fixed_p* (1.84×10^{-18}), Bin_p0 (2.95×10^{-9}) and *Bin_p1* (1.55×10^{-3}). The pathway enrichment analyses showed that the "chronic myelogenous leukemia" had adjusted p-value < 0.05 in five diabetes expression studies of adipose, arteries, blood and pancreatic tissues with effect = 2.87%–7.07% and the insulin response study of skeletal muscles with effect = 4.63% (Supplementary Table S4); the "Alzheimer's disease" had adjusted p-value < 0.05 in four diabetes studies of arteries, blood and pancreatic tissues with effect = 3.10%–4.06% (Supplementary Table S7). The joint analysis also showed that the p-values of *Fixed_p*, *Bin_p0* and *Bin_p1* were 9.46E−32, 1.12E−10 and 1.55E−03 for "UV response", 4.72E−29, 1.54E−15 and 1.55E−03 for "KLF1 targets", 1.11E−27, 6.28E−08 and 1.55E−03 for "SMARCA2 targets", and 2.93E−22, 1.12E−10 and 1.72E−04 for "stromal stem cells".

Table 3. Meta-analysis p-values of significant gene sets.

PID	GeneSet	Fixed_p	Bin_p0	Bin_p1
		Diabetes Studies		
5599	UV response	7.72E−17	4.01E−08	3.10E−03
4914	chronic myelogenous leukemia	1.45E−38	1.16E−09	1.97E−05
7922	KLF1 targets	3.35E−26	3.47E−13	3.10E−03
5947	SMARCA2 targets	1.95E−25	4.01E−08	3.10E−03
6442	Alzheimer's disease	1.65E−19	4.01E−08	2.87E−04
7145	stromal stem cells	1.88E−15	1.16E−09	2.87E−04
		Insulin Response Studies		
5599	UV response	1.48E−18	3.00E−05	0.023
4914	chronic myelogenous leukemia	4.27E−08	3.00E−05	0.023
7922	KLF1 targets	4.55E−05	3.00E−05	0.023
5947	SMARCA2 targets	1.12E−04	0.023	0.023
6442	Alzheimer's disease	0.042	1.16E−03	0.23
7145	stromal stem cells	2.14E−08	1.11E−03	0.023
		Joint Analysis		
5599	UV response	9.46E−32	1.12E−10	1.55E−03
4914	chronic myelogenous leukemia	3.91E−44	3.41E−12	1.52E−05
7922	KLF1 targets	4.72E−29	1.54E−15	1.55E−03
5947	SMARCA2 targets	1.11E−27	6.28E−08	1.55E−03
6442	Alzheimer's disease	1.84E−18	2.95E−09	1.55E−03
7145	stromal stem cells	2.93E−22	1.12E−10	1.72E−04

PID: ID of the significantly identified gene sets; GeneSet: name of the gene sets; *Fixed_p*: the unadjusted p-value by fixed-effect meta-analysis; *Bin_p0*: the unadjusted meta-analysis p-value by binomial test; *Bin_p1*: the adjusted meta-analysis p-value by binomial test.

The six significant gene sets were mainly observed in diabetes studies 1, 2, 3, 4, 8, 10 and 11 and insulin studies 3 and 4 (adjusted p-value < 0.05), involving tissues of adipose, arteries, blood, myotube, pancreatic tissues and skeletal muscles (Supplementary Tables S3–S8 and the Supplementary Figures S1–S6). Specifically, all six gene sets were significant at the diabetes study 3 of blood tissue (GDS3874/GDS3875); five gene sets except the "stromal stem cells" were significant in the diabetes study 2 of artery tissue (GDS3980); four gene sets except the "UV response" and the "stromal stem cells" were significant at the diabetes study 10 of pancreas (GDS3882); four gene sets except the "Alzheimer's disease" and the "stromal stem cells" were significant in the insulin study 4 of skeletal muscles (GDS3715); diabetes study 8 of myotube (GDS3681) and insulin study 3 of skeletal muscles (GDS3181) contained only a significant gene set of "stromal stem cells"; and diabetes study 4 of blood (GDS3963) had only a significant gene set of "chronic myelogenous leukemia".

3.4. Mapped KEGG Pathways for the Identified Gene Sets

To infer the potential functions of significant gene sets, mapping analysis was conducted to find their related KEGG pathways by mapping analysis. The most related KEGG pathways with estimated effects were shown at the Table 4. The "UV response" was mapped to the transforming growth factor beta 1 (TGF-beta) signaling pathway, presenting the effect of 20%, standard error (SE) of 0.03, p-value of 1.28E−9 and permutation-adjusted p-value < 0.0001. Similarly, we identified that the "chronic myelogenous leukemia", "KLF1 targets" and "SMARCA2 targets" were mapped to citrate cycle pathway (effect = 35% and adjusted p-value < 0.01), DNA replication (effect = 24% and adjusted p-value = 0.016) and nucleotide excision repair (effect = 11% and adjusted p-value = 0.034), respectively. The gene set of "Alzheimer's disease" was mapped to the pathways of "Oxidative phosphorylation" and "Parkinson's disease" with an effect of 35% and adjusted p-value < 0.001; and the "stromal stem cells" was mapped to the peroxisome proliferator-activated receptors (PPAR) and p53 signaling pathways with an effect of 10% and adjusted p-value of 0.029.

Table 4. The closest KEGG Pathway

Gene Set	KEGG Pathway	Size	Gene	Effect	SE	p	adj_p
UV response	TGF-beta signaling pathway	86	22	0.20	0.03	1.28E−09	<0.001
chronic myelogenous leukemia	The citrate cycle	32	15	0.35	0.06	1.11E−06	<0.001
KLF1 targets	DNA replication	36	13	0.24	0.05	1.54E−04	0.016
SMARCA2 targets	Nucleotide excision repair	44	6	0.11	0.02	4.29E−04	0.034
Alzheimer's disease	1. Oxidative phosphorylation	135	62	0.35	0.03	9.20E−27	<0.001
	2. Parkinson's disease	133	60	0.35	0.03	2.13E−25	<0.001
stromal stem cells	1. PPAR signaling pathway	69	9	0.10	0.02	3.02E−4	0.029
	2. p53 signaling pathway	69	9	0.10	0.02	3.02E−4	0.029

Size: the number of genes in the KEGG pathway; Gene: the number of overlapped genes between the gene set and the KEGG pathway; Effect: the higher probability for a gene of the gene set that belongs to the KEGG pathway than a random gene; SE: the standard error of the estimated effect; p: the unadjusted p-value; and adj_p: the adjusted p-value by permutation test.

3.5. Correlation and Independence of Gene and Pathway Expression Associations

Correlation and independent analyses of expression association profiles were conducted among the 13 diabetes studies, and a plot of the results was shown at the Figure 1. After adjustment for multiple testing, significantly correlated and independent association profiles of gene expression, respectively, accounted for 10 out of 78 analyses (or 12.8%) between studies (Supplementary Tables S9 and S10). However, for pathway expression, no studies were to be observed with significantly independent association, and, in contrast, 46 analyses (59.0%) were found to have significant correlation (Supplementary Table S11). The results showed that studies with correlated gene association profiles also tended to have correlated pathway association profiles. Study 2 of arteries and study 10 of the pancreas had the strongest gene correlation ($\rho = 0.074$ and p-value $= 7.07E−12$) and the second strongest pathway correlation ($\rho = 0.21$ and p-value $= 9.20E−104$), while studies 3 and 4 of blood had the second strongest gene correlation ($\rho = 0.065$ and p-value $= 6.65E−9$) and the strongest pathway correlation ($\rho = 0.24$ and p-value $= 5.17E−136$). The results suggested that different studies have both tissue specific and non-specific gene expression association with diabetes, and compared to the gene expression, the pathway expression tends to be tissue non-specific. Analyses of the insulin studies presented consistent conclusions.

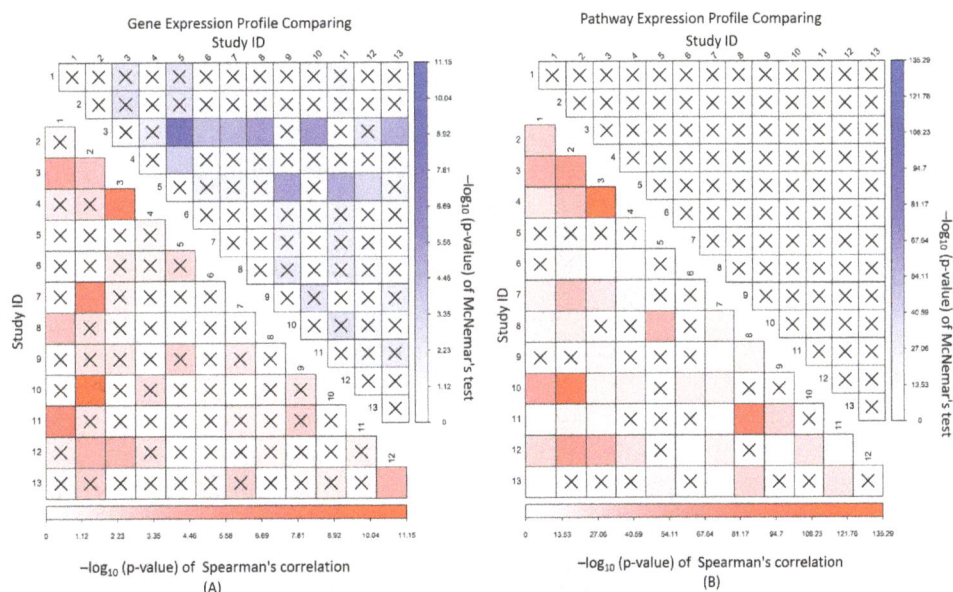

Figure 1. Correlation and independent analyses of gene (**A**) and pathway (**B**) expression association profiles between studies. The lower triangle is the correlation analysis and the upper triangle is the independent test. The 'X' indicates the analysis p-value is not significant.

4. Discussion

Although diabetes may have different etiology, they usually present some common clinical manifestations and share pathogenic mechanisms that are accompanied with pathochanges at different tissues. There was evidence in the animal studies showing that different tissues involve common genetic regulations in diabetes development. For example, GLUT4 heterozygous knockout in mice exhibited decreased expression in adipose tissue and muscle [35], and the knockouts in different tissues led to common observations of impaired whole-body glucose homeostasis and developed insulin resistance [36,37]. For this meta-analysis, we hypothesize that there exist tissue non-specific genetic regulations influencing human diabetes pathogenesis, and the study aim is to identify these genes and pathways based on measured gene expressions. Comparing to the original expression data submitted by the researcher, which is heterogeneous and may not be directly analyzed, our study is focused on the GEO Datasets that are curated by the NCBI and consisted of biologically and statistically comparable data [11].

We searched the GEO database and identified 27 gene expression datasets from different tissues, of which 14 datasets were related to 13 expression studies of diabetes states and 11 datasets were linked to five expression studies of insulin responses. The gene expressions were measured from different tissues including pancreas, skeletal muscles, liver, adipose and blood. For every study, we analyzed differential gene expressions to test gene association by an empirical Bayes approach that has robust behavior even for small sample size [16], and examined pathway expression association by hypergeometric test for enrichment of significant genes that provides parametric estimate of effect and calculation of p-value. We performed meta-analysis of measured genes and MSigDB gene sets over studies for identification of tissue non-specific genes and pathways. Our meta-analysis strategy consisted of tests for both genes and pathways. For pathway meta-analysis, two types of tests were also conducted to provide consistent evaluation of expression association: the binomial test was based on the number of significant studies, and the fixed-effect model was based on the sum of effects over studies.

Our meta-analysis showed that *PGRMC1*, *HADH*, *IRS1* and *MPST* were the four tissue non-specific genes presenting differential expression association with the diabetes or insulin response. These four genes are expressed in most tissues. For *PGRMC1* and *HADH*, their associations ranked at the top 5% (i.e., U-score ≤ 0.05) in the six diabetes studies and their best U-scores were 0.34% and 0.44% in the diabetes studies 9 and 11 of the pancreas, respectively (Table 2). Previous studies have indicated they are both related to the insulin secretion. The *PGRMC1* is located at the Chr X, encoding a progesterone steroid receptor. It interacts the glucagon-like peptide-1 (GLP-1) receptor, and its overexpression enhances GLP-1-induced insulin secretion [38]. The *HADH* is mapped to the Chr 4q22–26. It encodes an enzyme, which is crucial for β-oxidation of fatty acids and generation of acetyl-CoA and associated with ketogenesis. Downregulation of HADH mRNA and the gene mutations are associated with insulin secretion and hyperinsulinaemic hypoglycaemia [39].

The *IRS1*, located at the Chr 2q36, encodes a protein phosphorylated by insulin receptor tyrosine kinase, which is required for hormonal control of metabolism. The IRS1 protein is critical for insulin response, and impairment of insulin signaling by IRS1 is linked to insulin resistance [40]. The *MPST* gene, mapped to the Chr 22q13.1, encodes the 3-mercaptopyruvate sulfurtransferase. The enzyme is known to produce the hydrogen sulfide (H_2S) from cysteine and the increased H_2S in adipose tissues was observed to inhibit insulin-stimulated glucose metabolism and regulate insulin sensitivity [41]. Both IRS1 and MPST had U-score ≤ 0.05 in four out of five insulin studies. The IRS1 also showed a small U-score of 0.003% in the diabetes study 4 of blood. The MPST presented U-score ≤ 0.05 in four diabetes studies with the best U-score of 0.2% from study 4 of blood. The results suggested that both genes are related to insulin response, and their abnormal expression levels in the blood indicate the progression of diabetes.

Our pathway meta-analysis has identified six MSigDB gene sets with significant expression associations. However, genetic mechanisms of these gene sets and their biological functions related to diabetes remain unknown. We therefore proposed the mapping analysis and aimed to infer their roles underlying diabetes pathogenesis by the most related KEGG pathways. The six gene sets were significant in diabetes studies and joint analysis by all three p-values (*Fixed_p*, *Bin_p0* and *Bin_p1* at the Table 3): (1) the gene set of "UV response" are genes downregulated in fibroblasts after UV irradiation, and it was mapped to the TGF-beta signaling pathway, which regulates insulin gene transcription and β cell function [42]; (2) the gene set of "chronic myelogenous leukemia" (CML) is a collection of genes upregulated in the CD34+ cells of CML patients and previous study suggested that CML is connected to T2D [43]. The gene set was mapped to the pathway of the citrate cycle that is related to glucose metabolism and diabetes progression [44]; (3) the "KLF1 targets" is a collection of genes discovered to be downregulated in erythroid progenitor cells due to knockout of KLF1 gene, and the mapping study showed that it was related to the DNA replication pathway, suggesting its effects on diabetes potentially through pancreatic β-cell replication [45]; (4) the "SMARCA2 targets" presents genes positively correlated with the SMARCA2 gene, and its mapped pathway of nucleotide excision repair (NER) is responsible for recognizing and repairing bulky DNA damage that is commonly observed in diabetic patients [46]; (5) the "Alzheimer's disease" (AD) gene set lists genes that are downregulated in the brains of Alzheimer's patients, and it is mapped to pathways of oxidative phosphorylation that have important roles in causing diabetes [47] and Parkinson's disease that are known to have shared mechanisms with diabetes as AD [48]; and (6) the "stromal stem cells" gene set is identified as a group of genes upregulated in cultured stromal stem cells from adipose tissue, and the mapped PPAR and p53 signaling pathways are associated with insulin sensitivity [49] and insulin resistance [50], respectively.

To further evaluate the identified tissue non-specific genes and gene sets, we proposed the correlation and independent analyses for their expression association profile between different studies and tissues. The results showed that correlated association gene profiles accounted for 12.8% analyses: for example, diabetes study 2 of artery tissue and study 10 of the pancreas had rank correlation p-value of 7.07E−12, and study 1 of adipose tissue and study 11 of pancreas had rank correlation p-value of 5.59E−7 (Supplementary Table S9). The results also showed that 12.8% analyses had significantly independent gene association profile: for example, study 3 of EPC and study 5 of blood had the p-value of 7.47E−10. However, most studies did not present obviously correlated or independent profiles of gene expression association. In contrast, for pathway expression association, no studies showed a significantly independent profile, but 59.0% analyses had significant correlation. The results indicated that most tissues and studies have similar profiles of pathway expression associations with diabetes, and compared to genes, diabetes pathways tend to be tissue non-specific. For example, study 2 of arteries and study 10 of the pancreas had their rank correlation (p-value) of expression association profile changed from 0.074 (7.07E−12) for genes to 0.21 (9.20E−104) for pathways; and study 3 of EPC and study 5 of blood had independent p-value of expression association profile changed from 7.47E−10 for genes to >0.05 for pathways. The results suggested that a common pathway is mainly activated through tissue specific genes in different tissues to influence diabetes pathogenesis.

Our meta-analysis was performed on curated GDS of gene expressions identified from the GEO. However, these datasets have a few limitations: (1) most diabetes studies are for T2D and only two studies are for T1D; (2) all expression datasets have a relatively small sample size ($\leq$117); and (3) many tissues were collected in only 1–3 studies. These limitations can affect the statistical test and reduce the study power. Based on results of this analysis, it is worthwhile to conduct replication studies on more expression datasets with large sample size and different tissues in the next step. In addition, identification of tissue non-specific genes and pathways in the current study mainly relied on significantly statistical tests, which, however, had the limitation to provide direct evidence for their roles in diabetes pathogenesis. Therefore, in vivo biological studies of these genes and pathways in the future will play essential roles in understanding their genetic regulation mechanisms of diabetes.

5. Conclusions

In summary, we examined gene expression datasets from the GEO database that are related to the diabetes and insulin response, and performed meta-analysis with the aim to identify tissue non-specific genes and pathways. We also proposed the KEGG pathway mapping analysis to infer the function of MSigDB gene sets, and correlation and independent analysis of expression association profile between different studies and tissues. Our study successfully identified four and six tissue non-specific genes and gene sets, respectively. The results also suggested that effects of diabetes-related pathways are more likely tissue non-specific, compared to the effects of diabetes genes.

Acknowledgments: This study was supported in part by Grants N01-HC55021 and U01-HL096917 from National Institutes of Health (NIH)/NIH Heart, Lung and Blood Institute, and a Mississippi INBRE Grant P20GM103476. The funders had no role in study design, data collection and analysis, decision to publish, or preparation of the manuscript.

Author Contributions: Lianna Li, Shijian Liu and Hao Mei designed the study strategy and wrote the manuscript, Lianna Li and Hao Mei completed all analyses, Fan Jiang, Michael Griswold and Thomas Mosley contributed in study design, data interpretation and writing the manuscript.

References

1.   Alberti, K.G.; Zimmet, P.Z. Definition, diagnosis and classification of diabetes mellitus and its complications. Part 1: Diagnosis and classification of diabetes mellitus provisional report of a who consultation. *Diabet. Med. J. Br. Diabet. Assoc.* **1998**, *15*, 539–553. [CrossRef]

2.   DeFronzo, R.A.; Tripathy, D. Skeletal muscle insulin resistance is the primary defect in type 2 diabetes. *Diabetes Care* **2009**, *32*, S157–S163. [CrossRef] [PubMed]

3.   Reaven, G.M. Banting lecture 1988. Role of insulin resistance in human disease. *Diabetes* **1988**, *37*, 1595–1607. [CrossRef] [PubMed]

4.   Frederiksen, C.M.; Hojlund, K.; Hansen, L.; Oakeley, E.J.; Hemmings, B.; Abdallah, B.M.; Brusgaard, K.; Beck-Nielsen, H.; Gaster, M. Transcriptional profiling of myotubes from patients with type 2 diabetes: No evidence for a primary defect in oxidative phosphorylation genes. *Diabetologia* **2008**, *51*, 2068–2077. [CrossRef] [PubMed]

5.   Van Oostrom, O.; de Kleijn, D.P.; Fledderus, J.O.; Pescatori, M.; Stubbs, A.; Tuinenburg, A.; Lim, S.K.; Verhaar, M.C. Folic acid supplementation normalizes the endothelial progenitor cell transcriptome of patients with type 1 diabetes: A case-control pilot study. *Cardiovasc. Diabetol.* **2009**. [CrossRef] [PubMed]

6.   Cangemi, C.; Skov, V.; Poulsen, M.K.; Funder, J.; Twal, W.O.; Gall, M.A.; Hjortdal, V.; Jespersen, M.L.; Kruse, T.A.; Aagard, J.; et al. Fibulin-1 is a marker for arterial extracellular matrix alterations in type 2 diabetes. *Clin. Chem.* **2011**, *57*, 1556–1565. [CrossRef] [PubMed]

7.   Skov, V.; Knudsen, S.; Olesen, M.; Hansen, M.L.; Rasmussen, L.M. Global gene expression profiling displays a network of dysregulated genes in non-atherosclerotic arterial tissue from patients with type 2 diabetes. *Cardiovasc. Diabetol.* **2012**. [CrossRef] [PubMed]

8.   Misu, H.; Takamura, T.; Takayama, H.; Hayashi, H.; Matsuzawa-Nagata, N.; Kurita, S.; Ishikura, K.; Ando, H.; Takeshita, Y.; Ota, T.; et al. A liver-derived secretory protein, selenoprotein p, causes insulin resistance. *Cell Metab.* **2010**, *12*, 483–495. [CrossRef] [PubMed]

9.   Pihlajamaki, J.; Boes, T.; Kim, E.Y.; Dearie, F.; Kim, B.W.; Schroeder, J.; Mun, E.; Nasser, I.; Park, P.J.; Bianco, A.C.; et al. Thyroid hormone-related regulation of gene expression in human fatty liver. *J. Clin. Endocrinol. Metab.* **2009**, *94*, 3521–3529. [CrossRef] [PubMed]

10.   Kaizer, E.C.; Glaser, C.L.; Chaussabel, D.; Banchereau, J.; Pascual, V.; White, P.C. Gene expression in peripheral blood mononuclear cells from children with diabetes. *J. Clin. Endocrinol. Metab.* **2007**, *92*, 3705–3711. [CrossRef] [PubMed]

11.   Barrett, T.; Troup, D.B.; Wilhite, S.E.; Ledoux, P.; Evangelista, C.; Kim, I.F.; Tomashevsky, M.; Marshall, K.A.;

Phillippy, K.H.; Sherman, P.M.; et al. Ncbi geo: Archive for functional genomics data sets—10 years on. *Nucleic Acids Res.* **2011**, *39*, D1005–D1010. [CrossRef] [PubMed]

12. Davis, S.; Meltzer, P.S. Geoquery: A bridge between the gene expression omnibus (GEO) and bioconductor. *Bioinformatics* **2007**, *23*, 1846–1847. [CrossRef] [PubMed]

13. Chen, R.; Sigdel, T.K.; Li, L.; Kambham, N.; Dudley, J.T.; Hsieh, S.C.; Klassen, R.B.; Chen, A.; Caohuu, T.; Morgan, A.A.; et al. Differentially expressed rna from public microarray data identifies serum protein biomarkers for cross-organ transplant rejection and other conditions. *PLoS Comput. Biol.* **2010**. [CrossRef] [PubMed]

14. Bolstad, B.M.; Irizarry, R.A.; Astrand, M.; Speed, T.P. A comparison of normalization methods for high density oligonucleotide array data based on variance and bias. *Bioinformatics* **2003**, *19*, 185–193. [CrossRef] [PubMed]

15. Bolstad bm. Preprocesscore: A Collection of Pre-processing Functions. R Package Version 1.36.0. 2016. Available online: https://github.Com/bmbolstad/preprocesscore (accessed on 1 March 2006).

16. Smyth, G.K. Linear models and empirical bayes methods for assessing differential expression in microarray experiments. *Stat. Appl. Genet. Mol. Biol.* **2004**. [CrossRef] [PubMed]

17. Mei, H.; Li, L.; Liu, S.; Jiang, F.; Griswold, M.; Mosley, T. The uniform-score gene set analysis for identifying common pathways associated with different diabetes traits. *BMC Genom.* **2015**. [CrossRef] [PubMed]

18. Liberzon, A. A description of the molecular signatures database (MSigDb) web site. *Methods Mol. Biol.* **2014**, *1150*, 153–160. [PubMed]

19. Kanehisa, M.; Goto, S.; Sato, Y.; Furumichi, M.; Tanabe, M. Kegg for integration and interpretation of large-scale molecular data sets. *Nucleic Acids Res.* **2012**, *40*, D109–D114. [CrossRef] [PubMed]

20. Mei, H.; Li, L.; Jiang, F.; Simino, J.; Griswold, M.; Mosley, T.; Liu, S. snpGeneSets: An R package for genome-wide study annotation. *G3* **2016**, *6*, 4087–4095. [PubMed]

21. Viechtbauer, W. Conducting meta-analyses in R with the metafor package. *J. Stat. Softw.* **2010**, *36*, 1–48. [CrossRef]

22. Kanehisa, M.; Goto, S.; Furumichi, M.; Tanabe, M.; Hirakawa, M. Kegg for representation and analysis of molecular networks involving diseases and drugs. *Nucleic Acids Res.* **2010**, *38*, D355–D360. [CrossRef] [PubMed]

23. Karolina, D.S.; Armugam, A.; Tavintharan, S.; Wong, M.T.; Lim, S.C.; Sum, C.F.; Jeyaseelan, K. Microrna 144 impairs insulin signaling by inhibiting the expression of insulin receptor substrate 1 in type 2 diabetes mellitus. *PLoS ONE* **2011**, *6*, e22839. [CrossRef]

24. Marselli, L.; Thorne, J.; Dahiya, S.; Sgroi, D.C.; Sharma, A.; Bonner-Weir, S.; Marchetti, P.; Weir, G.C. Gene expression profiles of beta-cell enriched tissue obtained by laser capture microdissection from subjects with type 2 diabetes. *PLoS ONE* **2010**, *5*, e11499. [CrossRef] [PubMed]

25. Dominguez, V.; Raimondi, C.; Somanath, S.; Bugliani, M.; Loder, M.K.; Edling, C.E.; Divecha, N.; da Silva-Xavier, G.; Marselli, L.; Persaud, S.J.; et al. Class ii phosphoinositide 3-kinase regulates exocytosis of insulin granules in pancreatic beta cells. *J. Biol. Chem.* **2011**, *286*, 4216–4225. [CrossRef] [PubMed]

26. Taneera, J.; Lang, S.; Sharma, A.; Fadista, J.; Zhou, Y.; Ahlqvist, E.; Jonsson, A.; Lyssenko, V.; Vikman, P.; Hansson, O.; et al. A systems genetics approach identifies genes and pathways for type 2 diabetes in human islets. *Cell Metab.* **2012**, *16*, 122–134. [CrossRef] [PubMed]

27. Van Tienen, F.H.; Praet, S.F.; de Feyter, H.M.; van den Broek, N.M.; Lindsey, P.J.; Schoonderwoerd, K.G.; de Coo, I.F.; Nicolay, K.; Prompers, J.J.; Smeets, H.J.; et al. Physical activity is the key determinant of skeletal muscle mitochondrial function in type 2 diabetes. *J. Clin. Endocrinol. Metab.* **2012**, *97*, 3261–3269. [CrossRef] [PubMed]

28. Jin, W.; Goldfine, A.B.; Boes, T.; Henry, R.R.; Ciaraldi, T.P.; Kim, E.Y.; Emecan, M.; Fitzpatrick, C.; Sen, A.; Shah, A.; et al. Increased srf transcriptional activity in human and mouse skeletal muscle is a signature of insulin resistance. *J. Clin. Investig.* **2011**, *121*, 918–929. [CrossRef] [PubMed]

29. Yang, X.; Pratley, R.E.; Tokraks, S.; Bogardus, C.; Permana, P.A. Microarray profiling of skeletal muscle tissues from equally obese, non-diabetic insulin-sensitive and insulin-resistant pima indians. *Diabetologia* **2002**, *45*, 1584–1593. [PubMed]

30. Parikh, H.; Carlsson, E.; Chutkow, W.A.; Johansson, L.E.; Storgaard, H.; Poulsen, P.; Saxena, R.; Ladd, C.; Schulze, P.C.; Mazzini, M.J.; et al. Txnip regulates peripheral glucose metabolism in humans. *PLoS Med.* **2007**, *4*, e158. [CrossRef] [PubMed]

31. Coletta, D.K.; Balas, B.; Chavez, A.O.; Baig, M.; Abdul-Ghani, M.; Kashyap, S.R.; Folli, F.; Tripathy, D.; Mandarino, L.J.; Cornell, J.E.; et al. Effect of acute physiological hyperinsulinemia on gene expression in human skeletal muscle in vivo. *Am. J. Physiol. Endocrinol. Metab.* **2008**, *294*, E910–E917. [CrossRef] [PubMed]

32. Wu, X.; Wang, J.; Cui, X.; Maianu, L.; Rhees, B.; Rosinski, J.; So, W.V.; Willi, S.M.; Osier, M.V.; Hill, H.S.; et al. The effect of insulin on expression of genes and biochemical pathways in human skeletal muscle. *Endocrine* **2007**, *31*, 5–17. [CrossRef] [PubMed]

33. Wu, X.; Patki, A.; Lara-Castro, C.; Cui, X.; Zhang, K.; Walton, R.G.; Osier, M.V.; Gadbury, G.L.; Allison, D.B.; Martin, M.; et al. Genes and biochemical pathways in human skeletal muscle affecting resting energy expenditure and fuel partitioning. *J. Appl. Physiol.* **2011**, *110*, 746–755. [CrossRef] [PubMed]

34. Hardy, O.T.; Perugini, R.A.; Nicoloro, S.M.; Gallagher-Dorval, K.; Puri, V.; Straubhaar, J.; Czech, M.P. Body mass index-independent inflammation in omental adipose tissue associated with insulin resistance in morbid obesity. *Surg. Obes. Relat. Dis.* **2011**, *7*, 60–67. [CrossRef] [PubMed]

35. Stenbit, A.E.; Tsao, T.S.; Li, J.; Burcelin, R.; Geenen, D.L.; Factor, S.M.; Houseknecht, K.; Katz, E.B.; Charron, M.J. Glut4 heterozygous knockout mice develop muscle insulin resistance and diabetes. *Nat. Med.* **1997**, *3*, 1096–1101. [CrossRef] [PubMed]

36. Abel, E.D.; Peroni, O.; Kim, J.K.; Kim, Y.B.; Boss, O.; Hadro, E.; Minnemann, T.; Shulman, G.I.; Kahn, B.B. Adipose-selective targeting of the glut4 gene impairs insulin action in muscle and liver. *Nature* **2001**, *409*, 729–733. [CrossRef] [PubMed]

37. Zisman, A.; Peroni, O.D.; Abel, E.D.; Michael, M.D.; Mauvais-Jarvis, F.; Lowell, B.B.; Wojtaszewski, J.F.; Hirshman, M.F.; Virkamaki, A.; Goodyear, L.J.; et al. Targeted disruption of the glucose transporter 4 selectively in muscle causes insulin resistance and glucose intolerance. *Nat. Med.* **2000**, *6*, 924–928. [PubMed]

38. Zhang, M.; Robitaille, M.; Showalter, A.D.; Huang, X.; Liu, Y.; Bhattacharjee, A.; Willard, F.S.; Han, J.; Froese, S.; Wei, L.; et al. Progesterone receptor membrane component 1 is a functional part of the glucagon-like peptide-1 (GLP-1) receptor complex in pancreatic beta cells. *Mol. Cell Proteom.* **2014**, *13*, 3049–3062. [CrossRef] [PubMed]

39. Senniappan, S.; Shanti, B.; James, C.; Hussain, K. Hyperinsulinaemic hypoglycaemia: Genetic mechanisms, diagnosis and management. *J. Inherit. Metab. Dis.* **2012**, *35*, 589–601. [CrossRef] [PubMed]

40. Copps, K.D.; White, M.F. Regulation of insulin sensitivity by serine/threonine phosphorylation of insulin receptor substrate proteins IRS1 and IRS2. *Diabetologia* **2012**, *55*, 2565–2582. [CrossRef] [PubMed]

41. Zhang, Y.; Tang, Z.H.; Ren, Z.; Qu, S.L.; Liu, M.H.; Liu, L.S.; Jiang, Z.S. Hydrogen sulfide, the next potent preventive and therapeutic agent in aging and age-associated diseases. *Mol. Cell. Biol.* **2013**, *33*, 1104–1113. [CrossRef] [PubMed]

42. Lin, H.M.; Lee, J.H.; Yadav, H.; Kamaraju, A.K.; Liu, E.; Zhigang, D.; Vieira, A.; Kim, S.J.; Collins, H.; Matschinsky, F.; et al. Transforming growth factor-beta/smad3 signaling regulates insulin gene transcription and pancreatic islet beta-cell function. *J. Biol. Chem.* **2009**, *284*, 12246–12257. [CrossRef] [PubMed]

43. Veneri, D.; Franchini, M.; Bonora, E. Imatinib and regression of type 2 diabetes. *N. Engl. J. Med.* **2005**, *352*, 1049–1050. [CrossRef] [PubMed]

44. Brownlee, M. The pathobiology of diabetic complications: A unifying mechanism. *Diabetes* **2005**, *54*, 1615–1625. [CrossRef] [PubMed]

45. Bonner-Weir, S.; Li, W.C.; Ouziel-Yahalom, L.; Guo, L.; Weir, G.C.; Sharma, A. Beta-cell growth and regeneration: Replication is only part of the story. *Diabetes* **2010**, *59*, 2340–2348. [CrossRef] [PubMed]

46. Simone, S.; Gorin, Y.; Velagapudi, C.; Abboud, H.E.; Habib, S.L. Mechanism of oxidative DNA damage in diabetes: Tuberin inactivation and downregulation of DNA repair enzyme 8-oxo-7,8-dihydro-2′-deoxyguanosine-DNA glycosylase. *Diabetes* **2008**, *57*, 2626–2636. [CrossRef] [PubMed]

47. Lowell, B.B.; Shulman, G.I. Mitochondrial dysfunction and type 2 diabetes. *Science* **2005**, *307*, 384–387. [CrossRef] [PubMed]

48. Craft, S.; Watson, G.S. Insulin and neurodegenerative disease: Shared and specific mechanisms. *Lancet Neurol.* **2004**, *3*, 169–178. [CrossRef]

49. Ferre, P. The biology of peroxisome proliferator-activated receptors: Relationship with lipid metabolism and insulin sensitivity. *Diabetes* **2004**, *53*, S43–S50. [CrossRef] [PubMed]

50. Minamino, T.; Orimo, M.; Shimizu, I.; Kunieda, T.; Yokoyama, M.; Ito, T.; Nojima, A.; Nabetani, A.; Oike, Y.; Matsubara, H.; et al. A crucial role for adipose tissue p53 in the regulation of insulin resistance. *Nat. Med.* **2009**, *15*, 1082–1087. [CrossRef] [PubMed]

8

An Exploratory Study to Determine Whether *BRCA1* and *BRCA2* Mutation Carriers Have Higher Risk of Cardiac Toxicity

Monique Sajjad [1,2], Michael Fradley [2], Weihong Sun [1], Jongphil Kim [1], Xiuhua Zhao [1], Tuya Pal [1] and Roohi Ismail-Khan [1,*]

[1] H. Lee Moffitt Cancer Center and Research Institute, Tampa, FL 33612, USA;
moniquesajjad@gmail.com (M.S.); Weihong.sun@moffitt.org (W.S.); jongphil.kim@moffitt.org (J.K.);
xiuhua._zh@yahoo.com.org (X.Z.); tuya.pal@moffitt.org (T.P.)

[2] Division of Cardiovascular Medicine, University of South Florida, Tampa, FL 33620, USA;
Mfradley@health.usf.edu

* Correspondence: Roohi.Ismail-Khan@moffitt.org

Academic Editor: Nora L. Nock

Abstract: Anthracycline-based cardiotoxicity is concerning for women with breast cancer and portends a dose-dependent risk of developing left ventricular dysfunction. Overall, the prevalence of heart failure (HF) is ≈2% of the total US population; however, *BRCA*-deficient mice have shown increased HF. We evaluated for the inherent risk of HF in women with *BRCA* mutations to determine whether treatment with anthracycline-based therapy increased this risk. We obtained results on *BRCA* mutation carriers regarding cancer treatment and HF, identified through the *BRCA* patient advocacy organization Facing Our Risk for Cancer Empowered (FORCE) and the Moffitt-based Inherited Cancer Registry. In our patient group (232 *BRCA1* and 159 *BRCA2* patients; 10 with both mutations), 7.7% reported HF, with similar proportions in *BRCA1* versus *BRCA2* carriers (7.4% and 8.2%, respectively). These proportions are significantly higher than published rates ($p < 0.001$). There was no statistically significant difference in HF rates comparing anthracycline-treated versus anthracycline-naïve patients however (7.1% vs. 8.3%; $p = 0.67$). In addition, 9.1% of *BRCA1* carriers and 8.2% of *BRCA2* carriers reported arrhythmias. *BRCA* mutation carriers showed increased risk of cardiotoxicity versus the general population and an overall increased risk of cardiotoxicity from anthracycline-based therapy. Our study supports data that *BRCA* carriers have increased non-cancer mortality from cardiotoxicity. A prospective trial to determine HF and conduction abnormalities in this population is warranted.

Keywords: anthracycline-related cardiac toxicity; *BRCA1* mutation; *BRCA2* mutation; conduction abnormalities; heart failure

1. Introduction

Overall, approximately 5%–10% of breast and ovarian cancer cases are due to mutations in the high-penetrance genes *BRCA1* and *BRCA2* (*BRCA*) [1]. It is well-established that patients with mutations in the *BRCA* genes have increased mortality due to malignancy [2]; however, a recent study has also suggested a significant association between *BRCA* mutations and increased non-cancer mortality ($p = 0.024$) [3]. Mutation carriers have a 56%–87% risk of developing breast cancer by the age of 70 [2,4]. Lifetime risks for ovarian cancer are up to 44% and 27% for *BRCA1* and *BRCA2*, respectively [1].

Prevalence of heart failure (HF) in the general population of American women varies and increases with age. According to American Heart Association data, this risk varies between 0.3% in women under 30 and 11.8% in the elderly [5]. The overall risk for the general US population including males and females is estimated at 2% [5]. It is well-established that anthracycline use can lead to heart failure. The cardiotoxic effects of anthracycline therapy are seen in a dose-dependent fashion. The risk of heart failure due to adriamycin at 450 mg/m^2 is 3%–4% [6].

This exploratory study was performed to determine whether patients with *BRCA* mutations are at a higher inherent predisposition for cardiac disease. We also sought to determine whether mutation carriers were at a higher risk for toxicity from anthracycline therapy, which is one of the current standards of care for chemotherapy in this population. Third, we attempted to identify other possible causes responsible for increased mortality in these patients.

2. Materials and Methods

This study received Institutional Review Board approval (number Pro00004774; 20/06/2014). Patients with *BRCA1* or *BRCA2* mutation status were offered enrollment in the patient advocacy groups known as Facing Our Risk for Cancer Empowered (FORCE) and Inherited Cancer Registry (ICARE). FORCE is a nonprofit organization devoted to patients and family members with *BRCA* mutations. This organization meets annually and has over 13,000 members nationwide. We also identified patients using the ICARE initiative at the Moffitt Cancer Center (see Appendix A). Through FORCE and ICARE, the surveys were available to over 13,000 patients. Eligible patients included females with confirmed *BRCA1* or *BRCA2* mutations of any age and any treatment history. Patients were invited to participate in an optional online or paper survey (see Supplementary Survey S1) regarding general adverse effects surrounding their breast cancer treatment and overall health. Patients were consented by voluntarily completing the survey and had the option to give additional consent to be contacted for follow-up questions. Four hundred one surveys were completed, and all were included for analysis. The survey consisted of 57 questions, including a thorough patient-reported review of the cardiovascular system and disease states. Specific information regarding baseline cardiac risk factors and medical history prior to therapy was also obtained. Age, prior or current tobacco history, diabetic status, cholesterol levels, and history of cardiac disease were assessed. Targeted questions included Number 8 (cardiac)—"Did you experience any of the following: irregular rhythm, prior heart attack, heart failure"—Number 25—"Was your heart function normal before starting treatment?"—Number 26—"Is your heart function normal now?"—Number 42b—"Have you been diagnosed with heart attack?"—and Number 42c—"Have you been diagnosed with heart failure?" The survey also included detailed questions to determine preexisting conditions and comorbidities. Data of patients who underwent treatment at Moffitt Cancer Center were reviewed. Attempts were made to contact patients treated at outside centers, with questionnaire responses suggestive of cardiovascular complications.

The next part of the study included assessing the cardiac risk profiles of patients who had specifically been treated with anthracycline chemotherapy, comparing these values to established historical data. We contacted patients who answered "Yes" to any of the targeted questions (see above) and gave permission to do so. Of the 31 patients with heart failure, only 8 agreed to be contacted. They were all able to provide proof of HF diagnosis via clinical history or echo report. The survey, including targeted cardiac questions, is included in Supplementary Survey S1.

Population characteristics were summarized using descriptive statistics: frequency and proportion for categorical variables and median and range for continuous variables. The exact binomial distribution was used to compute the 95% confidence interval (CI) for proportions and to compare those with the proportions of a well-established population control. The association between two categorical variables was evaluated by the chi-square test.

3. Results

3.1. Population Characteristics

A total of 401 patients were entered into the study: 232 were *BRCA1* carriers, and 159 were *BRCA2* carriers (see Table 1 for patient characteristics). Ten patients carried both mutations. Patient accrual was initiated in April 2010 and concluded in November 2010 after 400 surveys were completed. All patients were female. Age ranged from 40 to 76 years. The patient cohort was generally healthy, and most patients with heart failure had 0–1 risk factor (see Figure 1).

Table 1. Patient characteristics.

	Number of Patients (%) ($n = 401$)
BRCA status	
BRCA1	232 (57.9)
BRCA2	159 (39.7)
BRCA1 and *BRCA2*	10 (2.5)
Hypertension	97 (24.2)
Diabetes	18 (4.5)
Hyperlipidemia	207 (51.6)
Tobacco use	37 (9.2)
Age group	
20–39 years	28 (7.0)
40–59 years	264 (65.8)
60–79 years	76 (19.0)
No answer	33 (8.2)

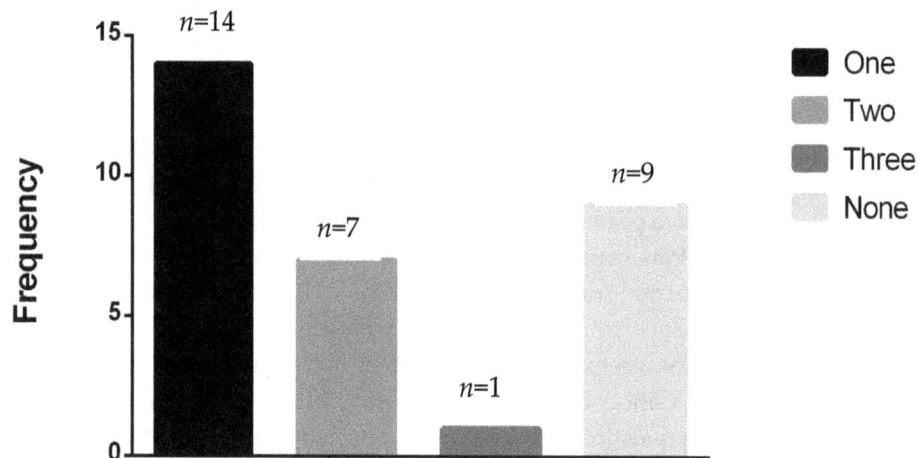

Figure 1. Number of confounding risk factors seen in the 31 *BRCA* patients with heart failure.

Thirty-one patients (7.3%) were diagnosed clinically with HF. Of these, 17 patients (7.3%) carried *BRCA1* mutations and 13 patients (8.2%) carried *BRCA2* mutations. One patient that carried both mutations was diagnosed with heart failure (Table 2).

Table 2. Patients diagnosed with heart failure.

Patient Group	Heart Failure, No. of Patients (%)	No Heart Failure	Total	95% Confidence Interval (%)	p Value Compared to 2%
BRCA1	17 (7.3%)	215	232	4.3–11.5	<0.0001
BRCA2	13 (8.2%)	146	159	4.4–13.6	<0.0001
BRCA1 and *BRCA2*	1 (10%)	9	10	0.3–44.5	0.37
Total	31 (7.7%)	370	401	5.3–10.8	<0.0001

When compared with rates of heart failure in the overall population, this proportion in our patient population was significantly higher (7.7% vs. 2%; $p < 0.001$). When adjusted for age, BRCA carriers in the 40- to 59-year-old age group (15/264 patients) had a 5.7% risk (CI, 3.2–9.2) versus 0.8% in the general population of women ($p < 0.0001$), and patients in the 60- to 79-year-old age group (12/76 patients) had a 15.8% risk (CI, 8.4–26.0) versus 5.4% in the general population ($p < 0.0001$). Four patients could not be included in age stratification because they did not include their age on the survey.

There was no difference between risk in *BRCA1* and *BRCA2* carriers who had 7.3% and 8.2% incidence of heart failure, respectively ($p = 0.76$). Of the 7.7% of *BRCA1* and *BRCA2* mutation carriers who developed heart failure, 8.3% were anthracycline naïve (statistically significant at $p \leq 0.001$) compared with the general population risk of 2%. This indicates that *BRCA* carriers may have a higher inherent risk of heart failure than the general population even when anthracycline use is excluded. In our patient group, 7.1% of mutation carriers developed heart failure after receiving anthracycline therapy with either adriamycin or epirubicin. When compared to the known risk of heart failure from anthracycline therapy of 3%, *BRCA* mutation carriers had a higher risk of cardiotoxicity from anthracycline therapy ($p = 0.008$) (Table 3). There was no statistically significant difference in the incidence of HF between the anthracycline-treated patients and those who were chemotherapy naïve (7.1% vs 8.3%; $p = 0.67$).

Table 3. Heart failure risk in BRCA patients with anthracyline therapy.

	Number of Patients				% HF on Anthracycline	% HF Chemotherapy Naïve
	Received Anthracycline	Chemotherapy Naïve	HF on Anthracycline	HF Chemotherapy Naïve		
BRCA1	106	126	6	11	5.7	8.7
BRCA2	70	89	6	7	8.6	7.9
Both	7	3	1	0	14.2	0
Total	183	218	13	18	7.1	8.3

Abbreviations: HF, heart failure.

In addition, 21 of 232 (9.1%) *BRCA1* carriers and 13 of 159 (8.2%) of *BRCA2* carriers reported arrhythmias. Overall, 37 patients of 391 (9.2%) reported arrhythmias (Table 4).

Table 4. Arrhythmias in BRCA patients.

Patient Group	Irregular Rhythm, *n*	No Irregular Rhythm or Unknown, *n*	Total, *n*	Irregular Rhythm, % (95% CI)
BRCA1	21	211	232	9.1 (5.7–13.5)
BRCA2	13	146	159	8.2 (4.4–13.6)
BRCA1 and *BRCA2*	3	7	10	30.0 (6.7–65.3)
Total	37	364	401	9.2 (6.6–12.5)

4. Discussion

There is clear evidence of increased non-malignant mortality in *BRCA* patients [3]. In our study, we sought to determine whether this risk was related to cardiac mortality, namely, heart failure. A recent study of 81 patients showed no change in ejection fraction, but this may be due to a small sample size and a short duration of follow-up [7].

BRCA1 is a human tumor suppressor gene that produces the breast cancer type 1 susceptibility protein. *BRCA1* is expressed in cells of the breast and other tissues, where it functions to repair or destroy damaged DNA and protect cells against oxidative and genotoxic stress. Mutations in this gene alter function and allow for tumors to arise [8]. Thus, mutations in this gene could lead to pathologic pathways, predisposing to physiologic dysfunction caused by oxidative stress, particularly cardiovascular disease. In support of this theory, murine studies conducted with cardiac myocytes revealed that wild-type *BRCA1* is essential to limiting apoptosis and improving cardiac function in response to genotoxic stress (doxorubicin) and oxidative stress (ischemia) [9]. Furthermore, heart-specific *BRCA1* deletion has been shown to promote severe systolic dysfunction and limited survival in mice [9]. *BRCA2* deletion has led to increased cardiomyocyte apoptosis after anthracycline therapy [10]. Follow-up studies reported an association between single nucleotide polymorphisms in *BRCA1*-associated protein and myocardial infarction risk in a large Japanese patient cohort with replication in additional Japanese and Taiwanese patient cohorts [11]. Even in women who do not receive anthracycline-based chemotherapy, BRCA protein deficiency can increase their susceptibility to any form of oxidative stress, whether it is from other types of cancer treatments, early surgical menopause, subclinical ischemia, or endothelial dysfunction from hypertension, hyperlipidemia, diabetes, or insulin resistance (which is known to be increased in the setting of BRCA protein deficiency) [12–14]. These mechanisms could play a role in the inherently higher risk for heart failure reported in those patients exposed to anthracyclines and those who were chemotherapy naïve.

Our study participants reported a significantly increased risk of heart failure after anthracycline therapy compared with the general population. The cardiotoxic effects of anthracyclines have been well-studied in cancer patients but have not been directly correlated to patients with *BRCA* mutations. Hypothesized mechanisms of anthracycline-induced cardiotoxicity include free radical damage to myocytes, which can occur non-enzymatically through direct interaction with iron [1] or enzymatically through interaction with cardiolipin. Anthracyclines act as reducing agents generating free radicals in both pathways and are shown to disrupt DNA synthesis by interference with topoisomerase, leading to apoptosis [1]. Because *BRCA1* is associated with DNA repair, mutations may portend a higher risk of damage by anthracycline therapy.

Our secondary goal of this study was to identify other possible mechanisms to explain the increased non-cancer-related mortality in this population. By means of the encompassing survey, we also found an alarming number of otherwise healthy patients who reported arrhythmias. There are no well-established statistics for the overall rates of arrhythmias in women in the United States; therefore, our question regarding arrhythmias should have been more specific, so that more detailed statistical analysis could have been performed. However, the most common conduction abnormalities are first-degree atrio-ventricular blocks, with a 3% risk in black women and a 1.3% risk in white women, and atrial fibrillation, which ranges from 6.6 per 100,000 women per year for 15- to 44-year-olds and 1203.7 per 100,000 women per year for those $\geq$85 years of age [15,16]. These rates are much lower than the percentages that we found in our study. Because our survey was not initially formatted with detailed questions regarding arrhythmias, we did not have data as to specific types, durations, and treatments required for the arrhythmias found in our patient cohort.

The major limitation of this exploratory study was the inability to contact all participants for further information that was not listed in the primary survey. This was an online survey reaching patients across multiple institutions. Because this was an anonymous survey, we were unable to further discuss specific survey responses and obtain objective data from a large percentage of participants, unless they provided consent and contact information for further communication. The data collected also consisted of patients' self-reporting their diagnoses and was limited to the questions addressed in the survey. Moreover, objective confirmatory data were unavailable in most cases. Of the contacted patients who reported heart failure, echocardiogram reports did confirm low ejection fraction when available. However, we considered the diagnosis valid since heart failure is a clinical diagnosis and the targeted questions were specific. Among the patients who reported arrhythmias, electrocardiograms confirmed the lack of sinus rhythm when available. An additional weakness was that these data were compared with historical controls, allowing different biases to be introduced, including selection bias given the inherent differences in the populations compared. Finally, the use of trastuzumab was not addressed in this study. However, the use of this therapy is less common in BRCA breast cancer patients, as most have triple negative tumors [17]. This may need to be addressed in the future if these data are confirmed with prospective studies.

5. Conclusions

Prior studies in humans have sought to determine why *BRCA* carriers have increased non-cancer mortality. Previous studies have indicated that this may be due to cardiotoxicity. The findings of our cross-sectional study design suggest this as well. We consider this an interesting hypothesis-generating observation; however, it remains important to conduct larger studies to test this hypothesis, ideally with prospective follow-up and objective record verification. A more in-depth review of cardiovascular effects in this patient population is warranted and currently underway. The prospective study will compare ejection fraction in *BRCA* patients with wild-type counterparts before and after treatment with anthracycline.

Acknowledgments: We thank Rasa Hamilton (Moffitt Cancer Center) for editorial assistance. We thank Sue Freidman, the director of FORCE, for posting our research survey on the FORCE website. Our study also received valuable assistance from the Survey Methods Core at the H. Lee Moffitt Cancer Center & Research Institute, an NCI-designated Comprehensive Cancer Center, supported under NIH grant P30-CA76292.

Author Contributions: M. Sajjad drafted the manuscript; M. Fradley reviewed and edited the manuscript; W. Sun performed data entry and survey collection; J. Kim participated in the design of the study and performed statistical analysis; X. Zhao performed statistical analysis; T. Pal provided access and assistance with genetic information through ICARE; R. Ismail-Khan conceived the study, participated in its design and coordination, and helped draft the manuscript; all authors read, approved, and provided input on all drafts, including the final draft.

Appendix A

ICARE: The ICARE initiative, based at Moffitt, is an inherited cancer registry and database (grant funded) and initiated by Dr. Tuya Pal. Christina Bittner, a Board-Certified Genetic Counselor (a member of the ICARE team), contacts patients with resources that may be of interest to the BRCA patient.

FORCE: The FORCE organization was established in 1999 and offers expert-reviewed research information, resources, advocacy, and peer support for individuals and families affected by hereditary cancers. FORCE efforts include an annual conference, print and electronic newsletters, a comprehensive website, a toll-free helpline, webinars and webcasts, regional support groups, educational brochures, and chat rooms through which individuals can access the latest information on treatment, risk-management, prevention, psychosocial and quality-of-life issues, and research studies, as well as discuss and share the cancer-related issues that they are facing.

FORCE is the only international, nonprofit organization devoted to improving the lives of individuals and families affected by hereditary breast and ovarian cancer. Their mission includes providing up-to-date information, resources, peer support, and advocacy, raising awareness, and promoting research specific to hereditary cancer. FORCE maintains a print newsletter mailing list of about 13,000 individuals and an e-mailing list of approximately 10,000 individuals, growing at a rate of approximately 100 contacts monthly. FORCE has collaborations with major cancer centers and cancer advocacy groups. FORCE is a partner member of the Ovarian Cancer National Alliance (OCNA) and the Allied Support Group for Ovarian Cancer (organized by SGO).

References

1. Claus, E.B.; Schildkraut, J.M.; Thompson, W.D.; Risch, N.J. The genetic attributable risk of breast and ovarian cancer. *Cancer* **1996**, *77*, 2318–2324. [CrossRef]
2. Struewing, J.P.; Hartge, P.; Wacholder, S.; Baker, S.M.; Berlin, M.; McAdams, M.; Timmerman, M.M.; Tucker, M.A. The risk of cancer associated with specific mutations of *BRCA1* and *BRCA2* among Ashkenazi Jews. *N. Engl. J. Med.* **1997**, *336*, 1401–1408. [CrossRef] [PubMed]

3.   Mai, P.L.; Chatterjee, N.; Hartge, P.; Tucker, M.; Brody, L.; Struewing, J.P.; Wacholder, S. Potential excess mortality in *BRCA1/2* mutation carriers beyond breast, ovarian, prostate, and pancreatic cancers, and melanoma. *PLoS ONE* **2009**, *4*, e4812. [CrossRef] [PubMed]

4.   Ford, D.; Easton, D.F.; Stratton, M.; Narod, S.; Goldgar, D.; Devilee, P.; Bishop, D.T.; Weber, B.; Lenoir, G.; Chang-Claude, J.; et al. Genetic heterogeneity and penetrance analysis of the *BRCA1* and *BRCA2* genes in breast cancer families. The Breast Cancer Linkage Consortium. *Am. J. Hum. Genet.* **1998**, *62*, 676–689. [CrossRef] [PubMed]

5.   Roger, V.L.; Go, A.S.; Lloyd-Jones, D.M.; Benjamin, E.J.; Berry, J.D.; Borden, W.B.; Bravata, D.M.; Dai, S.; Ford, E.S.; Fox, C.S.; et al. American Heart Association Statistics Committee and Stroke Statistics Subcommittee. Heart disease and stroke statistics—2012 update: A report from the American Heart Association. *Circulation* **2012**, *125*, e2–e220. [PubMed]

6.   Minow, R.A.; Benjamin, R.S.; Lee, E.T.; Gottlieb, J.A. Adriamycin cardiomyopathy—Risk factors. *Cancer* **1977**, *39*, 1397–1402. [CrossRef]

7.   Barac, A.; Lynce, F.; Smith, K.L.; Mete, M.; Shara, N.M.; Asch, F.M.; Nardacci, M.P.; Wray, L.; Herbolsheimer, P.; Nunes, R.A.; et al. Cardiac function in *BRCA1/2* mutation carriers with history of breast cancer treated with anthracyclines. *Breast. Cancer Res. Treat.* **2016**, *155*, 285–293. [CrossRef] [PubMed]

8.   Murphy, C.G.; Moynahan, M.E. *BRCA* gene structure and function in tumor suppression: A repair-centric perspective. *Cancer J.* **2010**, *16*, 39–47. [CrossRef] [PubMed]

9.   Shukla, P.C.; Singh, K.K.; Lovren, F.; Pan, Y.; Leong-Poi, H.; Erret, L.; Verma, S. *BRCA1* as an essential regulator of cardiac function. *Nat. Commun.* **2011**, *20*. [CrossRef]

10. Singh, K.K.; Shukla, P.C.; Quan, A.; Desjardins, J.F.; Lovren, F.; Pan, Y.; Garg, V.; Gosal, S.; Garg, A.; Szmitko, P.E.; et al. BRCA2 protein deficiency exaggerates doxorubicin-induced cardiomyocyte apoptosis and cardiac failure. *J. Biol. Chem.* **2012**, *287*, 6604–6614. [CrossRef] [PubMed]

11. Ozaki, K.; Sato, H.; Inoue, K.; Tsunoda, T.; Sakata, Y.; Mizuno, H.; Lin, T.H.; Miyamoto, Y.; Aoki, A.; Onouchi, Y.; et al. SNPs in BRAP associated with risk of myocardial infarction in Asian populations. *Nat. Genet.* **2009**, *41*, 329–333. [CrossRef] [PubMed]

12. Nozynski, J.K.; Konecka-Mrowka, D.; Zakliczynski, M.; Zembala-Nozynska, E.; Lange, D.; Zembala, M. BRCA1 Reflects Myocardial Adverse Remodeling in Idiopathic Dilated Cardiomyopathy. *Transplant. Proc.* **2016**, *48*, 1746–1750.

13. Bordeleau, L.; Lipscombe, L.; Lubinski, J.; Ghadirian, P.; Foulkes, W.D.; Neuhausen, S.; Ainsworth, P.; Pollak, M.; Sun, P.; Narod, S.A. Hereditary Breast Cancer Clinical Study Group. Diabetes and breast cancer among women with *BRCA1* and *BRCA2* mutations. *Cancer* **2011**, *117*, 1812–1818. [CrossRef] [PubMed]

14. van Westerop, L.L.; Arts-de Jong, M.; Hoogerbrugge, N.; de Hullu, J.A.; Maas, A.H. Cardiovascular risk of *BRCA1/2* mutation carriers: A review. *Maturitas* **2016**, *91*, 135–139. [CrossRef] [PubMed]

15. Vitelli, L.L.; Crow, R.S.; Shahar, E.; Hutchinson, R.G.; Rautaharju, P.M.; Folsom, A.R. Electrocardiographic findings in a healthy biracial population. Atherosclerosis Risk in Communities (ARIC) Study Investigators. *Am. J. Cardiol.* **1998**, *81*, 453–459. [CrossRef]

16. Miyasaka, Y.; Barnes, M.E.; Bailey, K.R.; Cha, S.S.; Gersh, B.J.; Seward, J.B.; Tsang, T.S. Mortality trends in patients diagnosed with first atrial fibrillation: A 21-year community-based study. *J. Am. Coll. Cardiol.* **2007**, *49*, 986–992. [CrossRef] [PubMed]

17. Lakhani, S.R.; Van De Vijver, M.J.; Jacquemier, J.; Anderson, T.J.; Osin, P.P.; McGuffog, L.; Easton, D.F. The pathology of familial breast cancer: predictive value of immunohistochemical markers estrogen receptor, progesterone receptor, HER-2, and p53 in patients with mutations in *BRCA1* and *BRCA2*. *J. Clin. Oncol.* **2002**, *20*, 2310–2318. [CrossRef] [PubMed]

9

In Vivo Imaging of Local Gene Expression Induced by Magnetic Hyperthermia

Olivier Sandre [1,*], Coralie Genevois [2], Eneko Garaio [3], Laurent Adumeau [4], Stéphane Mornet [4] and Franck Couillaud [2,*]

[1] Laboratory of Organic Polymer Chemistry, LCPO, UMR 5629 CNRS, University of Bordeaux, Bordeaux-INP, Pessac 33600, France
[2] Molecular Imaging and Innovative Therapies in Oncology, IMOTION, EA 7435, University of Bordeaux, 146 rue Léo Saignat, case 127, Bordeaux cedex 33076, France; coralie.genevois@u-bordeaux.fr
[3] Department of Electricity and Electronics, University of the Basque Country (UPV/EHU), P.K. 644, Leioa 48940, Spain; eneko.garayo@ehu.eus
[4] Institute for Condensed Matter Chemistry of Bordeaux, ICMCB, UPR 9048, CNRS, University of Bordeaux, Pessac F-33600 France; laurent.adumeau@gmail.com (L.A.); stephane.mornet@icmcb.cnrs.fr (S.M.)
* Correspondence: olivier.sandre@enscbp.fr (O.S.); franck.couillaud@u-bordeaux.fr (F.C.)

Academic Editor: Selvarangan Ponnazhagan

Abstract: The present work aims to demonstrate that colloidal dispersions of magnetic iron oxide nanoparticles stabilized with dextran macromolecules placed in an alternating magnetic field can not only produce heat, but also that these particles could be used in vivo for local and noninvasive deposition of a thermal dose sufficient to trigger thermo-induced gene expression. Iron oxide nanoparticles were first characterized in vitro on a bio-inspired setup, and then they were assayed in vivo using a transgenic mouse strain expressing the luciferase reporter gene under transcriptional control of a thermosensitive promoter. Iron oxide nanoparticles dispersions were applied topically on the mouse skin or injected subcutaneously with Matrigel™ to generate so-called pseudotumors. Temperature was monitored continuously with a feedback loop to control the power of the magnetic field generator and to avoid overheating. Thermo-induced luciferase expression was followed by bioluminescence imaging 6 h after heating. We showed that dextran-coated magnetic iron oxide nanoparticle dispersions were able to induce in vivo mild hyperthermia compatible with thermo-induced gene expression in surrounding tissues and without impairing cell viability. These data open new therapeutic perspectives for using mild magnetic hyperthermia as noninvasive modulation of tumor microenvironment by local thermo-induced gene expression or drug release.

Keywords: magnetic hyperthermia; gene therapies; heat shock protein promoter; in vivo optical imaging; magnetic polymer-coated nanoparticles

1. Introduction

Gene therapies are promising techniques for curing diseases either by repairing or replacing a defective gene or by expressing some therapeutic proteins or regulatory noncoding RNA. In spite of some well-known successes, such as therapy of crosslinked severe combined immunodeficiency (X-SCID) [1] and quite a lot of clinical trials [2], gene therapies are not widely used in clinical practices. Safe, specific, and efficient delivery of a therapeutic gene to an identified cell or organ still remains an important challenge but efficient control of transgene expression also requires considerable improvement.

Tight spatiotemporal regulation of gene expression, in the region where therapy is necessary and for the duration required to achieve a therapeutic effect, is very important for clinical applications of gene therapy and for minimizing systemic toxicity. Temporal control of gene expression may be achieved by several externally controlled, inducible gene promoters which respond to antibiotics [3] and other small molecules [4]. Spatial control is more frequently envisaged by using tissue-specific or disease-specific promoters, yet lacking features for both temporal and external controls. Furthermore, these specific promoters often exhibit a low level of therapeutic gene expression.

Hyperthermia in combination with the temperature-sensitive 70 kDa heat-shock protein (*HSP70*) promoter presents a unique approach, allowing noninvasive spatiotemporal control of transgene expression. We already demonstrated in vivo, by using transgenic mice and genetically modified cells, that focused-ultrasound (FUS) combined with real-time monitoring of the local temperature distribution by phase magnetic resonance imaging allows for a fine control of temperature increase and thus a good spatiotemporal control of local transgene expression by using thermosensitive promoters [5–7].

Magneto-activatable thermogenic nanoparticles (MTN) as heat sources also appeared as an attractive alternative, especially for deep-seated and poorly accessible tumors [8,9]. MTN injection into tumors and their subsequent heating using an alternating magnetic field has been developed as a cancer treatment for several decades, and after a phase II clinical trial at the end of 2011, it was authorized for use against severe brain cancer (glioblastoma) in combination with conventional radiotherapy and is currently proposed as a clinical tool. Depending on the temperature increase and the duration, tumor cell killing or increased susceptibility to concomitant radio- or chemotherapy has been observed [8–10]. MTN-based hyperthermia treatment also increases tumor sensitivity to natural killer cell-mediated lysis [11] and tumor-specific immune responses resulting from thermo-induction of heat-shock proteins (HSP) expression [12,13].

The present paper aims to demonstrate that MTNs could be used in vivo for local and noninvasive deposition of a sufficient thermal dose to trigger transgene expression by a thermosensitive promoter. Chosen MTNs are colloidal dispersions of magnetic iron oxide nanoparticles stabilized with dextran macromolecules. We first establish MTN heating properties on a bio-inspired in vitro setup, and then we move on to a thermosensitive transgenic mouse strain already characterized in vivo by bioluminescence imaging for its response to mild hyperthermia [14,15].

2. Materials and Methods

2.1. Animals and Animals Handling

Animal experiments were performed in agreement with European directives and approved by the local ethical comity (CEEA 50) under agreement A50120195. The double transgenic mice Hspa1b-LucF (+/+) Hspa1b-mPlum (+/+) [15] were housed at the University of Bordeaux facilities and maintained under 12 h light/dark cycle with water and food ad libitum. Animals were anesthetized with 2% isoflurane (Belamont, Nicholas Piramal Limited, London, UK) in air. Prior to experiments, mice were shaved with clippers and a depilatory cream was applied.

2.2. MTN Synthesis and Characterization

Magnetic nanoparticles (MNPs) synthesized by various chemical routes are studied for their potential use in medicine, in particular for magnetic hyperthermia [16,17]. In this work, MNPs were synthesized by alkaline coprecipitation of ferrous and ferric salts followed by a size-sorting procedure as previously described [18]. Briefly, magnetite (Fe_3O_4) MNPs, right after the coprecipitation reaction, were totally oxidized into maghemite (γ-Fe_2O_3) by boiling ferric nitrate [19] and dispersed in dilute nitric acid. Then, they were submitted to a size-sorting procedure based on the liquid–liquid phase-separation obtained by screening the electrostatic repulsions with an excess of electrolyte (HNO_3) concentration, yet below 0.4 M (pH > 0.4) in order to prevent MNP dissolution into ions. The principle of this sorting is that the concentrated phase is enriched with the larger MNPs, whereas the dilute phase contains the smaller MNPs [18,20,21]. After repeating several phase separations and washing steps (three cycles), the sample named C1C2C3 originated from the concentrated phase fractions corresponding to the concentrated phases obtained by adding HNO_3 electrolyte to the preceding pelleted fractions. The ions in excess were washed off with acetone and diethyl ether, resulting into a suspension in diluted HNO_3 at pH ~ 2 and at an iron oxide concentration of 110 g/L. These particles displayed a mean nanoparticle size of 12.4 nm (with standard deviation σ = 6.0 nm), as measured by automated particle counting on a transmission electronic microscope (TEM) image (Figure S1) This "ionic ferrofluid" was then coated by dextran T70 (Mw 70 kDa, Sigma-Aldrich, Saint-Quentin Fallavier, France) in order to be dispersible in aqueous buffer at neutral pH and also in a protein hydrogel matrix (called hereinafter Matrigel™). For this coating, a volume of 5 mL of ferrofluid at 110 g/L was added to 30 mL of dextran solution at 100 g/L and then incubated overnight under gentle shaking. This dispersion was precipitated in 120 mL of ethanol, and particles were separated magnetically from free (un-adsorbed) dextran macromolecular chains. This dextran precipitation not only allows removal of most of the polymer excess but also promotes the adhesion of glucose monomers onto the nanoparticle surface through hydrogen bonds. The pellet was dispersed in 40 mL of sterile ultrapure water (18 MΩ·cm) produced by using a SG-Labostar (7 TWF-UV system from Odemi, Grisy, France) and heated at 70 °C under vacuum in order to remove residual ethanol. Finally, free dextran chains were removed by several washes (two times with a dilution factor of about 10) with ultrapure water by tangential flow filtration (TFF) using 300 kDa cut-off ultrafiltration filters (Merck Millipore). At the end of the procedure, these magnetic nanoparticles contained 15.4% of dextran, determined by thermogravimetric analysis (TGA) (Figure S2). Finally, the so-prepared ferrofluid, thereafter denominated C1C2C3@dex, was concentrated by TFF in phosphate-buffered saline (PBS) to obtain an iron oxide concentration of 117 g/L as measured by a UV–vis spectrum (Figure S3). Their hydrodynamic diameter was measured by dynamic light scattering (DLS) after 200-times dilution in water with a Malvern™ NanoZS apparatus operating at 90° scattering angle (Figure S4), giving a Z-average diameter of 40.2 nm with low polydispersity (PDI = 0.215). Their specific heating power was characterized by the "specific absorption rate" (SAR) in the alternating magnetic field (AMF) conditions used (H = 10.2 kA/m at 755 kHz). A value for specific absorption rate (SAR) = 94 $\pm$ 1 W/g was found from the initial slope of the temperature profile within the first 3 s of AMF application (Figure S5). In previous work containing extensive magnetic characterization, we showed that this sample has a SAR verifying the linear response theory based on Néel and Brown relaxation processes of the magnetic moments, using independent magnetic measurements of the specific magnetization and of the magnetic anisotropy constant [18].

2.3. In Vitro MTN Experiments (Phantoms)

An aliquot of C1C2C3@dex nanoparticles (20, 25, 30, or 35 μL), diluted to 50 μL with water, was mixed with 50 μL of Matrigel™ (BD Matrigel™ Basement Membrane Matrix; Becton, Dickinson and Company, San José, CA, USA) at a temperature near 4 °C. Previously, it was checked that Matrigel™ diluted twice still jellified when raising the temperature to 37 °C. The mixture was poured into a cell culture insert (24 wells; ø pores: 0.8 μm) and covered by an agar plug (Figure 1A). The insert was placed into an agar gel (20 mL; 2% w/v) modeling the heat capacity of a mouse (body weight around 20 g) which was placed in a thermostatic double-wall chamber (Figure 1B). The system was then placed in the middle of the solenoid and maintained at 37 °C (Figure 1C) by the water circuit linked to a regulating bath (Huber Polystat CC, Offenburg, Germany) before the alternating magnetic field (AMF) was generated by the coil. An optical fiber thermo-probe (OTG-M420; Opsens™ Inc., Québec, QC, Canada) was introduced inside the insert through a catheter (20 G, TERUMO®, Terumo Medical Corporation, Somerset, NJ, USA) to reach the C1C2C3@dex nanoparticle/Matrigel™ mixture (Figure 1C).

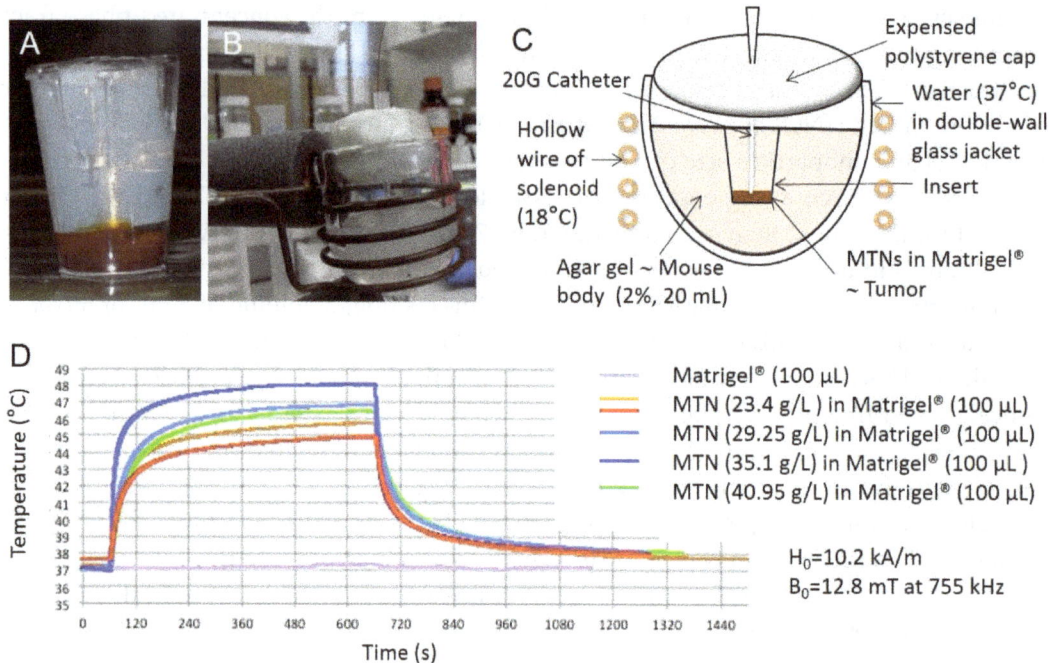

Figure 1. Heating performances of MTNs in a bio-inspired phantom. (**A**) Phantoms were made of 100 µL Matrigel™ containing magneto-activatable thermogenic nanoparticles (MTNs) at different concentrations and poured into a cell culture insert (24 wells; ø pores: 0.8 µm) to mimic a tumor; (**B**) Insert was covered by an agar plug and inserted into a 20 mL agar gel (2% w/v) of thermal inertia equivalent to the mouse body. Phantom was placed into a thermostatic chamber maintained at 37 °C through water circulation within double-walled glass water jacket to represent "active" thermal regulation by the mouse blood circulation, and located in the middle of the solenoid. The complete system reached thermal equilibrium at 37 °C before the alternating magnetic field was generated. An optical fiber thermal probe (Opsens®OTG-M420) was introduced into the insert up to reaching the MTN/Matrigel™ mixture through a catheter (20 G, TERUMO®); (**C**) Schematic representation of the experimental setup; (**D**) Time courses of temperature in the magnetic field of induction $B = 12.8$ mT at 755 kHz ($H = 10.2$ kA/m) according to different MTN concentrations.

2.4. In Vivo MTN Experiments (Animals)

MTNs were diluted in water and applied topically on the mouse skin. Diluted MTNs were also mixed with Matrigel™ at 4 °C and injected subcutaneously on the back of the mouse to generate so-called "pseudotumor". After 30 min, the Matrigel™ was gelled and the mouse was placed inside the solenoid. Optical fiber thermo-probe (OTG-M420; Opsens™, Québec, QC, Canada) was inserted into the pseudotumor using a Teflon™ catheter (20 G, TERUMO®) as guide. Room temperature around the coil was monitored during the experiment and maintained at 20 °C.

2.5. Hyperthermia Setup and Protocol

The electric generator (SEIT ELECTRONICA, Junior™ 3.5 kW model creates an alternating magnetic field (AMF) inside a 4-turn copper-ring solenoid of 55 mm outer diameter, 48 mm inner diameter, and 38 mm height. The solenoid is refrigerated by a cold (18 °C) water flux inside the 3.5 mm diameter, 0.4 mm wall thick, hollow wires of the solenoid. Magnetic field intensity is $H = 10.2$ kA/m, induction $B = 12.8$ mT and frequency 755 kHz, as determined by finite element modeling for magnetics (FEMM, http://www.femm.info), after measuring the high voltage (747 V) and deducing the current (234 Amps) from the calculated coil impedance ($R = 3.2$ Ω) and inductance ($L = 4.5 \times 10^{-9}$ H). The field value calculated by finite element modeling was also checked experimentally by measuring the electromotive force in a scout coil (1 turn of 17.5 mm diameter): 29.4 V peak to peak.

Homemade software was developed in LabVIEW™ allowing temperature regulation during the heating process according to predefined parameters (i.e., duration (in seconds), maximum temperature, and thermal dose above a threshold temperature). The thermal dose is defined as the integral of the temperature vs. time profile above a predetermined threshold, usually 42 °C. This definition is equivalent to the "equivalent total time at 43 °C" expressed in "degree-minute" (tdm$_{43}$) [22].

2.6. In Vivo Bioluminescence Imaging (BLI)

BLI was performed at the Vivoptic platform (Bordeaux University & CNRS, UMS 3767) using a NightOWL II LB 983 calibrated system equipped with an NC 100 CCD deep-cooled camera (Berthold Technologies™, Bad Wildbad, Germany). Mice were injected intraperitoneally with D-luciferin (2.9 mg in 100 µL PBS, Promega™, Madison, WI, USA,) and sedated 5 min later. Bioluminescence images (1 min exposure, 4 × 4 binning) and photographs (100 ms exposure) were taken 8 min after the luciferin injection. A low light emitting standard (Glowell, Lux Biotechnology Limited, Edinburgh, UK) was placed next to the animal during each image acquisition to provide a quality control. Pseudocolor images representing the spatial distribution of emitted photons were generated using IndiGO 2 software (Berthold Technologies™) and superposed to the corresponding photographs.

3. Results

3.1. Heating Performances of MTN In Vitro on Bio-Inspired Phantom

We first assayed MTN heating performances using an in vitro bio-inspired setup. MTNs were mixed with Matrigel™, a natural component of the tumor microenvironment, and the mixture was placed into a 20 mL agar gel to mimic a mouse weight of about 20 g. The tumor phantom model was designed with a volume of 100 µL in order to exceed the diameter of 1.1 mm proposed by Rabin as a minimum size of magnetic sample to exhibit macroscopic heating compared to environmental temperature [23]. The agar gel was actively thermo-regulated at 37 °C using a water flux. MTNs were mixed with Matrigel™ at a final iron oxide concentration of 23.4, 35.1, and 40.8 g/L. The maximum temperature that can be reached in Matrigel™ phantoms depends on their volume and on the concentration of their contained MTNs. Figure 1D shows the temperature profiles obtained for the maximum power of the generator providing magnetic field intensity of B = 12.8 mT at 755 kHz. The temperature, initially at 37 °C, rapidly increases and reaches a plateau that is maintained as long as the magnetic field is maintained, corresponding to a perfect balance between the heat flux created by the MTNs and the thermal losses at the interface between the magnetic phantom and the surrounding hydrogel mimicking the mouse body. Activation was repeated several times in order to check the reproducibility of the biomimetic system. Observation of the MTN-containing Matrigel™ after the experiment reveals a poor homogeneity of the spatial distribution of MTNs for an iron oxide concentration of 40.95 g/L. At a concentration of 35.1 g/L, the maximum temperature obtained is 48 °C while MTNs are homogeneously distributed in the Matrigel™ tumor phantom.

3.2. In Silico Modeling of Temperature Distribution in the Bio-Inspired Phantom

Numerical modeling of thermal distribution was used to predict MTN heating upon activation. The finite element software FEMM can perform numerical simulations not only of magnetics but also of the Fourier heat diffusion and convection equation. The grid model took into account all parameters of the experimental device, including dimensions of the various compartments, fixed temperature (37 °C) at the boundary of the glass jacket with convective conditions, and thermal characteristics of the Matrigel™ and agar media such as heat capacity and conductivity, taken as pure water values (C_P = 4.18 J/cm^3/K, k = 0.6 W/m/K). Figure 2 shows the modeling results for MTN concentration of 35 g/L in the 100 µL Matrigel™ phantom. The iron oxide mass in the phantom (3.5 mg) can irradiate a power $3.5 \times 10^{-3} \times 94 = 0.33$ W in a 100 µL volume, thus a power density of 3.3 W/cm^3. A somehow lower estimated value $p = 2\,W/cm^3$ was taken in the simulation to take into account the decrease of SAR when the MNPs are partially blocked within the gel matrix as compared to the homogeneous fluid (Brown relaxation mechanism is likely impeded). This 40% decrease of SAR is arbitrary, but it is quite in line with a systematic study of the local concentration's negative effect on SAR for similar hydrophilic polymer-coated MTNs [24]. The maximum temperature is predicted in the center of Matrigel™ insert, and temperature gradually decreases with distance from the center. The temperature difference is nearly 10 ° C between the Matrigel™ center and the border of the agar block containing the insert, yet with variation between 45 °C and 47 °C depending on the axial and lateral coordinates. The temperature predicted by the numerical model in the center rose to 48.5 °C, which is very close to the experimental value of 48 °C reported in Figure 1D for the same concentration (35 g/L). However, these curves show non-monotonous variation with concentration in the experiments, which might be ascribed to Matrigel™ heterogeneity or unprecise localization of the optical fiber temperature probe (in agreement with the thermal gradient calculated numerically).

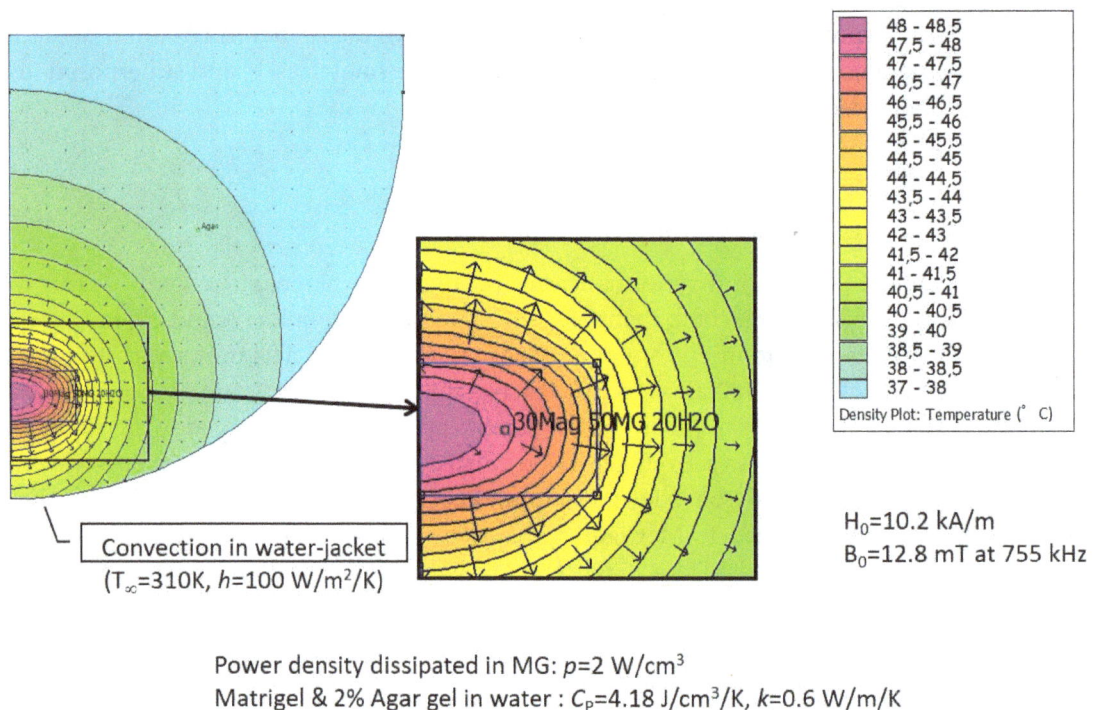

Figure 2. Numerical modeling of thermal distribution in the Matrigel™ phantom with 3709 nodes using finite element modeling for magnetics (FEMM) freeware (http://www.femm.info/). The power density was chosen to correspond to an MTN concentration of 35.1 g/L under an alternating magnetic field $B = 12.8\,mT$ at 755 kHz ($H = 10.2\,kA/m$). At thermal equilibrium (stationary state of the heat equation), maximal temperatures of 48 °C in the center of the Matrigel™ insert and of 44 °C at the insert periphery were predicted by the numerical model.

3.3. Topical Application of MTN Solution on Transgenic Mouse Skin

MTN aqueous suspension (69 g/L, 100 µL) was applied on the skin of *HSP70*–LucF transgenic mouse. Mouse was placed in the middle of the solenoid and an optical fiber thermoprobe was immersed in the MTN fluid droplet (Figure 3A). Temperature was monitored using LabVIEW™ homemade program to drive the generator and thus the magnetic field created by the solenoid (Figure 3B). The magnetic field generator was switched off when the temperature inside the MTN drop reached 45 °C and was switched on again at 44 °C, leading to a saw tooth temperature profile. Applied magnetic field was $B = 12.8\,mT$ at 755 kHz ($H = 10.2\,kA/m$). As shown in Figure 3C, heating was maintained for 10 min until the AMF was definitively switched off, with temperature decreasing rapidly to the baseline skin temperature (32 °C). BLI measurement was performed 6 h after magnetic heating according to previous data reporting *HSP70*-dependent LucF expression [15]. Figure 3D illustrates *HSP70*-dependent LucF expression as revealed by BLI. As controls, MTNs were applied on the mouse skin without magnetic field or the mouse was placed into the magnetic field for 10 min without MTNs; in these cases, no BLI signal was detected [25].

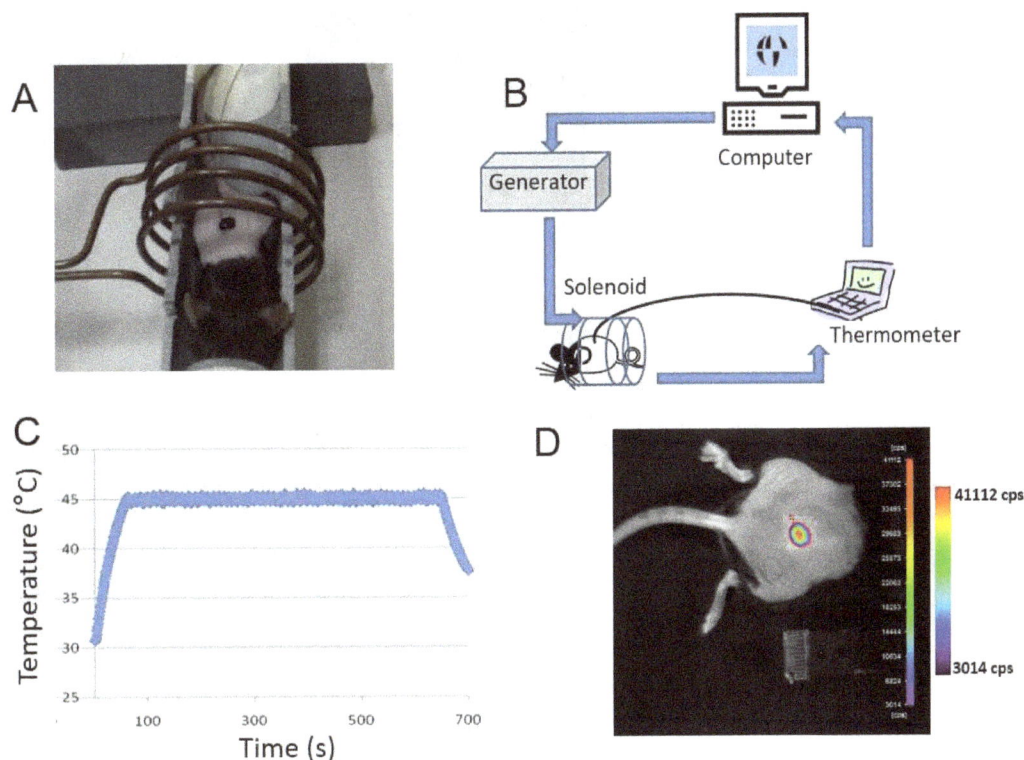

Figure 3. Imaging of heat-induced expression of luciferase by magnetic activation of MTN applied on the transgenic mouse skin. (**A**) An MTN drop (69 g/L, 100 μL water) was applied on *HSP70*–LucF transgenic mouse skin. The mouse was placed on a bed aligned with the central axis of the solenoid. An optical fiber thermoprobe was immersed in the MTN droplet; (**B**) On/off cycles of alternating magnetic field (B = 12.8 mT at 755 kHz (H = 10.2 kA/m)) were created by a computer-controlled generator according to the temperature recorded in the drop (T_{on} = 44°C, T_{off} = 45°C); (**C**) Examples of temperature measurements in the MTN solution; (**D**) Bioluminescence imaging of thermo-induced luciferase expression by the mouse skin, 6 h after magnetic hyperthermia.

3.4. Subcutaneous MTN-Containing Matrigel™ Pseudotumors in Mice

To test MTN-induced intratumoral hyperthermia with further realism towards therapeutic application, Matrigel™ pseudotumors containing MTNs (35.1 g/L) were created subcutaneously in HSP/LucF transgenic mice by injection of the mixture in the cold fluid state. A Teflon™ catheter was introduced into the pseudotumors to be used as a guide for the fiber optics temperature probe. Temperature was monitored using software developed under LabVIEW™ to drive the generator and thus the magnetic field inside the solenoid. Maximum temperature was fixed at 45 °C and running time was defined according to calculation of the predefined thermal dose expressed as tdm$_{43}$ (i.e., time interval equivalent to a thermal dose calculated above a threshold temperature of 42 °C). As shown in Figure 4, magnetic activation of Matrigel™ pseudotumors containing MTNs, heating results in LucF expression by mouse tissues as detected by BLI 6 h after heating. Different types of BLI patterns were obtained. Some mice exhibited a "Gaussian type" BLI pattern (Figure 4A; n = 5) whereas some other exhibited "ring shape" BLI patterns with no BLI signal in the middle of the heated zone (Figure 4B; n = 13). The pattern was not related to the predefined thermal dose, as illustrated in Figure 4.

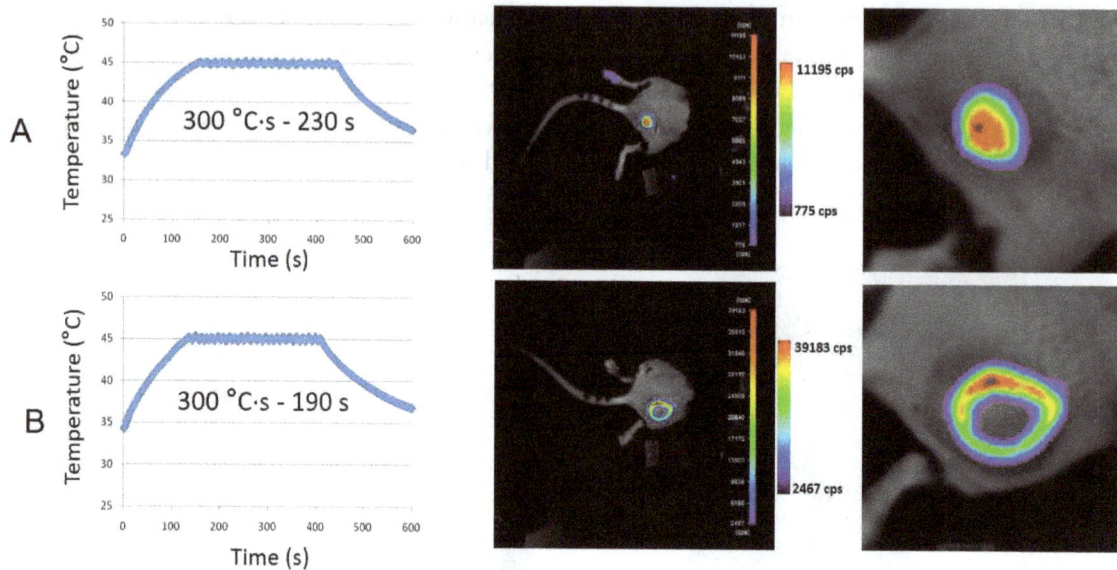

Figure 4. Imaging of heat-induced expression of luciferase by magnetic activation of Matrigel™ pseudotumors containing MTNs. Cold Matrigel™ (50 μL) added to MTNs (30 μL at 115 g/L, 20 μL water) leading to 100 μL at 34.5 g/L final MTN concentration were injected subcutaneously and allowed to solidify for 30 min at mouse body-temperature. The mouse is placed in the middle of the solenoid. An optical fiber thermoprobe (360 μm) was inserted in the Matrigel™ volume using a Teflon™ catheter. Magnetic field (B = 12.8 mT at 755 kHz (H = 10.2 kA/m)) is created by a PC-controlled generator (on/off cycles) according to temperature recorded in the Matrigel™ volume. Predefined thermal dose was 300 °C×s calculated above 42 °C, or an equivalent time at 43 °C, tdm$_{43}$ = 5 min, and upper limit temperature was set at 45 °C. Line (**A**) and (**B**) are BLI images of thermo-induced luciferase expression by the mouse tissues, 6 h after magnetic hyperthermia.

Soaking the leg of the transgenic mice in warm water (45 °C, 8 min) induces *HSP70* promoter activation and LucF expression, as detected by BLI 6 h later [14,15]. To determine the physiological status of the central area of the ring-shaped pattern, the mouse paw was immersed in a water bath (45 °C, 8 min, tdm$_{43}$ = 24 min). Six hours after this heating, the whole paw expresses luciferase except the central area of the ring. This lack of signal therefore corresponds to an area where cells have lost the ability to express luciferase. Transgenic mice with Matrigel™ pseudotumors containing MTNs without AMF application or with Matrigel™ pseudotumors alone and placed in the AMF did not exhibit BLI signal. Mice bearing pseudotumors with MTN concentrations lower than 35 g/L did not exhibit BLI signal (11 g/L, 15 min AMF on; 15 g/L 15 min AMF on; 18 g/L 15 min AMF on; n = 5; [25]).

3.5. In Silico Modeling of Temperature Distribution in MTN-Containing Pseudotumors

Using the same finite element modeling FEMM software as was used for bio-inspired phantoms, temperature-mapping in MTN-containing Matrigel™ pseudotumors was assayed upon activation with different tumor shapes that can arise after subcutaneous injection of a given volume of 100 μL. Thermal characteristics of the different tissues were obtained from the literature [26]. Figure 5 shows the modeling results for MTN concentration of 35 g/L in 100 μL Matrigel™ pseudotumors located below the skin. According to the pseudotumor shape, the estimated temperature on the skin surface facing the BLI camera increased from 46.2 °C (flat disk-like tumor) up to 57.2 °C (cylindrical tumor of larger thickness, thus lower surface area and less thermal loss to the surrounding tissues).

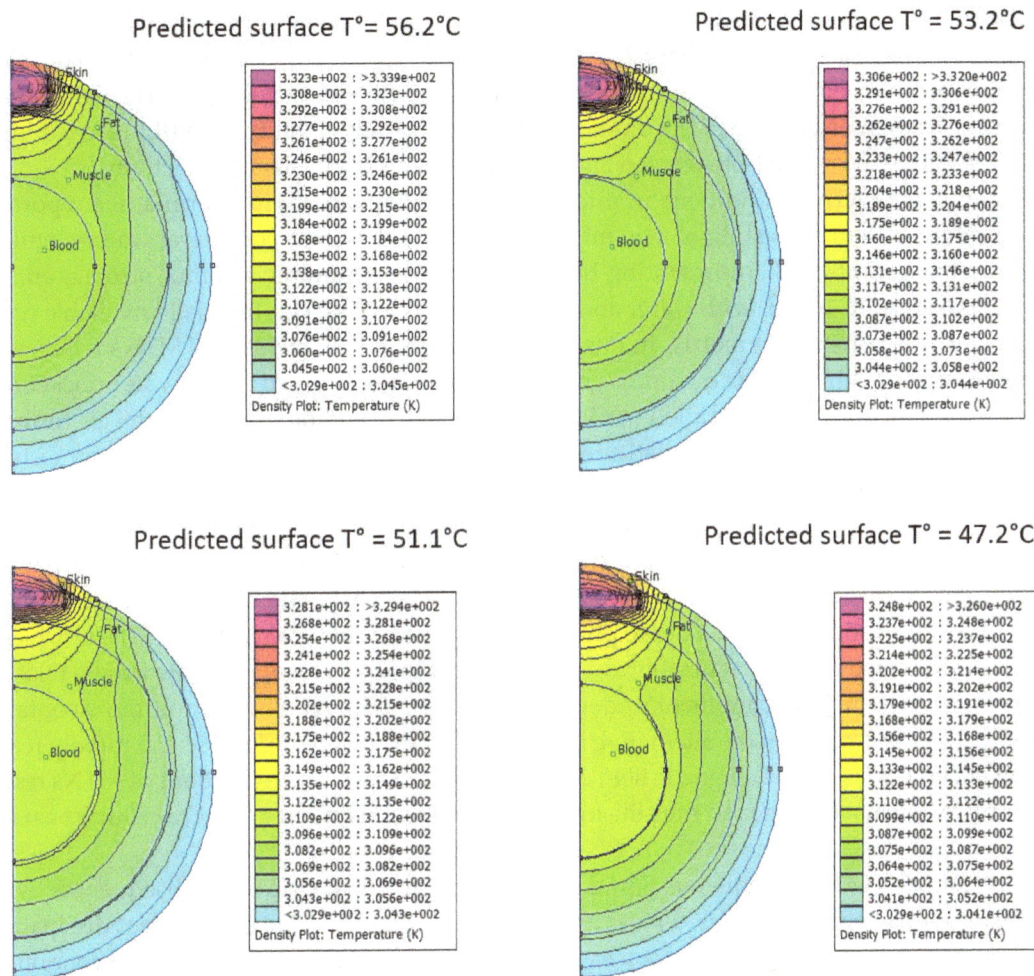

Figure 5. Numerical modeling (FEMM grid with ~7800 nodes) of temperature spatial distribution in and around a subcutaneous pseudotumor placed in the fat layer (3 mm thick) below skin (1 mm) and above muscles for different tumor shapes at constant 100 μL volume. Blood volume is 2.1 mL. MTNs are supposed to dissipate a heat power density $p = 2$ W/cm^3 under an alternating magnetic field of $B = 12.8$ mT at 755 kHz ($H = 10.2$ kA/m). Thermal parameters (heat capacity and conductivity) used are taken from literature [22]: ($C_P = 3.28$ J/cm^3/K; $k = 0.37$ W/m/K) for skin, ($C_P = 1.93$ J/cm^3/K; $k = 0.16$ W/m/K) for fat, ($C_P = 3.87$ J/cm^3/K; $k = 0.48$ W/m/K) for muscle, ($C_P = 4.18$ J/cm^3/K; $k = 0.6$ W/m/K) for blood pool, supposed isothermal at $T_0 = 310$ K. Predicted temperatures at the pseudotumor surface were 56.2, 53.2, 51.1, and 47.2 °C according to pseudotumor cylindrical shape heights of 2.6, 2, 1.34, and 1.04 mm, respectively (for the same 100 μL volume).

4. Discussion

In the present paper, we demonstrate the feasibility of inducing gene expression in vivo by magnetic hyperthermia. When MTNs were placed in an alternating magnetic field, they produced heat which dissipated into the surrounding environment. The challenge was to use MTN-based heating in physiological environment and to generate enough thermal energy to activate the thermosensitive promoter in cells of surrounding tissues. The transgenic mouse used is expressing the reporter gene *LucF* under transcriptional control of the *HSP70* thermosensitive promoter, and this mouse has been already fully characterized for its response to mild hyperthermia [14]. For temperature above 43 °C, temperature-induced *HSP70* promoter activation was modulated by both temperature as well as duration of hyperthermia and the reporter gene expression can be modeled by an Arrhenius analysis [14]. Typically, by soaking a mouse paw in a water bath at 45 °C for 8 min (calculated thermal dose = 1440 °C·s above 42 °C, thus tmd$_{43}$ = 24 min), a high BLI signal was expressed by skin [14,15].

The calculated average activation energy required for activation of *HSP70* promoter in this mouse was 559 kJ/mol leading to a K-factor of about 2, indicating that for temperatures above 43 °C, an increase in temperature of 1 °C is equivalent to decreasing the exposure time by a factor 2 [14]. The light emission after heating attests for both *HSP70* promoter activation and cell viability. The BLI readout indeed requires *LucF* translation and both oxygen and ATP-dependent LucF enzymatic activity.

To mimic the thermic conditions in mice exposed to magnetic hyperthermia, we reported at first a phantom system consisting of a 20 mL soft aqueous gel, with a peripheral water-circulating system thermalized at 37 °C representing blood circulation. To mimic a tumor stroma, an MTN droplet was diluted to 100 µL with water and Matrigel™, which is a commercially available natural component of the extracellular matrix. In this medium, the state of dispersion of MTNs is believed to be closed to mouse extracellular medium (i.e., they are partially blocked in the water pocket of the tight network. The upper concentration of MTNs in Matrigel™, without impeding the gel formation, is limited to about 35 g/L, which constituted the first limitation. Higher concentration results in MTN precipitation in Matrigel™ and impairs heating performance. As a second limitation, we found that in such conditions, heating capacity of MTNs is limited, lowered by a factor of around 40% (as estimated by the heat power density $p = 2$ W/cm^3 necessary for according experiments and thermic simulations, as compared to $p = 3.3$ W/cm^3 calculated from the SAR in liquid state). At a temperature of 37 °C, Matrigel™ indeed forms a dense protein network of 50–500 nm mesh-size [27], in which the polysaccharide-coated MNPs presumably are adsorbed and blocked, hampering the Brown relaxation mechanism of their magnetic moment under an AMF. When Matrigel™ pseudotumors containing MTNs (35 g/L) were placed in the alternative magnetic field, the temperature rose and rapidly reached a plateau at about 48 °C in the magnetic field conditions used. Lower concentrations of MTNs resulted in lower plateau temperatures, and thus the lower limit of 43 °C required to induce cellular heat stress was not reached [28].

Application of MTNs in aqueous liquid droplets on the mouse skin and subsequent activation by alternating magnetic field resulted in significant temperature increase, despite heat dissipation in air. To prevent any hazard and extensive tissues damage, the temperature was recorded continuously and limited to 45 °C. Such protocol ensures thermal dose calculation in real time and allowed for LucF expression. Although not as extensively studied in the present work as would be needed for a quantitative relationship, the BLI signal was well correlated with the thermal dose deposited on the mouse skin, defined as the integral of time–temperature curve above 42 °C with a coefficient of determination $R^2 = 0.999$ ($n = 3$: linear fit, Figure S6) and was consistent with previously reported data using warm water as heating source.

MTNs diluted in cold liquid Matrigel™ at 35 g/L and injected subcutaneously in mouse formed a solid mass when reaching physiological temperature of the animal. Placing the mouse into the alternating magnetic field resulted in MTN activation. The main result of this work was evidence that magnetically induced heat generation was sufficient to induce *HSP70*-dependent transgene expression in surrounding tissues. The temperature was recorded continuously by placing a thermal probe inside the Matrigel™ mass: Although the position of the 420 µm fiber in the millimeter-sized tumor phantom was not sufficiently precise to guarantee to be in the exact geometrical center of the pseudotumor, the temperature rise could be limited by software control of the AMF generation to 45 °C. However, certainly due to this imprecision of probe localization, BLI patterns of transgene activation appeared very variable and not well correlated to the recorded thermal dose. In particular, ring-shaped BLI patterns without a light-emitting area in the middle were observed in some cases. Ring-shaped activation patterns have been already reported and were resulting in local overheating, leading to central tissues necrosis and *HSP70* activation at the periphery [6]. This suggest that the recorded temperature was not representative of the overall temperature rise in the Matrigel™ mass and that the thermal probe may be not located in the warmer point of the Matrigel™ pseudotumor volume, but rather at its periphery, explaining the underestimation of the recorded temperature (and hence overheating and cell catabolism in the central part).

Modeling thermal exchanges by finite element modeling (FEM) simulation of the heat equation with biologically relevant parameters provided good correlation between simulations and experimental measurements. It is indeed the case that, when applied to the phantom setup, FEM provided consistent prediction of the maximum temperature reached at thermal equilibrium (i.e., when generated heat is balanced by thermal losses). For in vivo experiments using MTNs dispersed in Matrigel™, numerical modeling clearly illustrated that the temperature field is not homogeneous in the Matrigel™ volume. The higher temperature was calculated in the geometrical center of the Matrigel™ volume, and the predicted temperature profiles exhibited a centrifuge gradient. In vivo, perfect positioning of the thermal probe in the center of the Matrigel™ could neither be precisely achieved nor checked. Incorrect positioning provided false evaluation of the overall temperature in the Matrigel™ and further mistaking of the feedback loop software controlling the magnetic field. This may explain in part the ring-shaped BLI patterns ascribed to overheating and catabolism/necrosis of the cells in contact with the hottest spots. Modeling also illustrated changes in thermal profiles according to pseudotumor shapes. MTN/Matrigel™ mixtures were injected in a fluid state and solidified in various shapes, ranging from "almost spherical" to "flat disk". Modeling prediction revealed up to 9 °C of difference between expected temperatures on pseudotumor surface according to the different shapes, as ascribed to different surface-to-volume ratio and thus thermal losses to surrounding matrix, resulting in huge differences in the thermal dose delivered to tissues.

The *HSP70* promoter is activated not only by heat stress but also in response to a variety of stresses of both environmental and physiological origins [29]. Stressors include oxidative stress [30], toxic compounds [29], hypoxia, ischemia, acidosis, energy depletion, cytokines, and UV radiation [31]. In the current study, no BLI signal was detected after MTN topical application or MTN pseudotumor injection in absence of magnetic field activation. It is also noted that only female mice were used for the study to avoid BLI resulting from mechanical stresses and resulting scarring ascribed to fighting among males. In literature, it has been shown that intracellular magnetic field-activated MTN can activate the *HSP70* response in vitro [13] or induce toxicity on targeted cancer cells without a detectable temperature rise [10]. Intracellular internalization of MTNs is not expected to occur in our current experiments, as MTN suspension was deposited on skin for a very short time or the MTNs were embedded in Matrigel™. Activation of the *HSP70* response without consistent temperature increase was not observed. For instance, when recorded temperature did not reach a temperature above 43 °C or the thermal dose required for heat-induced HSP70 activation by using lower MTN concentration in Matrigel™ pseudotumors or shorter activation times, no BLI signal was detected. In other words, a macroscopic temperature rise was necessary to induce *HSP70* gene expression in this in vivo study.

The current work clearly revealed current practical limitations of the MTN-based strategy for controlling gene expression. As stated above, concentrations of the MTN required for *HSP70* promoter activation is very high, close to the limit of dispersibility of the MNPs in biological media. More efficient MNPs (exhibiting higher SAR) may be necessary to develop for future studies. Difficulties also occurred in controlling MTN homogeneity, which induced variations in the temperature field distribution. Finally, the major limitation is due to inadequate methods for temperature control in vivo and for determination of MTN distribution within the tumor microenvironment, resulting in unexpected events such as overheating. Possible amelioration could be to perform magnetic resonance imaging of the pseudotumor right before the AMF application, in order to check the exact position of the catheter and the fiber optics temperature probe.

5. Conclusions

In conclusion, we showed that polysaccharide-coated maghemite nanoparticles either deposited on the skin or injected into subcutaneous tumor phantoms were able to induce in vivo mild hyperthermia compatible with thermo-induced gene expression in surrounding tissues, without impairing cell viability. Although limitations still remain for finer temperature measurement and limitation, and thus, control of transgene expression, these data opened new therapeutic perspectives for using mild magnetic hyperthermia for noninvasive modulation of the tumor microenvironment by local thermo-induced gene expression or local drug release.

Acknowledgments: We thank Pauline Durand for technical assistance, Pierre Costet and Laetitia Medan (University of Bordeaux) for animal breading and care. TEM images were taken at the Bordeaux Imaging Center (BIC) of Bordeaux University with the acknowledged help of Sabrina Lacomme and Etienne Gontier. This work was supported by Emerging program of Cancéropôle Grand Sud-Ouest and 'Défi Nano' 2014 program from CNRS Interdisciplinary Mission. Authors also thank COST Action RADIOMAG (TD1402), supported by COST (European Cooperation in Science and Technology). Olivier Sandre thanks the Agence Nationale de la Recherche (grant ANR-13-BS08-0017 MagnetoChemoBlast). Work was done within LabEx TRAIL (ANR-10-LABX-57) community. Finally, the Vivoptic and BIC imaging platforms were both supported by France Life Imaging.

Author Contributions: Olivier Sandre, Laurent Adumeau and Stéphane Mornet performed MTN synthesis, polymer grafting and characterization, Eneko Garaio conceived and wrote the software, Olivier Sandre did thermal modeling; Coralie Genevois and Franck Couillaud performed mouse and imaging experiments. Olivier Sandre, Stéphane Mornet and Stéphane Mornet wrote the manuscript with input from all authors. All authors conceived and designed the experiments, read and approved the final manuscript.

References

1. Cavazzana-Calvo, M.; Hacein-Bey, S.; de Saint Basile, G.; Gross, F.; Yvon, E.; Nusbaum, P.; Selz, F.; Hue, C.; Certain, S.; Casanova, J.L.; et al. Gene therapy of human severe combined immunodeficiency (SCID)-X1 disease. *Science* **2000**, *288*, 669–672. [CrossRef] [PubMed]

2. Edelstein, M.L.; Abedi, M.R.; Wixon, J. Gene therapy clinical trials worldwide to 2007—An update. *J. Gene Med.* **2007**, *9*, 833–842. [CrossRef] [PubMed]

3. Iida, A.; Chen, S.T.; Friedmann, T.; Yee, J.K. Inducible gene expression by retrovirus-mediated transfer of a modified tetracycline-regulated system. *J. Virol.* **1996**, *70*, 6054–6059. [PubMed]

4. Clackson, T. Controlling mammalian gene expression with small molecules. *Curr. Opin. Chem. Biol.* **1997**, *1*, 210–218. [CrossRef]

5. Deckers, R.; Quesson, B.; Arsaut, J.; Eimer, S.; Couillaud, F.; Moonen, C.T. Image-guided, noninvasive, spatiotemporal control of gene expression. *Proc. Natl. Acad Sci. USA* **2009**, *106*, 1175–1180. [CrossRef] [PubMed]

6. Eker, O.F.; Quesson, B.; Rome, C.; Arsaut, J.; Deminière, C.; Moonen, C.T.; Grenier, N.; Couillaud, F. Combination of cell delivery and thermoinducible transcription for in vivo spatiotemporal control of gene expression: A feasibility study. *Radiology* **2011**, *258*, 496–504. [CrossRef] [PubMed]

7. Fortin, P.-Y.; Lepetit-Coiffé, M.; Genevois, C.; Debeissat, C.; Quesson, B.; Moonen, C.T.W.; Konsman, J.P.; Couillaud, F. Spatiotemporal control of gene expression in bone-marrow derived cells of the tumor microenvironment induced by MRI guided focused ultrasound. *Oncotarget* **2015**, *6*, 23417–23426. [CrossRef] [PubMed]

8. Thiesen, B.; Jordan, A. Clinical applications of magnetic nanoparticles for hyperthermia. *Int. J. Hyperth. Off. J. Eur. Soc. Hyperthermic Oncol. N. Am. Hyperth. Group* **2008**, *24*, 467–474. [CrossRef] [PubMed]

9. Kobayashi, T. Cancer hyperthermia using magnetic nanoparticles. *Biotechnol. J.* **2011**, *6*, 1342–1347. [CrossRef] [PubMed]

10. Creixell, M.; Bohórquez, A.C.; Torres-Lugo, M.; Rinaldi, C. EGFR-targeted magnetic nanoparticle heaters kill cancer cells without a perceptible temperature rise. *ACS Nano* **2011**, *5*, 7124–7129. [CrossRef] [PubMed]

11. Kubista, B.; Trieb, K.; Blahovec, H.; Kotz, R.; Micksche, M. Hyperthermia increases the susceptibility of chondro- and osteosarcoma cells to natural killer cell-mediated lysis. *Anticancer Res.* **2002**, *22*, 789–792. [PubMed]

12. Ito, A.; Matsuoka, F.; Honda, H.; Kobayashi, T. Antitumor effects of combined therapy of recombinant heat shock protein 70 and hyperthermia using magnetic nanoparticles in an experimental subcutaneous murine melanoma. *Cancer Immunol. Immunother.* **2004**, *53*, 26–32. [CrossRef] [PubMed]

13. Moros, M.; Ambrosone, A.; Stepien, G.; Fabozzi, F.; Marchesano, V.; Castaldi, A.; Tino, A.; de la Fuente, J.M.; Tortiglione, C. Deciphering intracellular events triggered by mild magnetic hyperthermia in vitro and in vivo. *Nanomed.* **2015**, *10*, 2167–2183. [CrossRef] [PubMed]

14. Deckers, R.; Debeissat, C.; Fortin, P.-Y.; Moonen, C.T.W.; Couillaud, F. Arrhenius analysis of the relationship between hyperthermia and Hsp70 promoter activation: A comparison between ex vivo and in vivo data. *Int. J. Hyperth. Off. J. Eur. Soc. Hyperthermic Oncol. N. Am. Hyperth. Group* **2012**, *28*, 441–450. [CrossRef] [PubMed]

15. Fortin, P.-Y.; Genevois, C.; Chapolard, M.; Santalucía, T.; Planas, A.M.; Couillaud, F. Dual-reporter in vivo imaging of transient and inducible heat-shock promoter activation. *Biomed. Opt. Express* **2014**, *5*, 457–467. [CrossRef] [PubMed]

16. Périgo, E.A.; Hemery, G.; Sandre, O.; Ortega, D.; Garaio, E.; Plazaola, F.; Teran, F.J. Fundamentals and advances in magnetic hyperthermia. *Appl. Phys. Rev.* **2015**. [CrossRef]

17. Duguet, E.; Vasseur, S.; Mornet, S.; Devoisselle, J.-M. Magnetic nanoparticles and their applications in medicine. *Nanomed.* **2006**, *1*, 157–168. [CrossRef] [PubMed]

18. Garaio, E.; Sandre, O.; Collantes, J.-M.; Garcia, J.A.; Mornet, S.; Plazaola, F. Specific absorption rate dependence on temperature in magnetic field hyperthermia measured by dynamic hysteresis losses (ac magnetometry). *Nanotechnology* **2015**. [CrossRef] [PubMed]

19. Tourinho, F. A.; Franck, R.; Massart, R. Aqueous ferrofluids based on manganese and cobalt ferrites. *J. Mater. Sci.* **1990**, *25*, 3249–3254. [CrossRef]

20. Massart, R.; Dubois, E.; Cabuil, V.; Hasmonay, E. Preparation and properties of monodisperse magnetic fluids. *J. Magn. Magn. Mater.* **1995**, *149*, 1–5. [CrossRef]

21. Arosio, P.; Thévenot, J.; Orlando, T.; Orsini, F.; Corti, M.; Mariani, M.; Bordonali, L.; Innocenti, C.; Sangregorio, C.; Oliveira, H.; et al. Hybrid iron oxide-copolymer micelles and vesicles as contrast agents for MRI: Impact of the nanostructure on the relaxometric properties. *J. Mater. Chem. B* **2013**, *1*, 5317–5328. [CrossRef]

22. Sapareto, S.A.; Dewey, W.C. Thermal dose determination in cancer therapy. *Int. J. Radiat. Oncol. Biol. Phys.* **1984**, *10*, 787–800. [CrossRef]

23. Rabin, Y. Is intracellular hyperthermia superior to extracellular hyperthermia in the thermal sense? *Int. J. Hyperthermia* **2002**, *18*, 194–202. [CrossRef] [PubMed]

24. Piñeiro-Redondo, Y.; Bañobre-López, M.; Pardiñas-Blanco, I.; Goya, G.; López-Quintela, M.A.; Rivas, J. The influence of colloidal parameters on the specific power absorption of PAA-coated magnetite nanoparticles. *Nanoscale Res. Lett.* **2011**. [CrossRef] [PubMed]

25. Sandre, O.; Genvois, C.; Garaio, E.; Adumeau, L.; Mornet, S.; Couillaud, F. University Bordeaux: Bordeaux, France, Data not shown. 2017.

26. Levy, A.; Dayan, A.; Ben-David, M.; Gannot, I. A new thermography-based approach to early detection of cancer utilizing magnetic nanoparticles theory simulation and in vitro validation. *Nanomedicine Nanotechnol. Biol. Med.* **2010**, *6*, 786–796. [CrossRef] [PubMed]

27. Poincloux, R.; Lizárraga, F.; Chavrier, P. Matrix invasion by tumour cells: A focus on MT1-MMP trafficking to invadopodia. *J. Cell Sci.* **2009**, *122*, 3015–3024. [CrossRef] [PubMed]

28. Van Rhoon, G.C.; Samaras, T.; Yarmolenko, P.S.; Dewhirst, M.W.; Neufeld, E.; Kuster, N. CEM43 °C thermal dose thresholds: A potential guide for magnetic resonance radiofrequency exposure levels? *Eur. Radiol.* **2013**, *23*, 2215–2227. [CrossRef] [PubMed]

29. Morimoto, R.I. Cells in stress: transcriptional activation of heat shock genes. *Science* **1993**, *259*, 1409–1410. [CrossRef] [PubMed]

30. Freeman, M.L.; Borrelli, M.J.; Syed, K.; Senisterra, G.; Stafford, D.M.; Lepock, J.R. Characterization of a signal generated by oxidation of protein thiols that activates the heat shock transcription factor. *J. Cell. Physiol.* **1995**, *164*, 356–366. [CrossRef] [PubMed]

31. Kregel, K.C. Heat shock proteins: Modifying factors in physiological stress responses and acquired thermotolerance. *J. Appl. Physiol. Bethesda Md 1985* **2002**, *92*, 2177–2186. [CrossRef] [PubMed]

10

Differential Binding of Three Major Human ADAR Isoforms to Coding and Long Non-Coding Transcripts

Josephine Galipon [1], **Rintaro Ishii** [2], **Yutaka Suzuki** [2], **Masaru Tomita** [1] and **Kumiko Ui-Tei** [2,3,*]

[1] Keio University Institute for Advanced Biosciences, Tsuruoka 997-0017, Japan; jgalipon@ttck.keio.ac.jp (J.G.); mt@sfc.keio.ac.jp (M.T.)

[2] Graduate School of Frontier Sciences, The University of Tokyo, Kashiwa 277-8562, Japan; rin-go2@jcom.home.ne.jp (R.I.); ysuzuki@k.u-tokyo.ac.jp (Y.S.)

[3] Graduate School of Science, The University of Tokyo, Tokyo 113-0032, Japan

* Correspondence: ktei@bs.s.u-tokyo.ac.jp

Academic Editor: H. Ulrich Göringer

Abstract: RNA editing by deamination of adenosine to inosine is an evolutionarily conserved process involved in many cellular pathways, from alternative splicing to miRNA targeting. In humans, it is carried out by no less than three major adenosine deaminases acting on RNA (ADARs): ADAR1-p150, ADAR1-p110, and ADAR2. However, the first two derive from alternative splicing, so that it is currently impossible to delete ADAR1-p110 without also knocking out ADAR1-p150 expression. Furthermore, the expression levels of ADARs varies wildly among cell types, and no study has systematically explored the effect of each of these isoforms on the cell transcriptome. In this study, RNA immunoprecipitation (RIP)-sequencing on overexpressed ADAR isoforms tagged with green fluorescent protein (GFP) shows that each ADAR is associated with a specific set of differentially expressed genes, and that they each bind to distinct set of RNA targets. Our results show a good overlap with known edited transcripts, establishing RIP-seq as a valid method for the investigation of RNA editing biology.

Keywords: ADAR1-p110; ADAR1-p150; ADAR3; RIP sequence; KEGG; GO

1. Introduction

The post-transcriptional modification of RNA is an evolutionarily conserved mechanism and a factor contributing to transcriptome diversity. In particular, the deamination of adenosine to inosine (A-to-I editing) on double-stranded RNA (dsRNA) was first identified for its ability to regulate the activity of neurotransmitter receptors [1]. The reaction is catalyzed by members of the ADAR family of dsRNA binding proteins (adenosine deaminases acting on RNA), which comprises of three constitutive isoforms in human, labeled ADAR (ADAR1), ADARB1 (ADAR2), and ADARB2 (ADAR3). While the deaminase domain of ADAR3 is not catalytically active, it is thought to act as a competitive inhibitor of ADAR1 and ADAR2 in the brain [2]. The generation of knockout mice for *ADAR1* and *ADAR2* further revealed that ADAR1 plays an essential role in cell survival and development, as *ADAR1−/−* embryos undergo massive apoptosis in early embryogenesis from E10.5 to E11.5 [3]. *ADAR2−/−* mice on the other hand are viable, but prone to epilepsy and their lifespan is shorter than wild-type [4]. The *ADAR1* locus codes for a constitutive 110 kDa isoform (ADAR1-p110), and the 150 kDa isoform (ADAR1-p150) is generated from the same locus during the interferon response [5].

The conversion of adenosine to inosine in the coding region is biologically meaningful, as inosine behaves as guanosine by favoring base-pairing with cytidine. In earlier years, studies on A-to-I editing primarily focused on modifications within coding sequences, as editing in those regions is susceptible to generate a protein product that differs by one amino-acid from the protein predicted by the original DNA sequence, which may modulate the protein activity [1]. Furthermore, the role of inosine in the formation of tRNA wobble base pairs has been described as early as the mid-1960s [6]. RNA editing has the further potential to regulate RNA structure via the modulation of base-pairing strength, which means it can virtually influence any cellular process requiring an interaction with RNA. While adenosine (A) typically binds to uridine (U) forming A:U pairs, inosine (I) forms a wobble interaction with cytosine (C), meaning that ADARs have the potential of either destabilizing (A:U to I/U) or stabilizing (A/C to I:C) dsRNA. This dual ability positions them as potential core players in the regulation of RNA activity. Meanwhile, the advent of RNA deep sequencing technologies (RNA-seq) has revealed that the majority of editing events occur on non-coding RNA [7]. However, there is presently no study that addresses the binding target preference of every ADAR active isoform, due to a lack of antibodies specific enough to distinguish ADAR1-p110 from ADAR1-p150 [8]. The transcription of ADAR1-p150 and ADAR1-p110 initiates from two promoters at the same gene locus: one interferon (IFN)-inducible promoter initiates the transcription of ADAR1-p150, while another constitutive promoter drives ADAR1-p110 [9]. Alternative splicing of exon 1 results in two sizes of ADAR1 proteins, approximately 150 kilodaltons (kDa) for IFN-inducible ADAR1-p150 and 110 kDa for the constitutively expressed ADAR1-p110. This makes it technically impossible to knock-out ADAR1-p110 specifically without knocking out ADAR1-p150 expression. Furthermore, popular methods for detecting genome-wide binding to RNA typically use UV crosslinking, which not only is inefficient in stabilizing interactions with perfect double-stranded RNAs [10], but prevents downstream analysis of editing sites due to the introduction of sequencing errors at the sites of crosslinking. Here, we present an RNA immunoprecipitation deep sequencing (RIP-seq) strategy that identifies targets for all known catalytically active ADAR isoforms: ADAR1-p150, ADAR1-p110, and ADAR2. We analyzed the effect of overexpressing green fluorescent protein (GFP)-tagged ADAR isoforms on the expression levels of size-selected long RNAs (>200 nt) in HeLa cells, and revealed ADAR isoform-specific binding target preferences.

2. Materials and Methods

2.1. Plasmid Construction

Total RNA was isolated from HeLa cells and cDNA was prepared by reverse transcription using random hexamers according to the manufacturer's protocol (Transcriptor High Fidelity cDNA Synthesis Kit, Roche Molecular Systems, Pleasanton, CA, USA). The genomic sequences for ADAR1-p150 (NM_001111), ADAR1-p110 (NM_001193495), and ADAR2 (NM_001112) were amplified by RT-PCR using the following primers (ADAR1-p150-F and ADAR1-R primers for ADAR1-p150, ADAR1-p110-F and ADAR1-R for ADAR1-p110, and ADAR2-F and ADAR2-R for ADAR2), each containing a restriction enzyme cleavage site (lower case letters):

ADAR1-p150-F: 5'-AAAGGGaagcttATGAATCCGCGGCAGGGGTATTCC-3' (HindIII),
ADAR1-p110-F: 5'-AAAGGGaagcttATGGCCGAGATCAAGGAGAAAATC-3' (HindIII),
ADAR1-R: 5'-AAAGGGtctagaCTATACTGGGCAGAGATAAAAGTTC-3' (XbaI),
ADAR2-F: 5'-AAAGGGgaattcATGGATATAGAAGATGAAGAAAACATG-3' (EcoRI),
ADAR2-R: 5'-AAAAGGAAAAgcggccgcTCAGGGCGTGAGTGAGAACTGGTC-3' (NotI).

The amplified fragments were each digested with the appropriate mixture of restriction enzymes (ADAR1-p150: HindIII+XbaI, ADAR1-p110: HindIII+XbaI, ADAR2: EcoRI+NotI). The pcDNA3.1(+) vector carrying pm-GFP-TNRC6A [11] was digested with HindIII+XbaI to remove the TNRC6A construct, and the digested ADAR1-p150 and -p110 fragments were inserted into pcDNA3.1(+) by ligation. The ADAR2 fragment was inserted into pcDNA3.1(+) after digestion with EcoRI+NotI.

2.2. Cell Culture and Transfection

HeLa cells (3×10^6) were inoculated on a 9-cm dish and incubated with Dulbecco's modified Eagle medium (DMEM) (Invitrogen, Carlsbad, CA, USA) supplemented with 10% heat-inactivated fetal bovine serum (Sigma-Aldrich, St. Louis, MO, USA) at 37 °C overnight. Cells were transfected at < 50% confluency with each construct (10 μg/dish) mixed with Lipofectamine 2000 (Invitrogen, Carlsbad, CA, USA) in Opti-MEM I (Invitrogen). Media was changed after 4 h and cells were sampled two days later.

2.3. RNA Immunoprecipitation and Illumina Library Preparation

HeLa cells transfected with expression constructs were washed with phosphate-buffered saline (PBS) (−) five times. Total RNA purification and immunoprecipitation (IP) were performed with the RiboCluster Profiler RIP-Assay Kit #RN1001 (MBL International Corporation, Woburn, MA, USA). Immunoprecipitation was carried out using anti-GFP antibody (Clonetech Laboratories, Mountain View, CA, USA) or normal rabbit immunoglobulin G (IgG) as a control. Total RNA was purified with DNase (TURBO™ DNase (0.88 U) (Invitrogen)) and fragmented using mRNA-Seq Sample Prep Kit (Illumina, San Diego, CA, USA). RNA sized 200–400 nt was excised from the acrylamide gel and subject to library preparation. Cluster amplification and single-end sequencing was performed using the Illumina TruSeq SBS Kit v5-GA #FC-104-5001 and Illumina Genome Analyzer GAIIx according to the manufacturer's protocol (read length: 36 nt). The sequence data was first converted to qseq format by CASAVA v1.8.2 (Illumina), and further converted to FASTQ format using the qseq2fastq converter provided by Kris Popendorf [12], and reads containing any base with a Phred quality score of less than 20 were filtered out using FASTX toolkit 0.0.13 [13] and custom python code to remove reads containing N's and homo-polymers consisting solely of one type of nucleotide.

2.4. DEG Analysis (Tuxedo Pipeline)

The latest version of the human genome (hg38) was downloaded in FASTA format from gencode release 25. Since the reads were short (36 nt) and unstranded, we opted for mapping reads using tophat 2.0.9 [14,15] in combination with bowtie 1.0.0 [16], and allowed up to one mismatch per read while always favoring the best alignment. The gene transfer format (GTF) annotation files provided with gencode release 25 (all genes + tRNAs) were supplied as one file to the tophat input to limit the splicing junctions search to known splicing junctions. Furthermore, the coverage-search option was activated as recommended for short (<45 nt) reads, with alignments reported exclusively across "GT–AG" introns. All IP and input samples were processed similarly. The command line used for mapping was as follows:

```
tophat -G annotation.gtf –no-novel-juncs –no-novel-indels -N 1 –read-gap-length 0
–read-edit-dist 1 –read-realign-edit-distance 0 –bowtie1 -o ./tophat_output_samplename/
–coverage-search bowtie_index input.fastq
```

For differential expression analysis, the four "input" samples were used as input to cuffdiff from the cufflinks 2.1.1-4 package [14]. Equal dispersion and variance was assumed among all four samples with the 'blind' dispersion method. Note that cuffdiff automatically switches to 'blind' mode if only one replicate per sample is provided. This method is expected to give a conservative estimate of the number of significant differentially expressed genes (DEGs). Results were visualized using R 3.3.1 "Bug in Your Hair" [17], Bioconductor 3.3 [18], cummeRbund 2.14.0 [14], as well as custom code. The command line used for differential expression analysis was as follows, with C2, C4, C6, C7, representing ADAR1-p150, ADAR1-p110, ADAR2, and GFP input samples, respectively. The BAM alignment files (accepted_hits.bam) produced by the above tophat pipeline were used as input, requiring prior installation of samtools 0.1.19-1 [19].

cuffdiff -p 8 -o ./cuffdiff_output_folder -b genome_bowtie.fa -L C2,C4,C6,C7 -u annotation.gtf ./tophat_output_C2/accepted_hits.bam ./tophat_output_C4/accepted_hits.bam ./tophat_output_C6/accepted_hits.bam ./tophat_output_C7/accepted_hits.bam

2.5. Identification of Binding Targets (RIPSeeker)

RIPSeeker is an R package that was designed specifically to detect significantly enriched peaks in RIP-seq data [20]. Bin size was first optimized per chromosome by testing bins ranging from 200 to 400 bp in steps of 5. Chromosome X was not included in the initial test due to lack of memory (when not using a fixed bin size, RIPSeeker may use up to several hundred gigabytes of RAM and is unfortunately not optimized for multicore). The optimal bin size for all chromosomes tested was between 200 and 250. We subsequently used a fixed bin size of 200 bp for all chromosomes to optimize memory performance. For future reference, it took RIPSeeker 12 h on average to analyse one BAM file containing approximately 100 million alignments (multiple hits are included). To run RIPSeeker, the following R script was run from the terminal using a bash script, assuming that all necessary R packages (BSgenome.Hsapiens.UCSC.hg38, biomaRt, RIPSeeker, among others) are loaded:

```
# read path to BAM files

hg38<-getBSgenome("hg38",masked=F) #load human genome version 38

extdata.dir<-system.file("tophat_out", package="RIPSeeker") #set location of tophat output

bamFiles<-list.files(extdata.dir, "
.bam$", recursive=T, full.names=T) #read filenames

outDir<-file.path("~/path/to/ripseeker_output") #set location of RIPSeeker output

file<-bamFiles[1] #set BAM file (replace 1 by number corresponding to desired file)

seqOut.file<-ripSeek(bamPath=file, genomeBuild="hg38", uniqueHit=T, assignMultihits=T,
rerunWithDisambiguatedMultihits=T, binSize=200, biomart="ensembl", biomaRt_dataset
="hsapiens_gene_ensembl", goAnno="org.Hs.eg.db", multicore=F, outDir=outDir)
#multicore should always be set to FALSE
```

2.6. Downstream Functional Analysis of Results

For the classification of ADAR-bound transcripts, the Kyoto Encyclopedia of Genes and Genomes (KEGG) database [21] was used to identify possible functional pathways. PANTHER overrepresentation test was performed on Gene Ontology (GO) biological processes with default settings (Bonferroni correction for multiple testing, $p < 0.05$) with a tool from the GO Consortium [22,23].

3. Results

3.1. RNA Immunoprecipitation (RIP)-Sequencing Experimental Setup

Catalytically-active members of the ADAR family comprise three major isoforms, the functional domain organization of which are summarized in Figure 1A. Both the constitutive isoforms ADAR1-p110 and ADAR2 are mostly nuclear and undergo shuttling in and out of the nucleolus, while ADAR1-p110 also undergoes nucleocytoplasmic shuttling [24]. ADAR2 is the shortest and possesses two dsRNA-binding domains; the activity of its deaminase domain is thought to be less sequence-specific [25]. ADAR1-p110 possesses three dsRNA-binding domains and a Z-DNA binding domain (β). Interferon-induced ADAR1-p150 comprises of the sequence of ADAR1-p110 with an additional Z-DNA binding domain (α), which is known to localize to cytoplasmic stress granules [26]. Figure 1B illustrates the dsRNA-binding activity of ADARs by showing the crystal structure of ADAR2 dsRNA binding domain in complex with a dsRNA helix [27]. In order to identify differentially-expressed genes and preferentially-bound RNA species, each ADAR isoform was overexpressed in HeLa cells as a protein fusion construct with myc-GFP in the N-terminal region, and myc-GFP alone was in turn expressed as a control. RNA immunoprecipitation (RNA-IP or RIP) was carried out with a monoclonal antibody against GFP, and RNA from both the input and IP was size-selected on an acrylamide gel for Illumina® library construction. For the study of mRNAs and long non-coding RNAs, total RNA was fragmented and fragments sized 200-400 nt were purified and subjected to next-generation sequencing by GAIIx (Figure 1C, Materials and Methods). In total, seven datasets were generated: GFP input, ADAR1-p110 IP and input, ADAR2 IP and input, as well as ADAR1-p150 IP and input. Each dataset was analyzed according to the computational workflow presented in Figure 1D. First, raw reads were pre-processed by converting to the standard FASTQ format and filtering out low-quality reads and artefacts. The quality filtering step yielded 22 to 30 million reads per sample, which were then mapped to the human genome. Differentially-expressed genes (DEGs) were identified using the tuxedo pipeline (tophat–cuffdiff) [14,15] optimized for short unstranded reads (Material and Methods) [28]. Finally, RNA binding targets were identified using RIPseeker, which uses hidden Markov models to accurately identify enriched transcripts from RIP-seq alignment files [20].

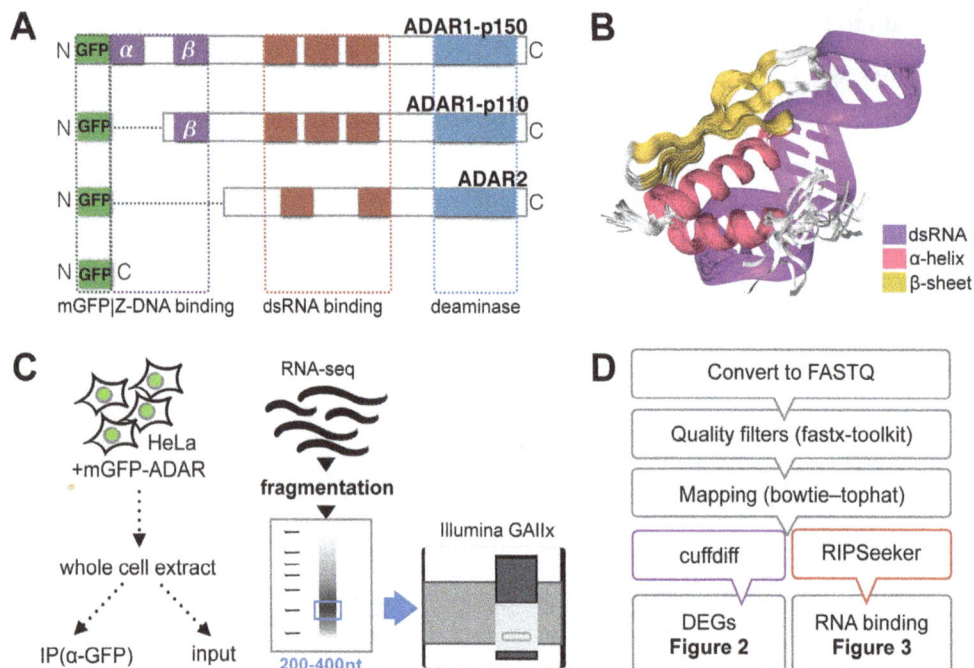

Figure 1. Experimental and computational workflow for RNA immunoprecipitation (RIP)-sequencing. (**A**) Schematic representation of known domains within adenosine deaminase acting on RNA (ADAR) enzyme isoforms with a catalytically-active deaminase domain. The final fusion proteins used in this experiment all harbor myc-green fluorescent protein (GFP) (mGFP) in the N-terminal region (left side in this figure). The dots within the protein sequence do not represent any real protein sequence, they were added in order to align and visually compare similar domains. Purple: Z-DNA binding domains; red: double-stranded RNA (dsRNA)-binding domains; blue: deaminase domain; green: mGFP; (**B**) the crystal structure of ADAR2 dsRNA-binding domain dsRBM1 bound to the free gluR-B R/G lower stem-loop (LSL) RNA rendered from PDB accession number (23LC). Different types of structures are represented in separate colors; (**C**) the experimental workflow for RIP-seq. HeLa cells are transfected with expression vectors for each mGFP-ADAR fusion protein. Part of the whole cell lysate was used for IP using anti-GFP antibody attached to magnetic beads, while the other part was used as a control (input). RNA was fragmented and material sized 200–400 nt was selected by gel electrophoresis and subjected to library preparation using mRNA-Seq Sample Prep Kit for Illumina GAIIx; and (**D**) the custom computational workflow used to obtain the results presented in this paper: differentially-expressed genes (DEGs) and ADAR-bound RNA targets.

3.2. Effect of ADAR Isoform Overexpression on Global Gene Expression

DEGs were identified by the expression levels in each ADAR input sample relative to the GFP input sample. To this end, the standard tuxedo pipeline described by Trapnell et al. [14] was adapted using available parameters most suitable for short unstranded reads (Materials and Methods). Second, the cuffdiff implementation for identifying DEGs goes beyond a traditional Poisson model for RNA-seq by providing several methods to estimate the dispersion present in a group of replicates. For instance, it provides a useful option for dispersion estimation in the case of only one replicate per sample, as is the case here. As shown in Supplementary Figure S1, the distribution of read counts and dispersion was fairly similar among samples. Therefore, we assumed equal variance among all samples, and treated all samples as replicates of a single condition. This trick is expected to produce a rather conservative estimation of DEGs [28]. The final output contained the level of significance and false discovery rate (FDR) for each gene. The overexpression of ADAR1-p150/-p110, and ADAR2 was successfully detected in each sample, respectively, with FDRs of 6.5%, 6.5%, and 1.9%, respectively, and levels of significance lower than 2.5×10^{-4} (Figure 2A). It has to be noted that, since DEG analysis reports gene expression changes per gene unit, it groups together the expression of isoforms originating from the same locus, such as ADAR1-p110 and –p150. Based on these results, DEGs that were up- or down-regulated more than two-fold relative to GFP input were kept only if their FDR and p-values were below the 6.5% and 2.5×10^{-4}, respectively (Figure 2B). Detailed fold-change values and the gene description for each DEG are presented in Supplementary Table S1.

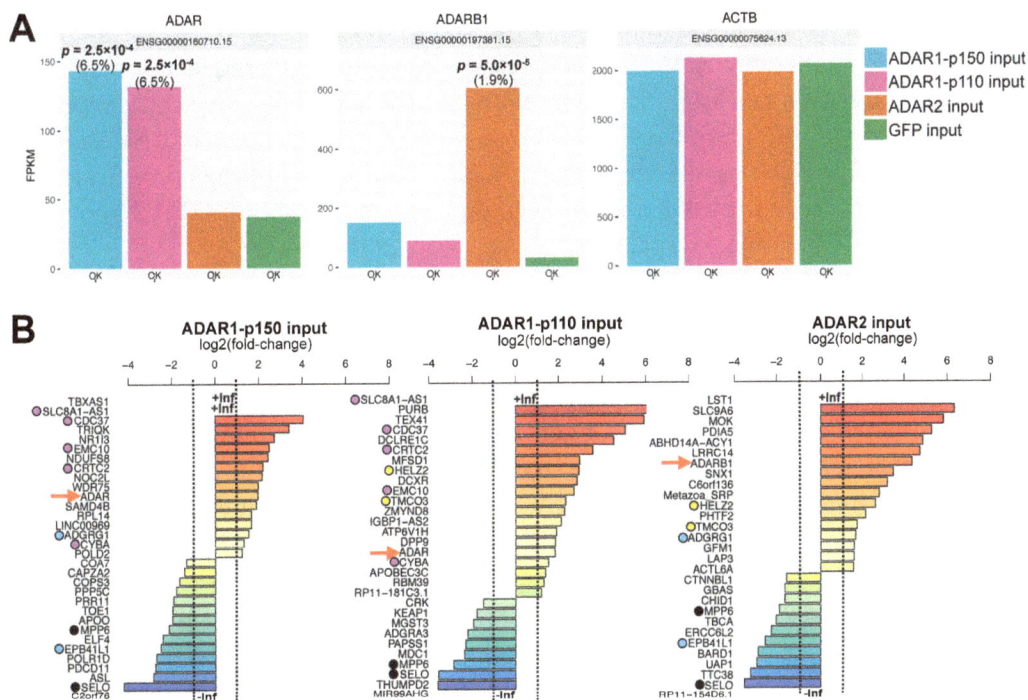

Figure 2. Differentially expressed genes upon specific ADAR isoform overexpression. (**A**) Coverage of the *ADAR* (*ADAR1*), *ADARB1* (*ADAR2*), and *ACTB* genes expressed in reads per kilobase per million reads (RPKM). ADAR codes for both the ADAR1-p150 and ADAR1-p110 isoforms, ADARB1 codes for ADAR2 isoform, and ACTB codes for Actin B. The *p*-value relative to GFP input is indicated in bold and the corresponding false discovery rate (FDR) in brackets; (**B**) Fold-change relative to GFP input for significant DEGs (FDR ≤ 6.5%; $p \le 2.5 \times 10^{-4}$) expressed on a logarithmic scale. The *ADAR* and *ADARB1* controls are highlighted with a red arrow. Black filled circles: DEGs common to all three ADAR isoforms; purple filled circles: common to ADAR1-p150 and ADAR1-p110; blue filled circles: common to ADAR1-p150 and ADAR2; yellow filled circles: common to ADAR1-p110 and ADAR2.

MPP6 and SELO transcripts were downregulated in all cases, meaning they are likely nonspecific. Few DEGs were in common between samples overexpressing different ADAR isoforms, except for ADAR1-p150 and ADAR1-p110, which both resulted in the significant upregulation of five common genes. Since ADAR1-p150 consists of the full amino acid sequence of ADAR1-p110 with an extension of N-terminal 295 amino acids [29], some overlap in function is to be expected. This included the up-regulation of two major phosphoinositide 3-kinases (PI3K)-Akt signaling pathway elements, cell division cycle 37 (CDC37) and mammalian target of rapamycin (mTOR)C2 (CRTC2), which are known to activate the RAC-alpha serine/threonine-protein kinase AKT by phosphorylation [30]. This pathway is overactive in a multitude of cancers, suggesting a tumorigenic role of excess ADAR1 expression. Consistent with this observation, strong inhibition of the mir-99a-let-7c cluster host gene MIR99AHG guiding the expression of let-7c, miR-99a and miR-125b was also observed in the ADAR1-p110 input sample compared to GFP input (Figure 2B, middle). Downregulation of this miRNA cluster was recently shown to induce tumorigenesis due to a loss of inhibition of key inflammatory cytokines involved in the IL-6/signal transducer and activator 3 (STAT3) pathway in cholangiocarcinoma [31]. An excess of ADAR1-p110 was also shown to be tumorigenic in the case of lung cancer [32]. The effect of different ADAR isoforms on miRNA maturation and expression levels remains to be investigated in an isoform-specific manner, and will be the subject of further study. Interestingly, the overexpression of ADAR1-p110 resulted in the up-regulation of another deaminase, APOBEC3C, which is localized in the nucleus and speculated to promote dC-to-dU DNA (and possibly also C-to-U RNA) editing [33].

In contrast, the genes influenced by *ADAR2* overexpression were less well documented. We observed up-regulation of MOK mitogen-activated protein (MAP)-kinase, and differential expression of proteins involved in vesicular and intracellular trafficking (SLC9A6, SNX1, GBAS) and metabolic enzymes (LAP3, UAP1). Notably, *ADAR2* overexpression led to the down-regulation of spliceosomal core component CTNNBL1, consistent with a recent report showing a negative correlation between splicing and editing by *ADAR2* [34]. Finally, we also observed down-regulation of DNA repair factors ERCC6L2 (Snf2 family of helicase-like proteins) and BRCA1-associated RING domain 1 (BARD1), a factor involved in the early steps of homologous recombination. Therefore, *ADAR2* overexpression may also affect the ability of cells to repair DNA.

3.3. ADAR Isoforms Bind to Distinct Targets Genome-Wide

RIP-seq analysis is typically carried out by simply comparing the coverage in the IP fraction with that of input samples, which gives an idea of the relative enrichment without providing confidence estimates. Quantification remains difficult for genomes containing a high number of repeats, as reads mapping to those regions may not be assigned accurately, and IP and input samples are not always directly comparable due to potentially divergent coverage of these repeat regions between IP and input. One solution is to perform peak calling on the IP sample, but many peak callers were designed for chromatin immunoprecipitation (ChIP-seq) data and, therefore, assume the presence of tandem peaks. RIPSeeker was designed specifically for RIP-seq, and uses machine learning on an alignment file to first model peak enrichment within the IP sample while taking into account only unique hits. Then, it reassigns multiple hits to their most likely location based on posterior probability, according to a "rich gets richer" model. Peak calling is then performed again on the unique and disambiguated multihits. Although computationally intensive, this method performs extremely well compared to other methods even when given an IP sample alignment file as sole input [20]. The parameters were optimized as described in Materials and Methods.

Peak calling with RIPSeeker was performed on each IP alignment file, which produced a list of peaks and their corresponding candidate bound transcripts for each ADAR isoform (Supplementary Table S2). Although RIPSeeker outputs neighboring gene features in the case a peak does not completely overlap with a gene annotation, we filtered these out and only kept peaks that were fully included inside a given annotation. This yielded 23, 890, and 290 unique gene identifiers (IDs) for candidate transcripts bound by ADAR1-p150, ADAR1-p110, and ADAR2, respectively. Although ADAR1-p150 bound relatively fewer targets, most of which were also bound by other ADAR isoforms, the general overlap was minimal, with only 11 transcripts out of 1144 (less than 1% of all candidates) bound by all three ADARs (Figure 3A). ADAR1-p110 and ADAR2 targets were especially clearly divided.

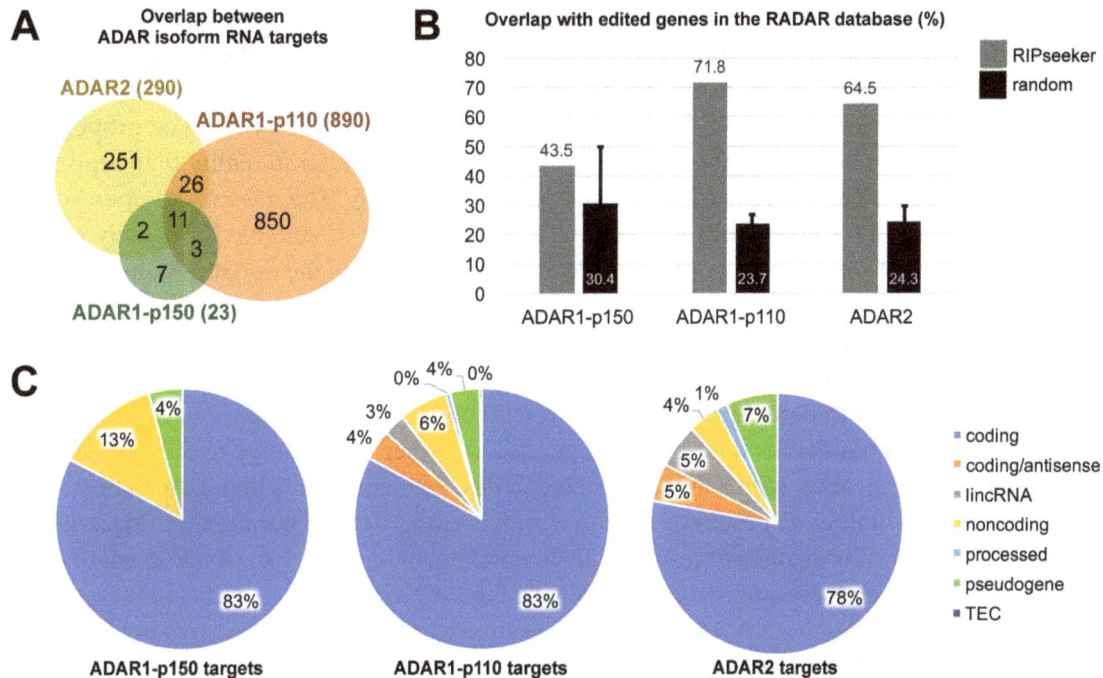

Figure 3. Identification of ADAR isoform-specific binding targets by RIPSeeker. (**A**) Venn diagram showing the overlap between the unique gene IDs of ADAR1-p150-, ADAR1-p110- and ADAR2-bound transcripts. (**B**) Overlap between unique gene IDs for each ADAR and unique gene IDs registered in the RADAR A-to-I editing site database (grey), compared to the mean overlap of 10,000 datasets chosen randomly from the human gene annotation file. Each random data set was similar in size to the corresponding ADAR; and (**C**) pie charts representing the proportion of RNA categories found in each ADAR-bound sample, respectively. Blue: coding transcripts; other colors: non-coding or uncharacterized transcripts.

ADAR binding does not necessarily entail A-to-I editing, but one may expect that ADAR-bound transcripts are more frequently edited than unbound transcripts. Indeed, as shown in Figure 3B, the transcripts detected in our study highly overlapped with those registered in the Rigorously Annotated Database of A-to-I RNA editing (RADAR) database (v2), that lists previously published A-to-I editing sites [35]. However, since this database currently lists more than 20,000 unique gene IDs, and the human genome annotation file contains a little above 60,000 entries, we needed to make sure that this high overlap was not merely due to chance. This was tested by generating 10,000 sets of gene IDs picked randomly from the gene annotation file, and calculating the percentage of overlap with transcripts in the RADAR database for each random set. The mean and standard deviation of this bootstrapping are presented as black bars in Figure 3B. Overall, these results show that the ADAR1-p110 and ADAR2-bound targets identified by this study are significantly enriched in edited transcripts.

Finally, the candidate ADAR-bound transcripts were manually curated into the following gene type categories based on information available on the Ensembl website: (1) coding, for protein-coding; (2) coding/antisense, when RIPSeeker could not determine which strand was relevant and both a coding transcript and its overlapping antisense RNA were detected ; (3) lincRNA, for long intergenic non-coding RNAs ; (4) non-coding, for any other type of long non-coding RNA, most of which were antisense RNAs; (5) processed, for transcripts indicated as such in the Ensembl database; (6) pseudogene, including known and unprocessed pseudogenes ; (7) TEC, for "to be experimentally confirmed", again when indicated as such (Figure 3C). Notably, ADAR1-p110 non-coding targets included small nucleolar RNA C/D box 3C (SNORD3C) and the miRNA precursor MIR568, which are both associated with the long non-coding RNA class (4). ADAR2 bound to a slightly higher proportion of non-coding transcripts than ADAR1-p110, although the difference was not quite significant (2×2 contingency table Fisher's exact test two-tailed p-value: 0.0738). When interpreting these results, one should keep in mind that coding transcripts might be overrepresented, as the genome annotation file available at GENCODE does not contain annotations for repeat elements such as Alu, short interspersed nuclear elements (SINEs), or long interspersed elements (LINEs). The proportion of non-coding transcripts bound by ADARs is expected to be much higher in reality than represented in Figure 3C, as Alu elements are heavily targeted by ADARs [36]. The question is open whether the reassignment of reads with multiple hits by RIPSeeker would be suitable for heavily repeated sequences, since there would be few unique reads available to perform the initial peak modeling step on those regions.

KEGG pathway analysis seem to indicate that ADAR1-p110 binds a great number of transcripts involved in Pathways in Cancer (ko05200), to which 14 genes corresponded out of the 159 ADAR1-p110-bound candidates currently registered (Table 1). The KEGG pathway search did not yield any convincing results for the transcripts bound by either of the other two ADAR isoforms, probably due to poor overlap with KEGG-registered genes. To increase functional prediction efficacy, GO enrichment for biological processes was performed using default parameters (Materials and Methods), and statistically significant results with an enrichment of more than two-fold relative to the expected number of genes are summarized in Table 2. Nothing statistically significant was found for ADAR2-bound targets, but both ADAR1 isoforms seemed to target more transcripts involved in the regulation of translation, mRNA degradation, and viral metabolism.

Table 1. ADAR1-p110-bound transcripts present in KEGG Pathways in Cancer (ko05200).

Gene ID	KEGG Name	KEGG ID	Description
APPL1	APPL	K08733	DCC-interacting protein 13 alpha
CTNNB1	β-catenin	K02105	Catenin beta 1
RHOA	Rho, Rac/Rho	K04513	Ras homolog gene family, member A
GSK3B	GSK-3β	K03083	Glycogen synthase kinase 3 beta
ITGB1	ITGB	K05719	integrin beta 1
GNB1	βγ	K04536	Guanine nucleotide-binding protein G(I)/G(S)/G(T) subunit beta-1
VHL	VHL	K03871	von Hippel-Lindau disease tumor suppressor
MLH1	hMLH1	K08734	DNA mismatch repair protein MutL homolog 1
TGFBR2	TGFβRII	K04388	Transforming growth factor (TGF)-beta receptor type 2
MITF	MITF	K09455	Melanogenesis associated transcription factor
RAF1	Raf	K04366	RAF proto-oncogene serine/threonine-protein kinase
TFG	TRK	K09292	Tyrosine kinase receptor (TRK)-fused gene
NCOA4	RET/PTC	K09289	Nuclear receptor coactivator 4
NFKB1	NFκB	K02580	Nuclear factor NF-kappa-B p105 subunit

Table 2. Positively enriched gene ontology (GO) biological processes for ADAR-bound transcripts.

Bound Isoform	GO Biological Process	Hits	Expected	Fold-Enrichment	p-Value
ADAR1-p150	SRP-dependent cotranslational protein targeting to membrane	4	0.09	45.99	1.34×10^{-2}
	Viral transcription	4	0.1	38.39	2.73×10^{-2}
	Nuclear-transcribed mRNA catabolic process, nonsense-mediated decay	4	0.11	36.79	3.23×10^{-2}
	rRNA processing	5	0.24	20.91	2.64×10^{-2}
ADAR1-p110	Nuclear-transcribed mRNA catabolic process	25	6.95	3.6	6.34×10^{-2}
	SRP-dependent cotranslational protein targeting to membrane	15	3.39	4.43	2.21×10^{-2}
	Viral life cycle	33	10.59	3.12	1.63×10^{-4}
	Translation	37	16.09	2.3	3.47×10^{-2}
ADAR2	No significantly enriched GO biological process				

4. Discussion

Enzyme isoforms in higher eukaryotes tend to specialize in their function, yet they can be difficult to distinguish experimentally, especially when these isoforms derive from alternative splicing. For instance, one can design ADAR1-p150-specific antibodies, but antibodies targeting ADAR1-p110 will inevitably target both ADAR1-p110 and -p150. This study attempts to address this issue of ADAR isoform specificity by overexpressing each major ADAR isoform with a GFP-tag in the N-terminal. Although the overexpression of a given gene may have secondary effects on global gene expression, at least the global distribution of gene expression values (RPKM) did not significantly change between samples (ADAR1-p150, -110, ADAR2, GFP inputs) (Supplementary Figure S1). One other issue that is specific to RNA editing is to preserve the ability to detect RNA editing sites on target transcripts. Other studies have tried crosslinking immunoprecipitation (CLIP)-seq, a method that stabilizes the interaction between a protein and its target RNA by UV-crosslinking [37]. Although this prevents the dissociation and re-association of ADAR on its target RNA, this introduces sequencing errors at the binding site due to irreversible effects of crosslinking, so that CLIP-seq can identify binding but not editing. Furthermore, traditional UV crosslinking is notably inefficient in the case of extended dsRNA targets [10], as is the case for ADARs, meaning that current CLIP-seq methods may introduce a bias for RNA targets containing more bulges and loops. One of the main goals of this paper was to present a crosslinking-free alternative by applying RIP-seq to the detection of ADAR-bound transcripts. The fact that we got a very significant overlap between ADAR-bound targets and transcripts registered in the major A-to-I editing database suggests that crosslinking is not absolutely necessary for the study of ADAR targets. The absence of crosslinking makes it theoretically possible to detect actual RNA editing sites within the IP samples. However, our current study was tentatively done on unstranded short reads from total RNA, which is not optimal for downstream detection of RNA editing sites. To this end, we would like to recommend (1) filtering out ribosomal RNA to maximize coverage; and (2) using strand-specific data with a longer read length to enable the study of bidirectional loci, as well as enhance signal-to-background ratio when detecting editing sites. Another challenge that arose was that the small number of candidate targets bound by ADAR1-p150 compared to other ADARs. Although this might reflect biological function, because ADAR1-p150 is the longest isoform, we speculate that the addition of a myc-GFP tag resulted in a protein so large that it may not have been overexpressed as efficiently as the other ADARs. We plan to address these issues in the near future. Meanwhile, the RIP-seq method presented here may be readily applied to other cell lines and various populations RNA may be enriched for the targeted study of longer mRNAs and lncRNAs, or smaller RNAs such as miRNAs.

Previously, ADAR1 overexpression was shown to have an inhibitory effect on iPS cell (iPSC) reprogramming, and the expression of ADAR1 in human embryonic stem cells (hESCs) resulted in the induction of differentiation-related genes [38]. Another study was unable to achieve overexpression of ADAR1-p110 in hESCs by traditional methods, suggesting that ADAR1 expression is tightly regulated in development [39]. Furthermore, it was also reported that iPSCs derived from cells in which ADAR1 was down-regulated exhibited the characteristics of cancer cells shortly after iPSC colony formation [38]. A-to-I editing is also reported to be altered in several cancers [31,39], and the results presented here are consistent with other studies suggesting a role of ADAR1 in cancer formation [40]. Indeed, the overexpression of ADAR1-p110 resulted in the up-regulation of CDC37 and mTORC2, which are involved in the PI3K-Akt signaling pathway (Figure 2C, Table 1). Furthermore, RIP-seq analysis revealed the binding by ADAR1-p110 of many transcripts involved in cancer, including major players such as β-catenin, transforming growth factor-beta (TGF-β) receptor, Raf, Rho, and nuclear factor-kappa B (NFκB) (Table 2). Further experimental validation is expected to confirm the molecular mechanism of ADAR oncogenicity.

5. Conclusions

This study is the first to present RIP-seq as a method to analyze the target specificity of all three major ADAR isoforms. We found that the overexpression of each ADAR isoform induces differential expression of distinct sets of genes, and that the genome-wide binding preferences of each isoform are clearly distinct, and in particular hint towards the mechanism of ADAR1-p110 in tumorigenesis. Furthermore, ADAR-bound targets substantially overlapped with transcripts for which at least one editing site is registered in the database of A-to-I editing sites. This shows that, contrary to current methods, such as CLIP-seq, RIP-seq may be more suitable for downstream detection of editing sites due to the absence of crosslinking.

Acknowledgments: This work was supported by grants from the Ministry of Education, Culture, Sports, Science and Technology of Japan (21310123, 21115004, 15H04319, and 16H14640), the grants provided from Suzuken Memorial Foundation to Kumiko Ui-Tei, and by a grant from the Nestlé Nutrition Council, Japan (NNCJ) to Josephine Galipon.

Author Contributions: Rintaro Ishii and Kumiko Ui-Tei conceived and designed the experiments; Rintaro Ishii performed the experiments; Yutaka Suzuki performed RNA sequencing; Josephine Galipon analyzed the data and wrote the manuscript, and Masaru Tomita and Kumiko Ui-Tei provided critical feedback that substantially improved the manuscript.

References

1. Higuchi, M.; Single, F.N.; Köhler, M.; Sommer, B.; Sprengel, R.; Seeburg, P.H. RNA editing of AMPA receptor subunit GluR-B: A base-paired intron-exon structure determines position and efficiency. *Cell* **1993**, *31*, 1361–1370. [CrossRef]

2. Chen, C.X.; Cho, D.S.; Wang, Q.; Lai, F.; Carter, K.C.; Nishikura, K. A third member of the RNA-specific adenosine deaminase gene family, ADAR3, contains both single- and double-stranded RNA binding domains. *RNA* **2000**, *6*, 755–767. [CrossRef] [PubMed]

3. Wang, Q.; Miyakoda, M.; Yang, W.; Khillan, J.; Stachura, D.L.; Weiss, M.J.; Nishikura, K. Stress-induced apoptosis associated with null mutation of *ADAR1* RNA editing deaminase gene. *J. Biol. Chem.* **2004**, *6*, 4952–4961. [CrossRef] [PubMed]

4. Higuchi, M.; Maas, S.; Single, F.N.; Hartner, J.; Rozov, A.; Burnashev, N.; Feldmeyer, D.; Sprengel, R.; Seeburg, P.H. Point mutation in an AMPA receptor gene rescues lethality in mice deficient in the RNA-editing enzyme ADAR2. *Nature* **2000**, *406*, 78–81. [PubMed]

5. Patterson, J.B.; Thomis, D.C.; Hans, S.L.; Samuel, C.E. Mechanism of interferon action: double-stranded RNA-specific adenosine deaminase from human cells is inducible by alpha and gamma interferons. *Virology* **1995**, *210*, 508–511. [CrossRef] [PubMed]

6. Crick, F.H. Codon-anticodon pairing: The wobble hypothesis. *J. Mol. Biol.* **1966**, *19*, 548–555. [CrossRef]

7. Li, J.B.; Levanon, E.Y.; Yoon, J.-K.; Aach, J.; Xie, B.; Leproust, E.; Zhang, K.; Gao, Y.; Church, G.M. Genome-wide identification of human RNA editing sites by parallel DNA capturing and sequencing. *Science* **2009**, *324*, 1210–1213. [CrossRef] [PubMed]

8. Nishikura, K. Functions and regulation of RNA editing by ADAR deaminases. *Annu. Rev. Biochem.* **2010**, *79*, 321–349. [CrossRef] [PubMed]

9. George, C.X.; Samuel, C.E. Human RNA-specific adenosine deaminase ADAR1 transcripts possess alternative exon 1 structures that initiate from different promoters, one constitutively active and the other interferon inducible. *Proc. Natl. Acad. Sci. USA* **1999**, *96*, 4621–4626. [CrossRef] [PubMed]

10. Liu, Z.R.; Wilkie, A.M.; Clemens, M.J.; Smith, C.W. Detection of double-stranded RNA-protein interactions by methylene blue-mediated photo-crosslinking. *RNA* **1996**, *2*, 611–621. [PubMed]

11. Nishi, K.; Nishi, A.; Nagasawa, T.; Ui-Tei, K. Human TNRC6A is an Argonaute-navigator protein for microRNA-mediated gene silencing in the nucleus. *RNA* **2013**, *19*, 17–35. [CrossRef]

12. Popendorf, K. qseq2fastq. Available online: www.dna.bio.keio.ac.jp/~krisp/qseq2fastq/ (accessed on 6 February 2017).

13. Hannon laboratory. Available online: http://hannonlab.cshl.edu/fastx_toolkit/ (accessed on 6 February 2017).

14. Trapnell, C.; Roberts, A.; Goff, L.; Pertea, G.; Kim, D.; Kelly, D.R.; Pimentel, H.; Salzberg, S.L.; Rinn, J.L.; Pachter, L. Differential gene and transcript expression analysis of RNA-seq experiments with TopHat and Cufflinks. *Nat. Protoc.* **2012**, *7*, 562–578. [CrossRef] [PubMed]

15. Kim, D.; Pertea, G.; Trapnell, C.; Pimentel, H.; Kelley, R.; Salzberg, S. TopHat2: Accurate alignment of transcriptomes in the presence of insertions, deletions and gene fusions. *Genome Biol.* **2013**, *14*, R36. [CrossRef] [PubMed]

16. Langmead, B.; Trapnell, C.; Pop, M.; Salzberg, S.L. Ultrafast and memory-efficient alignment of short DNA sequences to the human genome. *Genome Biol.* **2009**, *10*, R25. [CrossRef] [PubMed]

17. R Development Core Team. *R: A Language and Environment for Statistical Computing*; R Foundation for Statistical Computing: Vienna, Austria, 2008. Available online: http://www.R-project.org (accessed on 6 February 2017).

18. Huber, W.; Carey, V.J.; Gentleman, R.; Anders, S.; Carlson, M.; Carvalho, B.S.; Bravo, H.C.; Davis, S.; Gatto, L.; Girke, T.; et al. Orchestrating high-throughput genomic analysis with Bioconductor. *Nat. Methods* **2015**, *12*, 115–121. [CrossRef] [PubMed]

19. Li, H.; Handsaker, B.; Wysoker, A.; Fennell, T.; Ruan, J.; Homer, N.; Marth, G.; Abecasis, G.; Durbin, R.; 1000 Genome Project Data Processing Subgroup. The Sequence Alignment/Map format and SAMtools. *Bioinformatics* **2009**, *25*, 2078–2079. [CrossRef]

20. Li, Y.; Zhao, D.Y.; Greenblatt, J.F.; Zhang, Z. RIPSeeker: A statistical package for identifying protein-associated transcripts for RIP-seq experiments. *Nucleic Acids Res.* **2013**, *41*, e94. [CrossRef] [PubMed]

21. Kanehisa, M.; Sato, Y.; Kawashima, M.; Furumichi, M.; Tanabe, M. KEGG as a reference resource for gene and protein annotation. *Nucleic Acids Res.* **2016**, *44*, D457–D462. [CrossRef] [PubMed]

22. Thomas, P.D.; Campbell, M.J.; Kejariwal, A.; Mi, H.; Karlak, B.; Daverman, R.; Diemer, K.; Muruganujan, A.; Narechania, A. PANTHER: A library of protein families and subfamilies indexed by function. *Genome Res.* **2003**, *13*, 2129–2141. [CrossRef] [PubMed]

23. Mi, H.; Dong, Q.; Muruganujan, A.; Gaudet, P.; Lewis, S.; Thomas, P.D. PANTHER version 7: Improved phylogenetic trees, orthologs and collaboration with the Gene Ontology Consortium. *Nucleic Acids Res.* **2010**, *38*, D204–D210. [CrossRef] [PubMed]

24. Desterro, J.M.P.; Keegan, L.P.; Lafarga, M.; Berciano, M.T.; O'Connell, M.; Carmo-Fonseca, M. Dynamic association of RNA-editing enzymes with the nucleolus. *J. Cell Sci.* **2003**, *116*, 1805–1818. [CrossRef] [PubMed]

25. Barraud, P.; Allain, F.H.-T. ADAR proteins: Double-stranded RNA and Z-DNA binding domains. *Curr. Top. Microbiol. Immunol.* **2012**, *353*, 35–60. [PubMed]

26. Ng, S.K.; Weissbach, R.; Ronson, G.E.; Schadden, A.D. Proteins that contain a functional Z-DNA-binding domain localize to cytoplasmic stress granules. *Nucleic Acids Res.* **2013**, *41*, 9786–9899. [CrossRef] [PubMed]

27. Schwartz, T.; Behike, J.; Lowenhaupt, K.; Heinemann, U.; Rich, A. Structure of the DLM-1-Z-DNA complex reveals a conserved family of Z-DNA-binding proteins. *Nat. Struct. Biol.* **2001**, *8*, 761–765. [CrossRef] [PubMed]

28. Cufflinks Manual. Available online: http://cole-trapnell-lab.github.io/cufflinks/cuffdiff/#cross-replicate-dispersion-estimation-methods (accessed on 6 February 2017).

29. Samuel, C.E. Adenosine deaminases acting on RNA (ADARs) are both antiviral and proviral dependent upon the virus. *Virology* **2011**, *411*, 180–193. [CrossRef] [PubMed]

30. King, D.; Yeomanson, D.; Bryant, H.E. PI3King the Lock: Targeting the PI3K/Akt/mTOR Pathway as a Novel Therapeutic Strategy in Neuroblastoma. *J. Pediatr. Hematol. Oncol.* **2015**, *37*, 245–251. [CrossRef] [PubMed]

31. Lin, K.-Y.; Ye, H.; Han, B.-W.; Wang, W.-T.; Wei, P.-P.; He, B.; Li, X.-J.; Chen, Y.-Q. Genome-wide screen identified let-7c/miR-99a/miR-125b regulating tumor progression and stem-like properties in cholangiocarcinoma. *Oncogene* **2016**, *35*, 3376–3386. [CrossRef] [PubMed]

32. Anadón, C.; Guil, S.; Simó-Riudalbas, L.; Moutinho, C.; Setien, F.; Martínez-Cardús, A.; Moran, S.; Villanueva, A.; Calaf, M.; Vidal, A.; et al. Gene amplification-associated overexpression of the RNA editing enzyme ADAR1 enhances human lung tumorigenesis. *Oncogene* **2016**, *35*, 4407–4413. [CrossRef] [PubMed]

33. Jarmuz, A.; Chester, A.; Bayliss, J.; Gisbourne, J.; Dunham, I.; Scott, J.; Navaratnam, N. An anthropoid-specific locus of orphan C to U RNA-editing enzymes on chromosome 22. *Genomics* **2002**, *79*, 285–296. [CrossRef] [PubMed]

34. Licht, K.; Kapoor, U.; Mayrhofer, E.; Jantsch, M.F. Adenosine to inosine editing frequency controlled by splicing efficiency. *Nucleic Acids Res.* **2016**, *44*, 6398–6408. [CrossRef] [PubMed]

35. Ramaswami, G.; Li, J.B. RADAR: A rigorously annotated database of A-to-I editing. *Nucleic Acids Res.* **2014**, *42*, D109–D113. [CrossRef] [PubMed]

36. Kim, D.D.Y.; Kim, T.T.Y.; Walsh, T.; Kobayashi, Y.; Matise, T.C.; Byske, S.; Gabriel, A. Widespread RNA editing of embedded alu elements in the human transcriptome. *Genome Res.* **2004**, *14*, 1719–1725. [CrossRef] [PubMed]

37. Bahn, J.H.; Ahn, J.; Lin, X.; Zhang, Q.; Lee, J.H.; Civelek, M.; Xiao, X. Genomic analysis of ADAR1 binding and its involvement in multiple RNA processing pathways. *Nat. Commun.* **2015**, *6*, 6355. [CrossRef] [PubMed]

38. Germanguz, I.; Shtrichman, R.; Osenberg, S.; Ziskind, A.; Novak, A.; Domev, H.; Laevsky, I.; Jacob-Hirsch, J.; Feiler, Y.; Rechavi, G.; et al. ADAR1 is involved in the regulation of reprogramming human fibroblasts to induced pluripotent stem cells. *Stem Cells Dev.* **2014**, *23*, 443–456. [CrossRef] [PubMed]

39. Shtrichman, R.; Germanguz, I.; Mandel, R.; Ziskind, A.; Nahor, I.; Safran, M.; Osenberg, S.; Sherf, O.; Rechavi, G.; Itskovitz-Eldor, J. Altered A-to-I RNA editing in human embryogenesis. *PLoS ONE* **2012**, *7*, e41576. [CrossRef] [PubMed]

40. Han, L.; Diao, L.; Yu, S.; Xu, X.; Li, J.; Zhang, R.; Yang, Y.; Werner, H.M.; Eterovic, A.K.; Yuan, Y.; et al. The Genomic Landscape and Clinical Relevance of A-to-I RNA Editing in Human Cancers. *Cancer Cell* **2015**, *28*, 515–528. [CrossRef] [PubMed]

11

FTO Genotype and Type 2 Diabetes Mellitus: Spatial Analysis and Meta-Analysis of 62 Case-Control Studies from Different Regions

Ying Yang [1,†], Boyang Liu [2,†], Wei Xia [3,†], Jing Yan [4], Huan-Yu Liu [5], Ling Hu [6] and Song-Mei Liu [1,*]

[1] Center for Gene Diagnosis, Zhongnan Hospital of Wuhan University, Donghu Road 169#, Wuhan 430071, China; yangying0109@whu.edu.cn

[2] Department of Geography, Wilkeson Hall, State University of New York at Buffalo, Buffalo, NY 14261, USA; bliu24@buffalo.edu

[3] Department of Clinical Laboratory, Wuhan Children's Hospital (Wuhan Maternal and Child Healthcare Hospital), Tongji Medical College, Huazhong University of Science & Technology, Wuhan 430016, China; 18971319110@163.com

[4] Hubei Meteorological Information and Technology Support Center, Wuhan 430074, China; yanjing619@hotmail.com

[5] Department of Clinical Medicine, Hubei University of Medicine, Hubei 442000, China; gutentag95@sina.com

[6] Department of Neurology, Wuhan Children's Hospital (Wuhan Maternal and Child Healthcare Hospital), Tongji Medical College, Huazhong University of Science & Technology, Wuhan 430016, China; m18372622675@163.com

* Correspondence: smliu@whu.edu.cn

† These authors contributed equally to this work.

Academic Editor: Selvarangan Ponnazhagan

Abstract: Type 2 diabetes mellitus (T2DM) is a global health problem that results from the interaction of environmental factors with genetic variants. Although a number of studies have suggested that genetic polymorphisms in the fat mass and obesity-associated (*FTO*) gene are associated with T2DM risk, the results have been inconsistent. To investigate whether *FTO* polymorphisms associate with T2DM risk and whether this association is region-related, we performed this spatial analysis and meta-analysis. More than 60,000 T2DM patients and 90,000 controls from 62 case-control studies were included in this study. Odds ratios (ORs), 95% confidence intervals (CIs) and Moran's I statistic were used to estimate the association between *FTO* rs9939609, rs8050136, rs1421085, and rs17817499, and T2DM risk in different regions. rs9939609 (OR = 1.15, 95% CI 1.11–1.19) and rs8050136 (OR = 1.14, 95% CI 1.10–1.18) conferred a predisposition to T2DM. After adjustment for body mass index (BMI), the association remained statistically significant for rs9939609 (OR = 1.11, 95% CI 1.05–1.17) and rs8050136 (OR = 1.08, 95% CI 1.03–1.12). In the subgroup analysis of rs9939609 and rs8050136, similar results were observed in East Asia, while no association was found in North America. In South Asia, an association for rs9939609 was revealed but not for rs8050136. In addition, no relationship was found with rs1421085 or rs17817499 regardless of adjustment for BMI. Moran's I statistic showed that significant positive spatial autocorrelations existed in rs9939609 and rs8050136. Studies on rs9939609 and rs8050136 focused on East Asia and South Asia, whereas studies on rs1421085 and rs17817499 were distributed in North America and North Africa. Our data suggest that the associations between *FTO* rs9939609, rs8050136 and T2DM are region-related, and the two single-nucleotide polymorphisms contribute to an increased risk of T2DM. Future studies should investigate this issue in more regions.

Keywords: type 2 diabetes mellitus; T2DM; fat mass and obesity-associated; *FTO*; polymorphism(s); spatial analysis; meta-analysis

1. Introduction

Diabetes is a growing global health problem; more than 300 million people live with diabetes worldwide [1], and the prevalence of diabetes is estimated to rise [2]. Type 2 diabetes mellitus (T2DM) is the most common type of diabetes, as it accounts for more than 90% of diabetes cases [3]. Although the pathogenesis mechanisms of T2DM have not been clearly defined, a combination of genetic and environmental factors is believed to lead to the disease [4].

The fat mass and obesity-associated (*FTO*) gene is located on chromosome 16 (16 q12. 2), containing nine exons and several single-nucleotide polymorphisms (SNPs) [5]. In 2007, a genome-wide association study (GWAS) searching for type 2 diabetes-susceptibility genes confirmed a common variant (rs9939609) in the *FTO* gene that predisposes European populations to diabetes [6]. Since then, a large number of studies have focused on the association between *FTO* polymorphisms, expression and T2DM in different populations [7–10]. Meanwhile, some meta-analyses have been performed to elucidate the relationship between *FTO* polymorphisms and T2DM risk. For instance, a meta-analysis utilizing data from studies prior to 2010 identified an association between rs9939609 and T2DM in East and South Asians [11]. Additionally, a Norwegian population-based Nord-Trøndelag Health Study (HUNT study) [12], including three cohorts (HUNT, Malmö Diet and Cancer (MDC) and Malmö Preventive Project (MPP)), reported strong association between rs9939609 and T2DM risk in Scandinavians after adjustment for age, sex and body mass index (BMI). Another meta-analysis of association between obesity/BMI-associated loci and T2DM risk [13], using data from studies conducted between 2007 and 2012, revealed that *FTO* rs9939609 significantly associated with T2DM which also remained significant following adjustment for BMI; Analysis by Vasan et al. [14] has provided evidence that rs9939609 is associated with obesity and T2DM in Asian Indians, with modest attenuation observed when adjusting for BMI. These and the majority of other previous meta-analyses have focused on single population or one *FTO* loci without consideration of population-specific environmental influences among different regional subgroups. As such, the results of these meta-analyses cannot be generalized to the world.

More recently, geographic information systems (GIS) and spatial analysis are increasingly applied in the investigation of disease spatial pattern, including diabetes [15].

To more comprehensively clarify the association between *FTO* polymorphisms and T2DM risk, we performed this spatial analysis and meta-analysis to include most, if not all, eligible studies published before January 2017.

2. Materials and Methods

2.1. Search Strategy

Eligible articles were selected by searching up to January 2017 in PubMed and EMBASE using the following keywords: "*FTO* or fat mass and obesity-associated gene" and "variant or variation or polymorphism" and "type 2 diabetes or type 2 diabetes mellitus or T2D or T2DM". Articles obtained from the initial search were then screened based on the inclusion criteria described below. Only publications with English language were included. If more than one population was included in a given article, results were considered as separate studies.

2.2. Study Selection Criteria and Data Extraction

The selected studies met all of the following inclusion criteria. The studies had to: (1) evaluate the association between *FTO* polymorphisms and T2DM risk; (2) have a case–control or cohort design; and (3) provide odds ratios (OR) with a 95% confidence interval (CI) or sufficient data for calculation. From each study, the following information was collected: (1) name of the first author; (2) year of publication; (3) country of origin; (4) ethnicity of the samples; (5) sample size of cases and controls; (6) Hardy–Weinberg equilibrium (HWE) in control groups; and (7) data of SNPs. Data were independently extracted from eligible articles by two authors (YY and HYL) according to the criteria described. Discrepancies were resolved by discussion with a third reviewer (SML), and a consensus approach was used.

2.3. Spatial Analysis

The ArcGIS v10.3 software is a GIS tool that has become increasingly prevalent in public health research to understand the spatial pattern of diseases and genetic biodiversity [15]. This software was utilized to depict the geographic distribution of the association studies. R was used to calculate Moran's I, a statistic for evaluating the spatial autocorrelation [16,17]. By constructing the spatial weight matrix, Moran's I coefficient can be calculated as follows:

$$I = \frac{N}{\sum_i \sum_j w_{ij}} \frac{\sum_i \sum_j w_{ij}\left(X_i - \overline{X}\right)\left(X_j - \overline{X}\right)}{\sum_i \left(X_i - \overline{X}\right)^2}$$

N is the number of spatial units indexed by i and j; X is the variable of interest; $\overline{X}$ is the mean of X; and w_{ij} is an element of a matrix of spatial weights. In this study, we constructed the spatial weight matrix by making a distance threshold h. If the distance between point i and point j is smaller than h, w_{ij} will be 1. Otherwise, w_{ij} will be 0. It is worth noting that all diagonal elements of matrix w are all 0. Monte Carlo simulations were used to test for the significance of Moran's I.

2.4. Statistical Analysis

The strength of association between *FTO* SNPs and T2DM risk was expressed as a pooled OR and 95% CI. A z-test was performed to evaluate the significance of the pooled OR ($p < 0.05$ was considered statistically significant). The χ^2-test-based Q test and I^2 were performed to assess the heterogeneity of the studies. A value of $I^2(\%) > 50\%$ or $p \leq 0.10$ indicated significant heterogeneity. A random-effects model (DerSimonian–Laird method) [18] was used to determine the pooled OR in the presence of heterogeneity; otherwise a fixed-effects model (Mantel–Haenszel method) [19] was used. Subgroup analyses were performed by region. Sensitivity analyses were performed to assess the stability of the combined results by excluding the studies with unknown HWE in controls. Publication bias was evaluated by Begg's test [20] and Egger's test [21] ($p < 0.05$ was considered statistically significant). Data analyses were conducted using STATA 12.0 (Stata-Corp LP, College Station, TX, USA).

3. Results

3.1. Study Characteristics and Quality

A total of 202 potentially relevant papers were identified from PubMed and EMBASE. After reading the title and abstract, 148 articles were excluded because they addressed topics that did not match the inclusion criteria. The full texts of the remaining 54 articles were carefully screened. We excluded five meta-analyses or reviews, three articles that explored the association between *FTO* polymorphisms and gestational diabetes, two articles that did not include the full text, and three papers with insufficient data. In total, 41 articles met the inclusion criteria. A flow chart describing the article selection for our meta-analysis is shown in Figure 1. Of the articles included, 29 studies investigated rs9939609, 26 studies explored rs8050136, four studies investigated rs1421085 and three studies explored rs17817499. Other SNPs that were assessed in only one study were not analyzed. The detailed characteristics of the included studies are shown in Table 1.

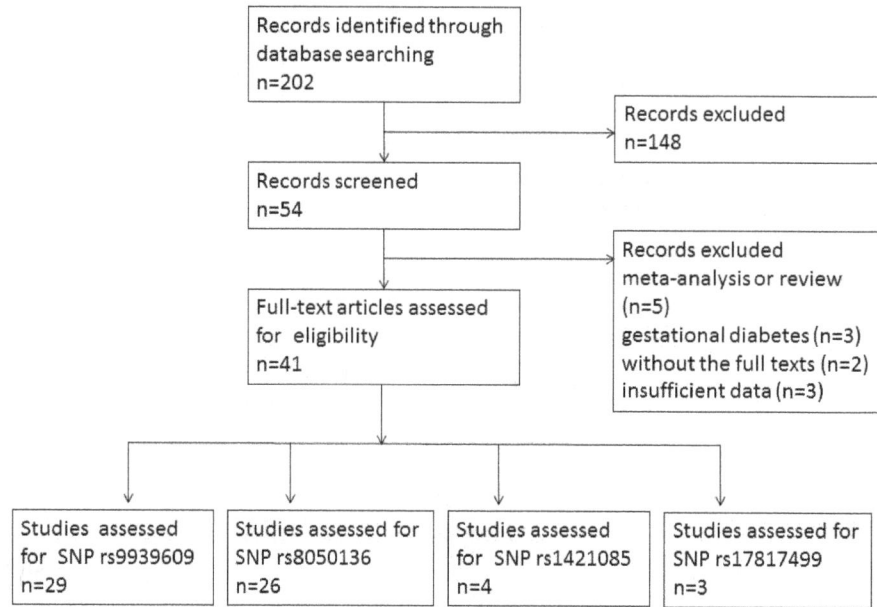

Figure 1. Study selection flow chart based on preferred reporting items for spatial analysis and meta-analysis.

Table 1. Characteristics of the included studies.

First Author	Year	Region	Sample Size		Risk Allele Frequency		HWE	Ref.
			T2DM	Control	T2DM	Control		
rs9939609								
Phani	2016	South Asia	518	518	0.54	0.59	NA	[7]
Xiao	2016	East Asia	879	895	0.341	0.295	yes	[22]
Xiao	2015	East Asia	849	873	0.336	0.292	yes	[8]
Shen	2015	East Asia	81	80	0.125	0.106	yes	[9]
Al-Sinani	2015	West Asia	992	294	0.48	0.435	yes	[23]
Fawwad	2015	South Asia	296	198	0.588	0.391	yes	[24]
Raza	2014	South Asia	101	97	0.406	0.376	NA	[25]
Bazzi	2014	South Asia	81	95	0.525	0.542	yes	[10]
Kalnina	2013	Europe	974	1075	0.501	0.438	yes	[26]
Ali	2013	South Asia	1583	1317	0.362	0.304	yes	[27]
Binh	2012	East Asia	98	251	0.255	0.181	yes	[28]
Iwata	2012	East Asia	722	758	0.206	0.182	yes	[29]
Rees(COBRA)	2011	South Asia	385	1281	0.336	0.294	yes	[30]
Rees(UKADS/DGP)	2011	South Asia	1568	1177	0.329	0.298	yes	[30]
Huang	2011	East Asia	591	1200	0.299	0.305	yes	[31]
Chauhan	2011	South Asia	2361	2755	0.35	0.34	yes	[32]
Cruz	2010	North America	519	547	0.252	0.212	yes	[33]
Bressler (African-American)	2010	North America	655	2685	0.463	0.483	yes	[34]
Bressler(white)	2010	North America	988	9915	0.465	0.443	yes	[34]
Liu	2010	East Asia	1774	1984	0.136	0.117	yes	[35]
Yajnik	2009	South Asia	1453	1361	0.353	0.3	yes	[36]
Legry	2009	Europe	283	2601	0.456	0.42	yes	[37]
Sanghera	2008	South Asia	513	353	0.363	0.31	yes	[38]
Horikawa	2008	East Asia	1849	1578	0.209	0.205	yes	[39]
Chang	2008	East Asia	735	726	0.132	0.127	yes	[40]
Omori	2008	East Asia	1621	1053	0.209	0.195	yes	[2]
Horikoshi	2007	East Asia	864	864	0.216	0.192	yes	[41]
Zeggini.	2007	Europe	5681	8284	0.435	0.394	yes	[42]
Frayling	2007	Europe	3757	5346	NA	NA	yes	[6]

Table 1. *Cont.*

First Author	Year	Region	Sample Size		Risk Allele Frequency		HWE	Ref.
			T2DM	Control	T2DM	Control		
rs8050136								
Xiao	2016	East Asia	879	895	0.313	0.275	yes	[22]
Xiao	2015	East Asia	849	873	0.308	0.274	yes	[8]
Shen	2015	East Asia	88	80	0.114	0.106	yes	[9]
Al-Sinani	2015	West Asia	992	294	0.458	0.425	yes	[23]
Chang	2014	East Asia	1502	1518	0.127	0.124	yes	[43]
Almawi	2013	West Asia	995	1195	0.487	0.551	yes	[44]
Qian	2013	East Asia	2898	3262	0.127	0.103	yes	[45]
Gamboa	2012	North America	1027	990	0.194	0.2	yes	[46]
Iwata	2012	East Asia	724	763	0.205	0.183	yes	[29]
Chauhan	2011	South Asia	1106	1800	0.35	0.34	yes	[32]
Ramya	2011	South Asia	1001	851	0.14	0.107	yes	[47]
Han	2010	East Asia	1007	995	0.13	0.11	yes	[48]
Bressler (African-American)	2010	North America	657	2728	0.425	0.44	yes	[34]
Bressler(White)	2010	North America	984	9873	0.444	0.402	yes	[34]
Wen	2010	East Asia	1165	1136	0.134	0.119	yes	[49]
Liu	2010	East Asia	1748	2015	0.139	0.117	yes	[35]
Hu	2009	East Asia	1849	1785	0.13	0.118	yes	[50]
Rong	2009	North America	1472	1825	0.151	0.136	yes	[51]
Lee	2008	East Asia	886	501	0.129	0.14	yes	[52]
Ng(HK)	2008	East Asia	1481	1530	0.156	0.136	yes	[53]
Ng(SNUH)	2008	East Asia	761	632	0.138	0.122	yes	[53]
Ng(KHGS)	2008	East Asia	799	1516	0.124	0.118	yes	[53]
Omori	2008	East Asia	1616	1060	0.208	0.194	yes	[2]
Horikoshi	2007	East Asia	857	861	0.238	0.2	yes	[41]
Zeggini	2007	Europe	4207	4111	0.44	0.39	yes	[42]
Scott	2007	Europe	2339	2401	0.406	0.381	yes	[54]
rs1421085								
Cauchi(Morocco)	2012	North Africa	1193	1095	0.395	0.356	yes	[55]
Cauchi(Tunisia)	2012	North Africa	1446	942	0.41	0.407	yes	[55]
Bressler (African-American)	2010	North America	657	2725	0.084	0.112	yes	[34]
Bressler(White)	2010	North America	989	9893	0.451	0.41	yes	[34]
rs17817499								
Almawi	2013	West Asia	995	1195	0.517	0.557	yes	[44]
Bressler (African-American)	2010	North America	653	2700	0.376	0.396	yes	[34]
Bressler(White)	2010	North America	986	9948	0.443	0.403	yes	[34]

T2DM, Type 2 diabetes mellitus; HWE, Hardy–Weinberg equilibrium; COBRA, Control of Blood Pressure and Risk Attenuation; UKADS/DGP, UK Asian Diabetes Study/Diabetes Genetics in Pakistan; HK, Hong Kong; SNUH, Seoul National University Hospital; KHGS, Korean Health and Genome Study.

3.2. Region-Related Associations Exist between rs8050136, rs9939609 and T2DM

For rs8050136, a total of 33,889 T2DM cases and 45,490 controls were included in the final data analysis. The overall results showed a significant association between rs8050136 and T2DM risk (OR = 1.14, 95% CI 1.10–1.18, p (z-test) < 0.001, I^2 = 37.4%) (Table 2, Figure 2a), with the association remaining statistically significant after adjustment for BMI (OR = 1.08, 95% CI 1.03–1.12, p (z-test) < 0.001, I^2 =27.1%) (Table 2, Figure 2b). To more clearly understand the association between rs8050136 and T2DM in different regions, we performed the subgroup analyses by region. Consequently, without BMI adjustment, a significant association between rs8050136 and T2DM was uncovered in East Asia (OR = 1.15, 95% CI 1.10–1.20), West Asia (OR = 1.17, 95% CI 1.05–1.29) and Europe (OR = 1.19, 95% CI 1.14–1.25) (Table 2, Figure 3a), with no such association in North America (OR = 1.06, 95% CI 0.93–1.19) or South Asia (OR = 1.19, 95% CI 0.91–1.48). After adjustment for BMI, significant association was only observed in East Asia (OR = 1.13, 95% CI 1.05–1.20) (Table 2,

Figure 3b). More importantly, as seen in Figure 4, the majority of studies on rs8050136 were distributed in East Asia. Several other studies were scattered throughout Europe, Northern America, South Asia and West Asia. More data for these regions may be required to detect an association.

Table 2. Meta-analysis of fat mass and obesity-associated (*FTO*) single-nucleotide polymorphisms (SNPs) and T2DM risk.

SNP	No. of Study (T2DM/Control)	Without BMI Adjustment				With BMI Adjustment			
		OR (95% CI)	p_z [a]	I^2 (%)	P_H [b]	OR (95% CI)	p_z [a]	I^2 (%)	P_H [b]
All									
rs9939609	29 (32771/50161)	1.15 (1.11–1.19)	0	53.2	0	1.11 (1.05–1.17)	0	56.1	0.003
rs8050136	26 (33889/45490)	1.14 (1.10–1.18)	0	37.4	0.032	1.08 (1.03–1.12)	0	27.1	0.151
rs1421085	4 (4285/16279)	1.05 (0.91–1.21))	0.48	80.6	0.001	1.02 (0.88–1.19)	0.755	78.2	0.003
rs17817499	3 (2634/15482)	1.09 (0.93–1.28)	0.271	82.7	0.003	1.05 (0.90–1.23)	0.539	80	0.007
East Asia									
rs9939609	11 (10063/10262)	1.11 (1.05–1.17)	0	19.5	0.257	1.11 (1.02–1.20)	0	0	0.535
rs8050136	15 (19109/19422)	1.15 (1.10–1.20)	0	0	0.789	1.13 (1.05–1.20)	0	0	0.531
North America									
rs9939609	3 (2162/14790)	1.11 (0.89–1.32)	0	85.4	0.001	1.02 (0.81–1.22)	0	85.7	0.008
rs8050136	4 (4140/17082)	1.06 (0.93–1.19)	0	74.1	0.009	1.03 (0.97–1.10)	0	69.9	0.019
Europe									
rs9939609	4 (10695/17306)	1.18 (1.14–1.22)	0	0	0.49	1.11 (0.93–1.29)	0	75.6	0.043
rs8050136	2 (8020/10685)	1.19 (1.14–1.25)	0	46.5	0.172	NA	NA	NA	NA
South Asia									
rs9939609	10 (8859/9152)	1.19 (1.10–1.29)	0	58.6	0.01	1.19 (1.06–1.31)	0	69.7	0.01
rs8050136	2 (2107/2651)	1.19 (0.91–1.48)	0	68	0.077	1.06 (0.94–1.18)	0	0	0.808
West Asia									
rs8050136	2 (1987/1489)	1.17 (1.05–1.29)	0	0	0.76	1.12 (0.98–1.25)	0	0	0.369

[a] p value for z-test; [b] p value for χ^2-test based Q test; BMI, body mass index; OR, odds ratio; CI, confidence interval; NA, not available.

Study ID		OR (95% CI)	% Weight
Xiao (2016)		1.20 (1.03, 1.38)	3.58
Xiao (2015)		1.18 (1.11, 1.37)	5.27
Shen (2015)		1.07 (0.54, 2.14)	0.23
Al-Sinani (2015)		1.14 (0.95, 1.38)	2.58
Chang (2014)		1.02 (0.88, 1.20)	4.05
Almawi (2013)		1.18 (1.07, 1.35)	4.79
Qian (2013)		1.27 (1.03, 1.41)	3.13
Gamboa (2012)		0.96 (0.82, 1.12)	4.38
Iwata (2012)		1.15 (0.95, 1.38)	2.58
Chauhan (2011)		1.08 (0.97, 1.21)	5.75
Ramya (2011)		1.38 (1.10, 1.72)	1.38
Han (2010)		1.24 (1.01, 1.52)	1.95
Bressler(African-Americans) (2010)		0.94 (0.83, 1.06)	6.02
Bressler(White) (2010)		1.19 (1.08, 1.31)	6.02
Liu (2010)		1.22 (1.07, 1.40)	3.85
Hu (2009)		1.13 (0.98, 1.29)	4.08
Rong (2009)		1.13 (0.99, 1.30)	4.19
Lee (2008)		0.91 (0.73, 1.40)	1.20
Ng(HK) (2008)		1.18 (1.02, 1.37)	3.54
Ng(SNUH) (2008)		1.15 (0.92, 1.44)	1.88
Ng(KHGS) (2008)		1.06 (0.88, 1.28)	2.89
Omori (2008)		1.09 (0.95, 1.25)	4.38
Horikoshi (2007)		1.22 (1.03, 1.46)	2.58
Zeggini (2007)		1.23 (1.18, 1.32)	9.08
Scott (2007)		1.17 (1.12, 1.22)	10.63
Overall (I-squared = 37.4%, p = 0.032)		1.14 (1.10, 1.18)	100.00

NOTE: Weights are from random effects analysis

-2.14   0   2.14

(a)

Study ID		OR (95% CI)	% Weight
Xiao (2016)		1.11 (0.77, 1.36)	2.19
Al-Sinani (2015)		1.03 (0.83, 1.29)	3.58
Chang (2014)		1.02 (0.87, 1.92)	0.69
Almawi (2013)		1.16 (1.01, 1.34)	6.98
Qian (2013)		1.17 (1.03, 1.32)	9.04
Gamboa (2012)		0.90 (0.74, 1.09)	6.21
Iwata (2012)		1.14 (0.94, 1.38)	3.93
Chauhan (2011)		1.05 (0.93, 1.19)	11.25
Ramya (2011)		1.09 (0.83, 1.42)	2.19
Bressler(African-Americans) (2010)		0.94 (0.83, 1.06)	14.38
Bressler(White) (2010)		1.12 (1.01, 1.23)	15.72
Wen (2010)		1.15 (0.96, 1.38)	4.31
Liu (2010)		1.22 (1.05, 1.41)	5.87
Rong (2009)		1.20 (1.02, 1.41)	5.00
Lee (2008)		0.89 (0.70, 1.14)	3.93
Omori (2008)		1.11 (0.93, 1.33)	4.73
Overall (I-squared = 27.1%, p = 0.151)		1.08 (1.03, 1.12)	100.00

-1.92   0   1.92

(b)

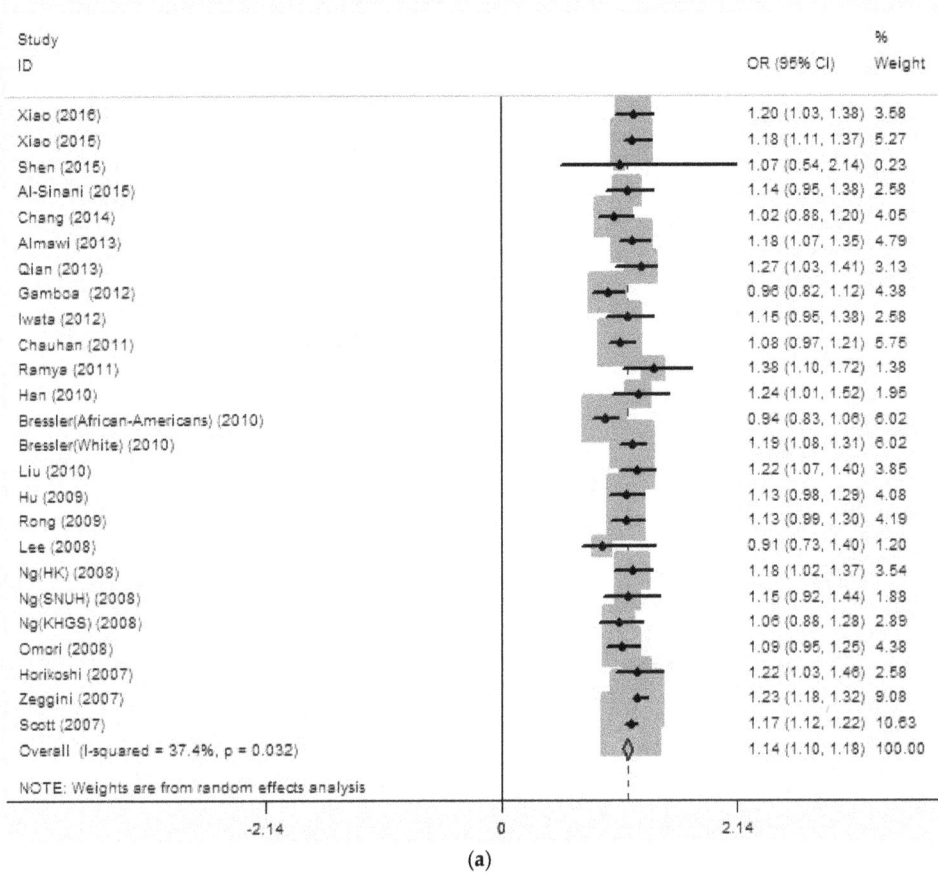

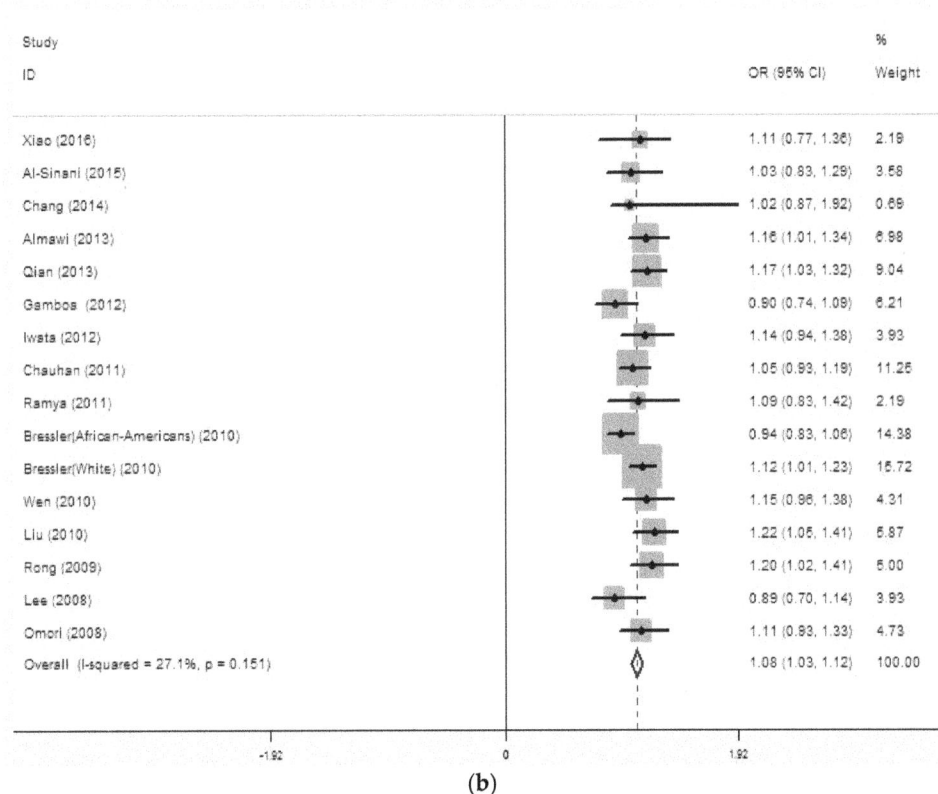

Figure 2. Meta-analysis for the associations between rs8050136 and Type 2 diabetes mellitus (T2DM) risk: (**a**) without; and (**b**) with adjustment for body mass index (BMI).

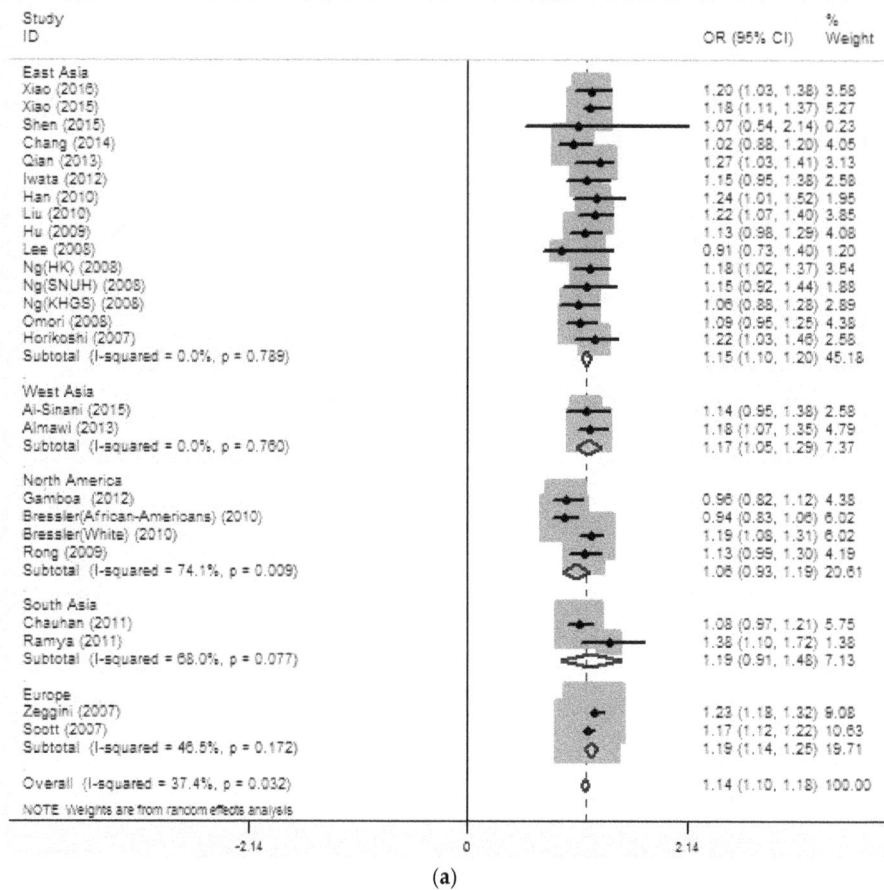

(a)

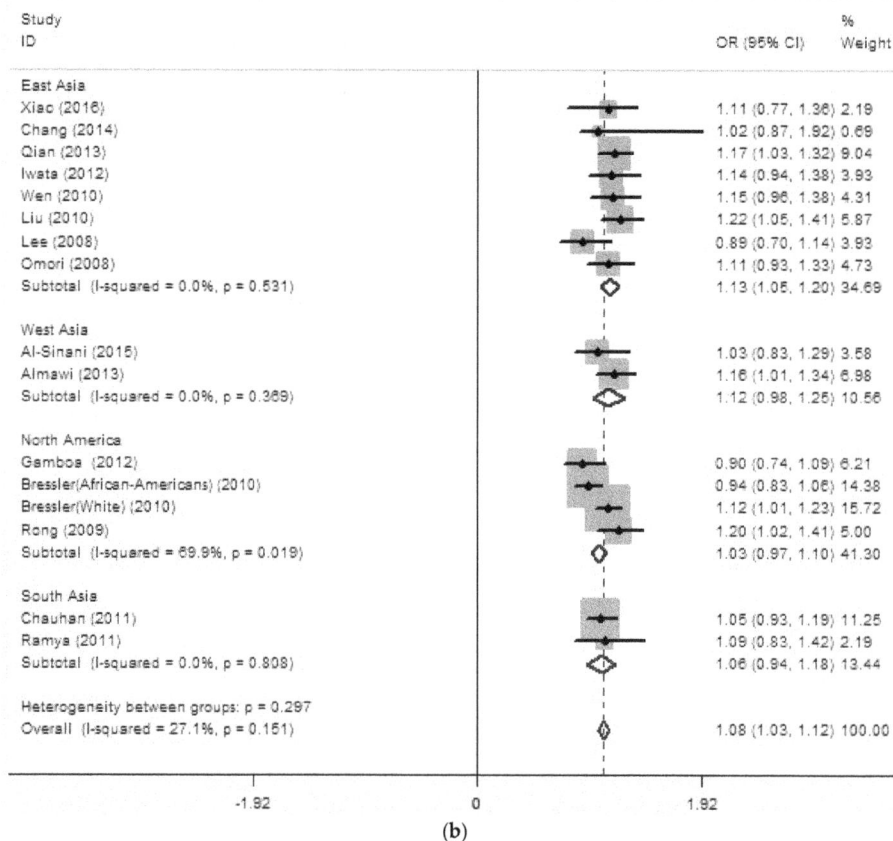

(b)

Figure 3. The stratified analysis results of rs8050136 grouped by region: (**a**) without; and (**b**) with adjustment for BMI.

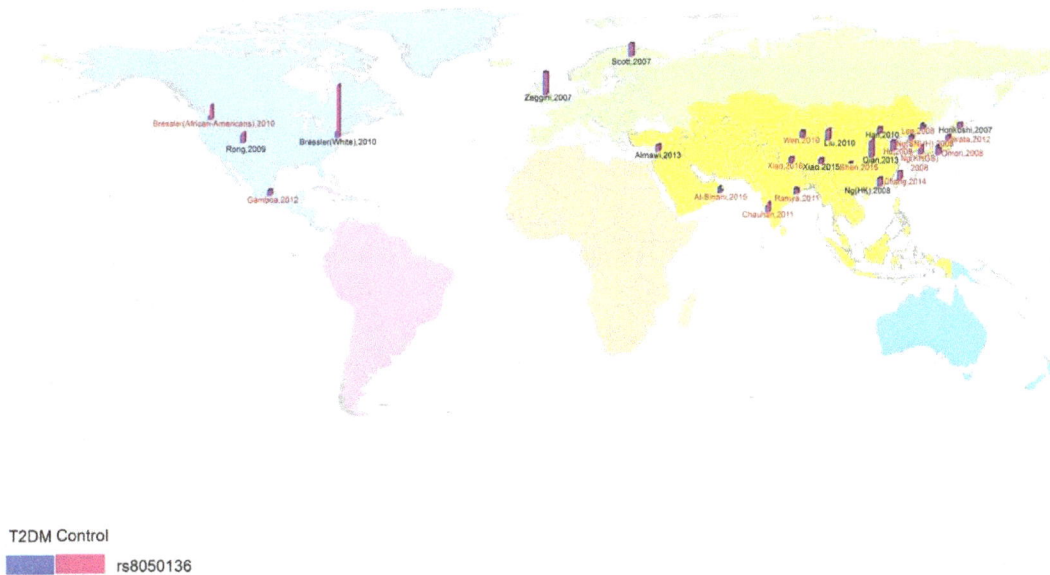

T2DM Control

rs8050136

Figure 4. Geographic distribution of selected studies exploring the association between rs8050136 and T2DM risk. Blue bars indicate T2DM patients while pink bars indicate controls; the height of bars is proportional to sample size. Studies in black text represent those that showed a significant association between the SNP and T2DM risk. Studies in red text indicate no significant association.

For rs9939609, a total of 32,771 T2DM cases and 50,161 controls were included in the meta-analysis. The overall results indicated that rs9939609 was significantly associated with an increased risk of T2DM (OR = 1.15, 95% CI 1.11–1.19, p (z-test) < 0.001, I^2 = 53.2%) (Table 2, Figure S1a). After adjustment for BMI, the association remained statistically significant (OR = 1.11, 95% CI 1.05–1.17, p (z-test) < 0.001, I^2 = 56.1%) (Table 2, Figure S1b). Due to the heterogeneity that existed between studies, we performed stratified analyses grouped by region. In the subgroup analyses, similar results were found in East Asia (without BMI adjustment: OR = 1.11, 95% CI 1.05–1.17; with BMI adjustment: OR = 1.11, 95% CI 1.02–1.20) and South Asia (without BMI adjustment: OR = 1.19, 95% CI 1.10–1.29; with BMI adjustment: OR = 1.19, 95% CI 1.06–1.31), whereas no such association existed between rs9939609 and T2DM in North America (without BMI adjustment: OR = 1.11, 95% CI 0.89–1.32; with BMI adjustment: OR = 1.02, 95% CI 0.81–1.22) (Table 2, Figure S2). Additionally, in Europe, a significant association between rs9939609 and T2DM was observed without BMI adjustment (OR = 1.18, 95% CI 1.14–1.22), whereas no association was uncovered with BMI adjustment (OR = 1.11, 95% CI 0.93–1.29). Similar to the distributions of rs8050136 studies, the geographic distribution of researches on rs9939609 were concentrated in East Asia and South Asia, where the association was found to be significant.

As illustrated in Figure 5, when the spatial scale was smaller than 1,000,000 meters, there was significant positive spatial autocorrelation in terms of both rs9969309 and rs8050136. It turned out that in relative small spatial scale (h < 1,000,000 meters), the studies with significant correlations tended to be clustered, which indicated that the correlation between rs9969309 and rs8050136, and T2DM risk was strongly associated with the geographic factors. With the h increasing, Moran's I showed no positive spatial autocorrelation of these two SNPs and T2DM risk, which meant we cannot reject the null hypothesis of completed spatial randomness. Our results follow Tobler's first law of geography: "Everything is related to everything else, but near things are more related than distant things" (pp.236, [56]). It seemed that in Asia, there was a strong positive-positive (significant-significant) spatial autocorrelation while in Europe there may be some negative-negative (non-significant-non-significant) spatial autocorrelation. In North America, the spatial autocorrelation was not significant, maintaining a relatively random spatial pattern.

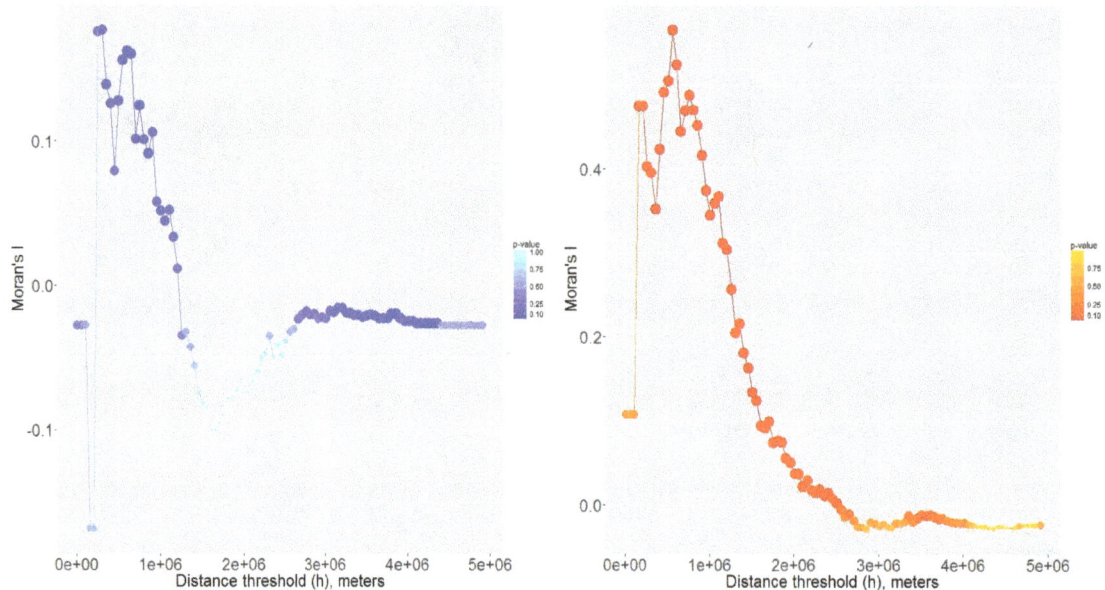

Figure 5. Spatial autocorrelation analysis of rs8050136, rs9939609 and T2DM by Moran's I. Blue indicates the results of rs8050136 and red indicates the results of rs9939609; the size and shade of the circles are proportional to the significance of Moran's I.

For rs1421085 and rs17817499, a total of 4,285 T2DM cases with 16,279 controls and 2,634 T2DM cases with 15,482 controls, respectively, were identified for data analysis. The results indicated that neither rs1421085 nor rs17817499 were associated with T2DM, independent of BMI adjustment (Table 2, Figures S3 and S4). Compared with rs9939609 and rs8050136, studies that focused on rs1421085 and rs17817499 were relatively fewer and were distributed in North America and North Africa.

3.3. Sensitivity Analyses

To assess the stability of the combined results obtained by excluding studies of unknown HWE in controls [7,25], a sensitivity analysis was conducted (Figure S5). The analysis confirmed that the rs9939609 polymorphism conferred a predisposition to T2DM.

3.4. Assessment of Publication Bias

To evaluate the publication bias, we performed Begg's test and Egger's test. The results showed that there was no publication bias for the associations between the four *FTO* polymorphisms and T2DM risk ($p > 0.05$ for Begg's test and Egger's test) (Table S1).

4. Discussion

Our meta-analysis and spatial analysis are based on a large sample size, including over 60,000 and 90,000 subjects for rs9939609 and rs8050136, respectively, spanning regions across Asia, Europe and Northern America. In line with previous meta-analyses of Asian populations [14,36,45], we further demonstrated a strong association between rs9939609 and rs8050136, and T2DM regardless of adjustment for BMI (Table 2, Figures 2 and 3, Figures S1 and S2). Notably, the associations are region-related.

Indeed, some statistics such as Moran's I [16,17], and local indicators of spatial autocorrelation (LISA) [57] can be used to quantitatively study spatial autocorrelation. However, due to obstacles including the modifiable areal unit problem (MAUP) (i.e., some papers only provide a country location while some papers have the city location) and the low data volume, it is difficult to perform spatial statistics for rs1421085 and rs17817499 to further explore the spatial pattern. Nevertheless, our data still indicate the geographic factor may play an important role in the correlations between T2DM risk and rs8050136 (Figures 4 and 5), rs9939609 (Figure 5).

Initially, the articles we reviewed contained more than 10 types of *FTO* SNPs in T2DM patients and controls, but we eventually chose the four most common SNPs, namely rs9939609, rs8050136, rs1421085 and rs17817499. All four SNPs are located in intron 1 of the *FTO* gene, a region of strong linkage disequilibrium [40]. Some studies have found no direct connection between the variants and *FTO* expression or function [9], while other studies have suggested that variants of *FTO* play an important role in regulating body weight and fat mass by influencing food intake [6]. A recent report revealed that SNPs in *FTO* could influence obesity by altering the expression of the adjacent genes *IRX3* and *RPGRIP1L* [58]. Although mechanisms regarding how these noncoding variants affect T2DM are not yet clear, Smemo et al. have demonstrated that variants within *FTO* can form long-range functional connections with *IRX3*, representing a determinant of body mass and composition [59]. Additionally, recent studies have suggested hepatic FTO contributes to glucose homeostasis [60–62], indicating that FTO may play a role in the regulation of carbohydrate metabolism.

Of note, the overall heterogeneity of rs9939609 increased slightly after BMI adjustment ($I^2 = 53.2\%$, $p < 0.001$ without BMI adjustment vs. $I^2 = 56.1\%$, $p = 0.003$ with BMI adjustment) (Table 2), suggesting that BMI may not primarily account for heterogeneity. To this end, we performed additional subgroup analyses by region and found that heterogeneity still existed in the group of North America and South Asia independent of BMI adjustment. We then excluded each study in South Asia and North America and performed subgroup analyses, respectively. When omitting studies by Fawwad et al. or Chauhan et al. in South Asia, as well as Bressler et al. (African-Americans) in North America [24,32,34], the heterogeneity disappeared in the South Asian ($I^2 = 34.6\%$, $p = 0.141$ and $I^2 = 37.2\%$, $p = 0.121$) and North American ($I^2 = 0.0\%$, $p = 0.667$) subgroups, respectively, without BMI adjustment (Table S2). Of note, the heterogeneity showed no change by removing other studies in South Asian or North American subgroup. Alternatively, only removing the study by Ali et al. [27], heterogeneity in the South Asian subgroup also attenuated sharply ($I^2 = 20.3\%$, $p = 0.288$) after adjustment for BMI (Table S2). These results demonstrated that these studies mentioned above were the main source of heterogeneity in South Asia and North America. Unlike rs9939609, owing to the low data volume of the studies, the heterogeneity in rs1421085 and rs17817499 showed no change by subgroup analyses.

BMI is widely considered as a confounder of T2DM risk. In this study, the overall associations between the four SNPs and T2DM risk were not affected by BMI adjustment. (Table 2), indicating that the overall associations were BMI-independent. Nevertheless, in Europe for rs9939609 and West Asia for rs8050136, the BMI adjustment altered the associations (Table 2). In agreement with previous reports [11,12], our data showed that rs9939609 was also associated with T2DM risk somewhat independently of BMI in East and South Asia as well as in Europe. Interestingly, different regions showed different associations between rs9939609 and rs8050136, and T2DM risk, demonstrating that the associations were region-dependent. Generally, a race/ethnicity population might live in the same region in most of the non-immigrant countries. Thus, our results might reflect the influence of different races/ethnicities to some extent.

The rs9939609 was the first SNP discovered within the *FTO* gene that showed a strong association with BMI and as such is the most widely investigated SNP of *FTO* [63]. Additionally, the A allele of rs9939609 is known to indicate a predisposition to obesity, T2DM, polycystic ovary syndrome (PCOS) and some cancers [41,64,65]. Our results of rs9939609 are not only consistent with earlier reports [11–14], but also include more recent studies with greater geographical coverage [7–9,22,23,34] (Table 2, Figure S2), providing stronger evidence for these associations. Similarly, rs8050136 was also found to function as a susceptible SNP to rs9939609-related diseases. Unlike rs9939609 and rs8050136, studies on rs1421085 and rs17817499 are scarce, and have limited regional coverage; lack of association maybe due to smaller sample size and less studies involved.

The study we present here still possesses several limitations. First, a large proportion of the studies focused on Asian populations, with European and Northern American populations only accounting for a small part. Second, there were relatively few studies on rs1421085 and rs17817499, which may lead to bias in negative results (Table 2, Figures S3 and S4). Lastly, except for BMI, we used genotype data without considering other possible confounders (such as age and sex) or gene-gene and gene–environment interactions. Although BMI is widely used to measure obesity, it has been suggested that different criteria (not necessarily > 30) may be used in different ethnic populations. Adiposity (or specific distribution of fat) rather than body weight (or BMI) may play a critical role in the regulation of insulin sensitivity and the development of diabetes. This may lead to an inconsistency in the effect of BMI on the association between *FTO* variants and T2DM risk. Therefore, further studies that adjust for more concomitant factors and cover more regions should be conducted.

5. Conclusions

The spatial analysis and meta-analysis showed that the associations between genetic polymorphisms in *FTO* and T2DM are region-related and that shedding light on spatial variations can provide new insights into well-established relationships. The rs9939609 and rs8050136 SNPs contributed to an increased risk of T2DM, which could provide new solutions for T2DM prevention and therapy. This study presented an initial step in spatial analysis for genetic and regional factors in the development of diabetes, although more work remains to be done before we can understand the impact of genetics, environment, geography, BMI and fat distribution on diabetes as well as how these associations may vary across space.

Acknowledgments: This work was supported by the National Natural Science Foundation of China (81472023, and 81271919), the National Basic Research Program of China (973 Program) (2012CB720605) and Research Project Foundation of Health and Family Planning Commission of Wuhan City (WX15C22). We thank Roy Morgan for careful reviewing of this manuscript (Committee on Genetics, Genomics and Systems Biology, The University of Chicago, USA).

Author Contributions: Y.Y. and S.M.L. designed research; Y.Y., B.Y.L., J.Y. and H.Y.L. conducted research; Y.Y., W.X., L.H. and H.Y.L. selected articles and performed meta-analysis; B.Y.L. and J.Y. conducted spatial analysis; Y.Y. and B.Y.L. prepared the initial manuscript draft; SML edited and revised subsequent drafts; and S.M.L. had primary responsibility for final content.

References

1. Kleinberger, J.W.; Pollin, T.I. Personalized medicine in diabetes mellitus: Current opportunities and future prospects. *Ann. N. Y. Acad. Sci.* **2015**, *1346*, 45–56. [CrossRef] [PubMed]
2. Omori, S.; Tanaka, Y.; Takahashi, A.; Hirose, H.; Kashiwagi, A.; Kaku, K.; Kawamori, R.; Nakamura, Y.; Maeda, S. Association of *CDKAL1*, *IGF2BP2*, *CDKN2A/B*, *HHEX*, *SLC30A8*, and *KCNJ11* with susceptibility to type 2 diabetes in a Japanese population. *Diabetes* **2008**, *57*, 791–795. [CrossRef] [PubMed]
3. American Diabetes Association. Classification and diagnosis of diabetes. Sec. 2. In standards of medical care in diabetes—2015. *Diabetes Care* **2015**, *38* (Suppl. S1), S8–S16.
4. O'Rahilly, S.; Barroso, I.; Wareham, N.J. Genetic factors in type 2 diabetes: The end of the beginning? *Science* **2005**, *307*, 370–373. [CrossRef] [PubMed]
5. Dina, C.; Meyre, D.; Gallina, S.; Durand, E.; Korner, A.; Jacobson, P.; Carlsson, L.M.; Kiess, W.; Vatin, V.; Lecoeur, C.; et al. Variation in *FTO* contributes to childhood obesity and severe adult obesity. *Nat. Genet.* **2007**, *39*, 724–726. [CrossRef] [PubMed]
6. Frayling, T.M.; Timpson, N.J.; Weedon, M.N.; Zeggini, E.; Freathy, R.M.; Lindgren, C.M.; Perry, J.R.;

Elliott, K.S.; Lango, H.; Rayner, N.W.; et al. A common variant in the *FTO* gene is associated with body mass index and predisposes to childhood and adult obesity. *Science* **2007**, *316*, 889–894. [CrossRef] [PubMed]

7. Phani, N.M.; Vohra, M.; Rajesh, S.; Adhikari, P.; Nagri, S.K.; D'Souza, S.C.; Satyamoorthy, K.; Rai, P.S. Implications of critical *PPARγ2*, *ADIPOQ* and *FTO* gene polymorphisms in type 2 diabetes and obesity-mediated susceptibility to type 2 diabetes in an Indian population. *Mol. Genet. Genomics* **2016**, *291*, 193–204. [CrossRef] [PubMed]

8. Xiao, S.; Zeng, X.; Quan, L.; Zhu, J. Correlation between polymorphism of *FTO* gene and type 2 diabetes mellitus in Uygur people from Northwest China. *Int. J. Clin. Exp. Med.* **2015**, *8*, 9744–9750. [PubMed]

9. Shen, F.; Huang, W.; Huang, J.T.; Xiong, J.; Yang, Y.; Wu, K.; Jia, G.F.; Chen, J.; Feng, Y.Q.; Yuan, B.F.; et al. Decreased N^6-methyladenosine in peripheral blood rna from diabetic patients is associated with *FTO* expression rather than *ALKBH5*. *J. Clin. Endocrinol. Metab.* **2015**, *100*, E148–E154. [CrossRef] [PubMed]

10. Bazzi, M.D.; Nasr, F.A.; Alanazi, M.S.; Alamri, A.; Turjoman, A.A.; Moustafa, A.S.; Alfadda, A.A.; Pathan, A.A.; Parine, N.R. Association between *FTO*, *MC4R*, *SLC30A8*, and *KCNQ1* gene variants and type 2 diabetes in Saudi population. *Genet. Mol. Res.* **2014**, *13*, 10194–10203. [CrossRef] [PubMed]

11. Li, H.; Kilpelainen, T.O.; Liu, C.; Zhu, J.; Liu, Y.; Hu, C.; Yang, Z.; Zhang, W.; Bao, W.; Cha, S.; et al. Association of genetic variation in *FTO* with risk of obesity and type 2 diabetes with data from 96,551 East and South Asians. *Diabetologia* **2012**, *55*, 981–995. [CrossRef] [PubMed]

12. Hertel, J.K.; Johansson, S.; Sonestedt, E.; Jonsson, A.; Lie, R.T.; Platou, C.G.; Nilsson, P.M.; Rukh, G.; Midthjell, K.; Hveem, K.; et al. *FTO*, type 2 diabetes, and weight gain throughout adult life: A meta-analysis of 41,504 subjects from the Scandinavian HUNT, MDC, and MPP studies. *Diabetes* **2011**, *60*, 1637–1644. [CrossRef] [PubMed]

13. Xi, B.; Takeuchi, F.; Meirhaeghe, A.; Kato, N.; Chambers, J.C.; Morris, A.P.; Cho, Y.S.; Zhang, W.; Mohlke, K.L.; Kooner, J.S.; et al. Associations of genetic variants in/near body mass index-associated genes with type 2 diabetes: A systematic meta-analysis. *Clin. Endocrinol.* **2014**, *81*, 702–710. [CrossRef] [PubMed]

14. Vasan, S.K.; Karpe, F.; Gu, H.F.; Brismar, K.; Fall, C.H.; Ingelsson, E.; Fall, T. *FTO* genetic variants and risk of obesity and type 2 diabetes: A meta-analysis of 28,394 Indians. *Obesity* **2014**, *22*, 964–970. [CrossRef] [PubMed]

15. Hipp, J.A.; Chalise, N. Spatial analysis and correlates of county-level diabetes prevalence, 2009–2010. *Prev. Chronic Dis.* **2015**, *12*, E08. [CrossRef] [PubMed]

16. Moran, P.A.P. The interpretation of statistical maps. *J. R. Stat. Soc.* **1947**, *10*, 243–251.

17. Moran, P.A. A test for the serial independence of residuals. *Biometrika* **1950**, *37*, 178–181. [CrossRef] [PubMed]

18. DerSimonian, R.; Laird, N. Meta-analysis in clinical trials. *Control. Clin. Trials* **1986**, *7*, 177–188. [CrossRef]

19. Mantel, N.; Haenszel, W. Statistical aspects of the analysis of data from retrospective studies of disease. *J. Natl. Cancer Inst.* **1959**, *22*, 719–748. [PubMed]

20. Begg, C.B.; Mazumdar, M. Operating characteristics of a rank correlation test for publication bias. *Biometrics* **1994**, *50*, 1088–1101. [CrossRef] [PubMed]

21. Egger, M.; Davey Smith, G.; Schneider, M.; Minder, C. Bias in meta-analysis detected by a simple, graphical test. *BMJ* **1997**, *315*, 629–634. [CrossRef] [PubMed]

22. Xiao, S.; Zeng, X.; Fan, Y.; Su, Y.; Ma, Q.; Zhu, J.; Yao, H. Gene polymorphism association with type 2 diabetes and related gene-gene and gene-environment interactions in a Uyghur population. *Med. Sci. Monit.* **2016**, *22*, 474–487. [PubMed]

23. Al-Sinani, S.; Woodhouse, N.; Al-Mamari, A.; Al-Shafie, O.; Al-Shafaee, M.; Al-Yahyaee, S.; Hassan, M.; Jaju, D.; Al-Hashmi, K.; Al-Abri, M.; et al. Association of gene variants with susceptibility to type 2 diabetes among Omanis. *World J. Diabetes* **2015**, *6*, 358–366. [CrossRef] [PubMed]

24. Fawwad, A.; Siddiqui, I.A.; Zeeshan, N.F.; Shahid, S.M.; Basit, A. Association of snp rs9939609 in *FTO* gene with metabolic syndrome in type 2 diabetic subjects, rectruited from a tertiary care unit of Karachi, Pakistan. *Pak. J. Med. Sci.* **2015**, *31*, 140–145. [PubMed]

25. Raza, S.T.; Abbas, S.; Ahmad, A.; Ahmed, F.; Zaidi, Z.H.; Mahdi, F. Association of glutathione-s-transferase (GSTM1 and GSTT1) and FTO gene polymorphisms with type 2 diabetes mellitus cases in Northern India. *Balk. J. Med. Genet.* **2014**, *17*, 47–54.

26. Kalnina, I.; Zaharenko, L.; Vaivade, I.; Rovite, V.; Nikitina-Zake, L.; Peculis, R.; Fridmanis, D.; Geldnere, K.; Jacobsson, J.A.; Almen, M.S.; et al. Polymorphisms in *FTO* and near *TMEM18* associate with type 2 diabetes and predispose to younger age at diagnosis of diabetes. *Gene* **2013**, *527*, 462–468. [CrossRef] [PubMed]

27. Ali, S.; Chopra, R.; Manvati, S.; Singh, Y.P.; Kaul, N.; Behura, A.; Mahajan, A.; Sehajpal, P.; Gupta, S.; Dhar, M.K.; et al. Replication of type 2 diabetes candidate genes variations in three geographically unrelated Indian population groups. *PLoS ONE* **2013**, *8*, e58881. [CrossRef] [PubMed]

28. Binh, T.Q.; Phuong, P.T.; Nhung, B.T.; Thoang, D.D.; Lien, H.T.; Thanh, D.V. Association of the common *FTO*-rs9939609 polymorphism with type 2 diabetes, independent of obesity-related traits in a Vietnamese population. *Gene* **2013**, *513*, 31–35. [CrossRef] [PubMed]

29. Iwata, M.; Maeda, S.; Kamura, Y.; Takano, A.; Kato, H.; Murakami, S.; Higuchi, K.; Takahashi, A.; Fujita, H.; Hara, K.; et al. Genetic risk score constructed using 14 susceptibility alleles for type 2 diabetes is associated with the early onset of diabetes and may predict the future requirement of insulin injections among Japanese individuals. *Diabetes Care* **2012**, *35*, 1763–1770. [CrossRef] [PubMed]

30. Rees, S.D.; Islam, M.; Hydrie, M.Z.I.; Chaudhary, B.; Bellary, S.; Hashmi, S.; O'Hare, J.P.; Kumar, S.; Sanghera, D.K.; Chaturvedi, N.; et al. An *FTO* variant is associated with type 2 diabetes in South Asian populations after accounting for body mass index and waist circumference. *Diabet. Med.* **2011**, *28*, 673–680. [CrossRef] [PubMed]

31. Huang, W.; Sun, Y.; Sun, J. Combined effects of *FTO* rs9939609 and *MC4R* rs17782313 on obesity and BMI in Chinese Han populations. *Endocrine* **2011**, *39*, 69–74. [CrossRef] [PubMed]

32. Chauhan, G.; Tabassum, R.; Mahajan, A.; Dwivedi, O.P.; Mahendran, Y.; Kaur, I.; Nigam, S.; Dubey, H.; Varma, B.; Madhu, S.V.; et al. Common variants of *FTO* and the risk of obesity and type 2 diabetes in Indians. *J. Hum. Genet.* **2011**, *56*, 720–726. [CrossRef] [PubMed]

33. Cruz, M.; Valladares-Salgado, A.; Garcia-Mena, J.; Ross, K.; Edwards, M.; Angeles-Martinez, J.; Ortega-Camarillo, C.; de la Pena, J.E.; Burguete-Garcia, A.I.; Wacher-Rodarte, N.; et al. Candidate gene association study conditioning on individual ancestry in patients with type 2 diabetes and metabolic syndrome from Mexico City. *Diabetes Metab. Res. Rev.* **2010**, *26*, 261–270. [CrossRef] [PubMed]

34. Bressler, J.; Kao, W.H.; Pankow, J.S.; Boerwinkle, E. Risk of type 2 diabetes and obesity is differentially associated with variation in *FTO* in Whites and African-Americans in the ARIC study. *PLoS ONE* **2010**, *5*, e10521. [CrossRef] [PubMed]

35. Liu, Y.; Liu, Z.; Song, Y.; Zhou, D.; Zhang, D.; Zhao, T.; Chen, Z.; Yu, L.; Yang, Y.; Feng, G.; et al. Meta-analysis added power to identify variants in *FTO* associated with type 2 diabetes and obesity in the Asian population. *Obesity* **2010**, *18*, 1619–1624. [CrossRef] [PubMed]

36. Yajnik, C.S.; Janipalli, C.S.; Bhaskar, S.; Kulkarni, S.R.; Freathy, R.M.; Prakash, S.; Mani, K.R.; Weedon, M.N.; Kale, S.D.; Deshpande, J.; et al. *FTO* gene variants are strongly associated with type 2 diabetes in South Asian Indians. *Diabetologia* **2009**, *52*, 247–252. [CrossRef] [PubMed]

37. Legry, V.; Cottel, D.; Ferrieres, J.; Arveiler, D.; Andrieux, N.; Bingham, A.; Wagner, A.; Ruidavets, J.B.; Ducimetiere, P.; Amouyel, P.; et al. Effect of an *FTO* polymorphism on fat mass, obesity, and type 2 diabetes mellitus in the French MONICA study. *Metabolism* **2009**, *58*, 971–975. [CrossRef] [PubMed]

38. Sanghera, D.K.; Ortega, L.; Han, S.; Singh, J.; Ralhan, S.K.; Wander, G.S.; Mehra, N.K.; Mulvihill, J.J.; Ferrell, R.E.; Nath, S.K.; et al. Impact of nine common type 2 diabetes risk polymorphisms in Asian Indian Sikhs: *PPARG2* (*Pro12Ala*), *IGF2BP2*, *TCF7L2* and *FTO* variants confer a significant risk. *BMC Med. Genet.* **2008**, *9*, 59. [CrossRef] [PubMed]

39. Horikawa, Y.; Miyake, K.; Yasuda, K.; Enya, M.; Hirota, Y.; Yamagata, K.; Hinokio, Y.; Oka, Y.; Iwasaki, N.; Iwamoto, Y.; et al. Replication of genome-wide association studies of type 2 diabetes susceptibility in Japan. *J. Clin. Endocrinol. Metab.* **2008**, *93*, 3136–3141. [CrossRef] [PubMed]

40. Chang, Y.C.; Liu, P.H.; Lee, W.J.; Chang, T.J.; Jiang, Y.D.; Li, H.Y.; Kuo, S.S.; Lee, K.C.; Chuang, L.M. Common variation in the fat mass and obesity-associated (*FTO*) gene confers risk of obesity and modulates BMI in the Chinese population. *Diabetes* **2008**, *57*, 2245–2252. [CrossRef] [PubMed]

41. Horikoshi, M.; Hara, K.; Ito, C.; Shojima, N.; Nagai, R.; Ueki, K.; Froguel, P.; Kadowaki, T. Variations in the *HHEX* gene are associated with increased risk of type 2 diabetes in the Japanese population. *Diabetologia* **2007**, *50*, 2461–2466. [CrossRef] [PubMed]

42. Zeggini, E.; Weedon, M.N.; Lindgren, C.M.; Frayling, T.M.; Elliott, K.S.; Lango, H.; Timpson, N.J.; Perry, J.R.; Rayner, N.W.; Freathy, R.M.; et al. Replication of genome-wide association signals in UK samples reveals risk loci for type 2 diabetes. *Science* **2007**, *316*, 1336–1341. [CrossRef] [PubMed]

43. Chang, Y.C.; Liu, P.H.; Yu, Y.H.; Kuo, S.S.; Chang, T.J.; Jiang, Y.D.; Nong, J.Y.; Hwang, J.J.; Chuang, L.M. Validation of type 2 diabetes risk variants identified by genome-wide association studies in Han Chinese population: A replication study and meta-analysis. *PLoS ONE* **2014**, *9*, e95045. [CrossRef] [PubMed]

44. Almawi, W.Y.; Nemr, R.; Keleshian, S.H.; Echtay, A.; Saldanha, F.L.; AlDoseri, F.A.; Racoubian, E. A replication study of 19 GWAS-validated type 2 diabetes at-risk variants in the Lebanese population. *Diabetes Res. Clin. Pract.* **2013**, *102*, 117–122. [CrossRef] [PubMed]

45. Qian, Y.; Liu, S.J.; Lu, F.; Li, H.Z.; Dong, M.H.; Lin, Y.D.; Du, J.B.; Lin, Y.; Gong, J.H.; Jin, G.F.; et al. Genetic variant in fat mass and obesity-associated gene associated with type 2 diabetes risk in Han Chinese. *BMC Genet.* **2013**, *14*, 86. [CrossRef] [PubMed]

46. Gamboa-Melendez, M.A.; Huerta-Chagoya, A.; Moreno-Macias, H.; Vazquez-Cardenas, P.; Ordonez-Sanchez, M.L.; Rodriguez-Guillen, R.; Riba, L.; Rodriguez-Torres, M.; Guerra-Garcia, M.T.; Guillen-Pineda, L.E.; et al. Contribution of common genetic variation to the risk of type 2 diabetes in the Mexican Mestizo population. *Diabetes* **2012**, *61*, 3314–3321. [CrossRef] [PubMed]

47. Ramya, K.; Radha, V.; Ghosh, S.; Majumder, P.P.; Mohan, V. Genetic variations in the *FTO* gene are associated with type 2 diabetes and obesity in South Indians (cures-79). *Diabetes Technol. Ther.* **2011**, *13*, 33–42. [CrossRef] [PubMed]

48. Han, X.; Luo, Y.; Ren, Q.; Zhang, X.; Wang, F.; Sun, X.; Zhou, X.; Ji, L. Implication of genetic variants near *SLC30A8*, *HHEX*, *CDKAL1*, *CDKN2A/B*, *IGF2BP2*, *FTO*, *TCF2*, *KCNQ1*, and *WFS1* in type 2 diabetes in a Chinese population. *BMC Med. Genet.* **2010**, *11*, 81. [CrossRef] [PubMed]

49. Wen, J.; Ronn, T.; Olsson, A.; Yang, Z.; Lu, B.; Du, Y.; Groop, L.; Ling, C.; Hu, R. Investigation of type 2 diabetes risk alleles support *CDKN2A/B*, *CDKAL1*, and *TCF7L2* as susceptibility genes in a Han Chinese cohort. *PLoS ONE* **2010**, *5*, e9153. [CrossRef] [PubMed]

50. Hu, C.; Zhang, R.; Wang, C.; Wang, J.; Ma, X.; Lu, J.; Qin, W.; Hou, X.; Wang, C.; Bao, Y.; et al. *PPARG*, *KCNJ11*, *CDKAL1*, *CDKN2A-CDKN2B*, *IDE-KIF11-HHEX*, *IGF2BP2* and *SLC30A8* are associated with type 2 diabetes in a Chinese population. *PLoS ONE* **2009**, *4*, e7643. [CrossRef] [PubMed]

51. Rong, R.; Hanson, R.L.; Ortiz, D.; Wiedrich, C.; Kobes, S.; Knowler, W.C.; Bogardus, C.; Baier, L.J. Association analysis of variation in/near *FTO*, *CDKAL1*, *SLC30A8*, *HHEX*, *EXT2*, *IGF2BP2*, *LOC387761*, and *CDKN2B* with type 2 diabetes and related quantitative traits in Pima Indians. *Diabetes* **2009**, *58*, 478–488. [CrossRef] [PubMed]

52. Lee, Y.H.; Kang, E.S.; Kim, S.H.; Han, S.J.; Kim, C.H.; Kim, H.J.; Ahn, C.W.; Cha, B.S.; Nam, M.; Nam, C.M.; et al. Association between polymorphisms in *SLC30A8*, *HHEX*, *CDKN2A/B*, *IGF2BP2*, *FTO*, *WFS1*, *CDKAL1*, *KCNQ1* and type 2 diabetes in the Korean population. *J. Hum. Genet.* **2008**, *53*, 991–998. [CrossRef] [PubMed]

53. Ng, M.C.Y.; Park, K.S.; Oh, B.; Tam, C.H.T.; Cho, Y.M.; Shin, H.D.; Lam, V.K.L.; Ma, R.C.W.; So, W.Y.; Cho, Y.S.; et al. Implication of genetic variants near *TCF7L2*, *SLC30A8*, *HHEX*, *CDKAL1*, *CDKN2A/B*, *IGF2BP2*, and *FTO* in type 2 diabetes and obesity in 6719 Asians. *Diabetes* **2008**, *57*, 2226–2233. [CrossRef] [PubMed]

54. Scott, L.J.; Mohlke, K.L.; Bonnycastle, L.L.; Willer, C.J.; Li, Y.; Duren, W.L.; Erdos, M.R.; Stringham, H.M.; Chines, P.S.; Jackson, A.U.; et al. A genome-wide association study of type 2 diabetes in Finns detects multiple susceptibility variants. *Science* **2007**, *316*, 1341–1345. [CrossRef] [PubMed]

55. Cauchi, S.; Ezzidi, I.; El Achhab, Y.; Mtiraoui, N.; Chaieb, L.; Salah, D.; Nejjari, C.; Labrune, Y.; Yengo, L.; Beury, D.; et al. European genetic variants associated with type 2 diabetes in North African Arabs. *Diabetes Metab.* **2012**, *38*, 316–323. [CrossRef] [PubMed]

56. Tobler, A.W.R. A computer movie simulation urban growth in Detroit region. *Econ. Geogr.* **1970**, *46*, 234–240.

57. Anselin, L. Local indicators of spatial association LISA. *Geogr. Anal.* **1995**, *27*, 93–115. [CrossRef]

58. Merkestein, M.; Laber, S.; McMurray, F.; Andrew, D.; Sachse, G.; Sanderson, J.; Li, M.; Usher, S.; Sellayah, D.; Ashcroft, F.M.; et al. FTO influences adipogenesis by regulating mitotic clonal expansion. *Nat. Commun.* **2015**, *6*, 6792. [CrossRef] [PubMed]

59. Smemo, S.; Tena, J.J.; Kim, K.H.; Gamazon, E.R.; Sakabe, N.J.; Gomez-Marin, C.; Aneas, I.; Credidio, F.L.; Sobreira, D.R.; Wasserman, N.F.; et al. Obesity-associated variants within *FTO* form long-range functional connections with *IRX3*. *Nature* **2014**, *507*, 371–375. [CrossRef] [PubMed]

60.   Bravard, A.; Vial, G.; Chauvin, M.A.; Rouille, Y.; Bailleul, B.; Vidal, H.; Rieusset, J. FTO contributes to hepatic metabolism regulation through regulation of leptin action and SAT3 signalling in liver. *Cell Commun. Signal.* **2014**, *12*, 4. [CrossRef] [PubMed]

61.   Guo, F.; Zhang, Y.; Zhang, C.; Wang, S.; Ni, Y.; Zhao, R. Fatmass and obesity associated (FTO) gene regulates gluconeogenesis in chicken embryo fibroblast cells. *Comp. Biochem. Physiol. A Mol. Integr. Physiol.* **2015**, *179*, 149–156. [CrossRef] [PubMed]

62.   Mizuno, T.M.; Lew, P.S.; Luo, Y.; Leckstrom, A. Negative regulation of hepatic fat mass and obesity associated (Fto) gene expression by insulin. *Life Sci.* **2017**, *170*, 50–55. [CrossRef] [PubMed]

63.   Speakman, J.R. FTO effect on energy demand versus food intake. *Nature* **2010**, *464*, E1. [CrossRef] [PubMed]

64.   Cai, X.; Liu, C.; Mou, S. Association between fat mass- and obesity-associated (FTO) gene polymorphism and polycystic ovary syndrome: A meta-analysis. *PLoS ONE* **2014**, *9*, e86972. [CrossRef] [PubMed]

65.   Li, G.; Chen, Q.; Wang, L.; Ke, D.; Yuan, Z. Association between *FTO* gene polymorphism and cancer risk: Evidence from 16,277 cases and 31,153 controls. *Tumour. Biol.* **2012**, *33*, 1237–1243. [CrossRef] [PubMed]

12

Systematic Identification and Assessment of Therapeutic Targets for Breast Cancer based on Genome-Wide RNA Interference Transcriptomes

Yang Liu [1,†], Xiaoyao Yin [2,†], Jing Zhong [3,†], Naiyang Guan [2], Zhigang Luo [2], Lishan Min [3], Xing Yao [3], Xiaochen Bo [4], Licheng Dai [3,*] and Hui Bai [4,5,*]

[1]   Research Center for Clinical & Translational Medicine, Beijing 302 Hospital, Beijing 100039, China; liuyang@bmi.ac.cn
[2]   Science and technology on Parallel and Distributed Processing Laboratory, National University of Defense Technology, Changsha 410073, China; yinxy1992@sina.com (X.Y.); ny_guan@nudt.edu.cn (N.G.); zgluo@nudt.edu.cn (Z.L.)
[3]   Huzhou Key Laboratory of Molecular Medicine, Huzhou Central Hospital, Huzhou 313000, China; zhongjing1003@126.com (J.Z.); malisha362@126.com (L.M.); yaoy333@126.com (X.Y.)
[4]   Beijing Institute of Radiation Medicine, Beijing 100850, China; boxc@bmi.ac.cn
[5]   No. 451 Hospital of PLA, Xi'an 710054, China
[*]   Correspondence: dlc21@126.com (L.D.); huibai13@hotmail.com (H.B.)

[†]   These authors contributed equally to this work

Academic Editor: Wenyi Gu

Abstract: With accumulating public omics data, great efforts have been made to characterize the genetic heterogeneity of breast cancer. However, identifying novel targets and selecting the best from the sizeable lists of candidate targets is still a key challenge for targeted therapy, largely owing to the lack of economical, efficient and systematic discovery and assessment to prioritize potential therapeutic targets. Here, we describe an approach that combines the computational evaluation and objective, multifaceted assessment to systematically identify and prioritize targets for biological validation and therapeutic exploration. We first establish the reference gene expression profiles from breast cancer cell line MCF7 upon genome-wide RNA interference (RNAi) of a total of 3689 genes, and the breast cancer query signatures using RNA-seq data generated from tissue samples of clinical breast cancer patients in the Cancer Genome Atlas (TCGA). Based on gene set enrichment analysis, we identified a set of 510 genes that when knocked down could significantly reverse the transcriptome of breast cancer state. We then perform multifaceted assessment to analyze the gene set to prioritize potential targets for gene therapy. We also propose drug repurposing opportunities and identify potentially druggable proteins that have been poorly explored with regard to the discovery of small-molecule modulators. Finally, we obtained a small list of candidate therapeutic targets for four major breast cancer subtypes, i.e., luminal A, luminal B, HER2+ and triple negative breast cancer. This RNAi transcriptome-based approach can be a helpful paradigm for relevant researches to identify and prioritize candidate targets for experimental validation.

Keywords: breast cancer; library of integrated network-based cellular signatures; gene set enrichment analysis; drug target; DNA methylation; Cancer Gene Census

1. Introduction

As a heterogeneous disease, breast cancer is the most frequently diagnosed cancer in women and the second leading cause of cancer death among females worldwide, accounting for 25% of all cancer cases and 15% of all cancer deaths [1,2]. Traditionally, breast cancer prognosis and classification have relied on analysis of tumor morphology and expression of three markers, i.e., estrogen receptor (ER), progesterone receptor (PR) and human epidermal growth factor receptor2 (HER2). These proteins also serve as targets for specific treatment [3]. Triple-negative breast cancers (TNBC), which refers to an absence of the expression of ER, PR and HER2, accounts for approximately 15%–20% of all diagnosed breast cancer cases. TNBC is more likely to affect younger women, African-Americans, Hispanics, and/or those with a BRCA1 gene mutation [4–6]. In addition, TNBC is more aggressive than other types of breast cancer and is associated with poorer survival than non-TNBC, owing to more frequent relapse in TNBC patients with residual disease [7,8].

Despite major advances in ER-positive or HER2-amplified breast cancers, which can be targeted by drugs such as tamoxifen and trastuzumab (Herceptin), there is no targeted therapy currently available for TNBC. Cytotoxic chemotherapy [4] is the mainstay of treatment since pathological complete responses after chemotherapy are more likely in TNBC than in non-TNBC [7,8]. However, research shows only 31% of TNBC patients experience pathological complete responses after chemotherapy [9], emphasizing the urgent need to explore potential gene targets for development of effective therapeutics and gene therapies.

Large-scale genomics initiatives such as The Cancer Genome Atlas (TCGA), and other complementary omics data are providing a growing list of genes that are causally involved or playing biologically and pathologically compelling roles in breast cancer, especially in TNBC [10–15]. Well studied therapeutic targets for TNBC include those proteins in proliferative and survival-dependent pathways, such as epidermal growth factor receptor (EGFR) [16,17], vascular endothelial growth factor receptor (VEGFR) [18], JAK2/STAT3 [19] and PI3K [20–23], and those control the cell cycle and the DNA damage responses such as Poly-ADP Ribose Polymerase (PARPs) [24,25]. However, the initial results of candidate compounds targeting these proteins from clinical trials [3,26–30] remain modest, highlighting the importance of systematic target discovering, assessing and prioritizing for biological validation and therapeutic exploitation.

RNA interference (RNAi) is an endogenous process that regulates expression of genes and corresponding proteins to maintain homeostasis in diverse organisms. In addition, RNAi therapy is emerging rapidly for personalized cancer treatment [31,32]. RNAi has also been invaluable research tool for unraveling critical genes and pathways involved in cancer development, growth and metastasis and has identified critical tumor-type specific gene targets for chemotherapy [33]. Importantly, RNAi can be combined with transcriptional profiling and used in applications such as identifying molecular biomarkers and discovering new potential targets for drug treatment [34]. However, existing studies mainly focus on knocking down a small set of genes from single cancer cell line of interest to observe typical phenotype changes. In addition, those using transcriptome analysis generally make inferences directly from the statistically significant dysregulated gene sets without taking the complex interactions into consideration.

Fortunately, the emerging of the Library of Integrated Network-Based Cellular Signatures (LINCS) data brought us with new opportunities to evaluate the cellular status from the aspect of whole genomic expression profiles after being perturbed by all kinds of perturbagens including RNAi reagents. Based on the L1000 technology [35], LINCS has recently provided open access to 1,328,098 gene expression profiles generated from 77 different cellular contexts including human primary and cancer cell lines, upon genetic knockdown of a total of 4372 genes. Ever since, the data landscape has changed substantially, making it possible for the first time to address comprehensive, objective and in depth target discovery and assessment. Therefore, there is a clear need for a systematic, objective, data-driven gene target discovery and further assessment of such gene lists, based on information integrated from different disciplines and sources, with the goal of prioritizing genes for further detailed biological validation studies.

Here, we demonstrate such a systematic, unbiased and objective computational approach that combines the RNAi transcriptome based target identification and multifaceted assessment. We applied our approach to identify and assess the candidate gene targets for different breast cancer subtypes, i.e., luminal A, luminal B, HER2+ and TNBC. The workflow with the computational identification, annotation scheme and assessment criteria is summarized in Figure 1 (Further information on the data sets and analysis is provided in Materials and Methods). We first establish the reference gene expression profiles from breast cancer cell line MCF7 upon genome-wide RNAi of a total of 3689 genes, and identified breast cancer query signatures using RNA-seq data generated from tissue samples of clinical breast cancer patients in the Cancer Genome Atlas (TCGA). Based on gene set enrichment analysis, we identify genes that when knocked down could significantly reverse the transcriptome of breast cancer state. By carrying out unbiased multifaceted assessment, including gene expression pattern characterization, survival analysis, DNA methylation, Cancer Gene Census, and chemoinformatics, we have been able to extrapolate and prioritize the most therapeutically promising targets within the gene list for further experimental work. In addition, we suggest potential alternative repurposing indications for known drugs and identify potentially novel proteins that have been poorly explored with regard to the discovery of small-molecule modulators.

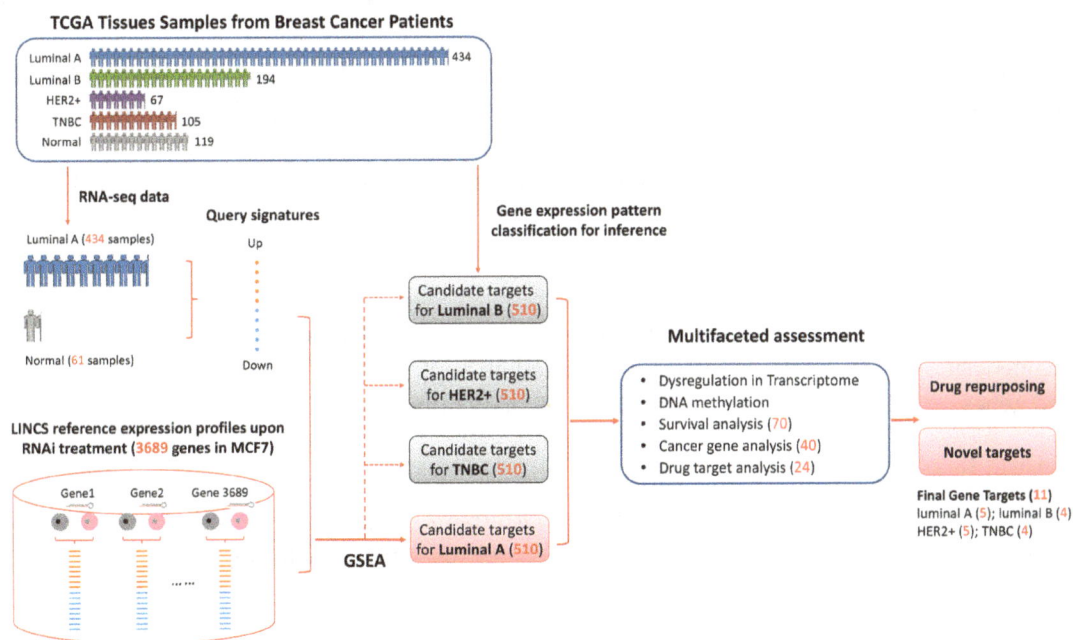

Figure 1. Workflow of candidate therapeutic target identification and multifaceted assessment. Luminal A phenotype specific signatures were calculated using TCGA gene expression data of luminal A breast cancer and corresponding normal samples. The signatures were then queried against Library of Integrated Network-Based Cellular Signatures (LINCS) MCF7 RNAi gene expression profile of 3689 genes using gene set enrichment analysis. Genes negatively connected to the phenotype were considered as candidate targets for luminal A. Then the targets were inferred to other three subtypes (i.e., luminal B, HER2+, TNBC) based on gene expression pattern analysis and further validated with transcriptome analysis, methylation analysis, cancer gene analysis and drug target analysis. Detailed information is provided in Materials and Methods.

2. Materials and Methods

2.1. Data Source

2.1.1. Gene Expression and Methylation Data

mRNA expression data and methylation data for breast cancer patients were downloaded from UCSC Cancer Genomics Browser [36]. The dataset (including TCGA_BRCA_exp_HiSeqV2-2015-02-24 for mRNA expression data and TCGA_BRCA_hMethyl27-2015-02-24, TCGA_BRCA_hMethyl450-2015-02-24 for methylation data) contains 1241 samples composed of 142 basal like, 434 luminal A, 194 luminal B, 67 HER2+, 119 normal tissue and 285 undefined under the taxonomy of PAM50 [37] for breast cancers. Of all the TCGA data, 120 samples have negative expression level of ER, PR and HER2. Samples in basal like and undefined group were selected, resulting in 105 samples as TNBC (Supplementary materials Table S1). A total of 919 samples were kept for final analysis, including 434 luminal A, 194 luminal B, 67 HER2+, 105 TNBC and 119 normal tissues.

2.1.2. Clinical Survival Data

Survival data for the breast cancer patients were downloaded from UCSC Cancer Genomics Browser [36]. Overall survival (OS) (in days) and vital status (LIVING or DECEASED) were provided in the dataset.

2.1.3. LINCS Dataset

Publicly available LINCS L1000 level three data (q2norm) were downloaded from lincscloud (http://www.lincscloud.org) with permission. Gene knockdown expression data generated from breast cancer cell line MCF7 were used for correspondence. This dataset contains 53,763 gene expression profiles involving 12,792 RNAi reagents targeting 3689 genes and 2922 untreated profiles on 165 plates as control (Supplementary Materials Table S2). The L1000 technology originally measures expressions of 978 landmark genes to infer the entire transcriptome via linear regression. Each profile in LINCS contains 22,268 probe sets, which roughly correspond to the Affymetrix Human Genome U133A Array (HG-U133A).

2.1.4. Cancer Gene Census

Cancer Gene Census [38] dataset aims to identify the mutated genes that are causally implicated in oncogenesis. The latest released dataset (accessed on Sept. 30th, 2016), which contains 602 cancer related genes roughly accounting for more than 1% of all human genes, were downloaded.

2.1.5. Drug Target

Drug target information was obtained from DrugBank 5.0 [39]. The latest released DrugBank database (accessed on Sept. 1st, 2016) contains 8206 drug entries including 1991 FDA-approved small molecule drugs, 207 FDA-approved biotech (protein/peptide) drugs, 93 nutraceuticals and over 6000 experimental drugs. In total, 4253 protein targets were included in our analysis.

2.2. Candidate Target Idenfication Pipeline

Connectivity map uses Gene Set Enrichment Analysis (GSEA) [40] to connect a signature with reference database. To generate phenotype specific gene signatures for luminal A, hierarchical clustering was first performed to exclude outlier using 1000 most-variable genes as determined by variation. Then differential analysis was conducted with R/Bioconductor package limma [41]. Phenotypic gene signature was generated by selecting top up- and down-regulated 250 genes.

To generate the reference database, the LINCS RNAi data were processed by merging the replicates of the same shRNA reagents and calculating the log fold change via subtracting the control untreated gene expression profiles. shRNA reagent with best knocking-down effect (lowest gene expression of corresponding gene) was selected to represent the silencing effect of this gene. The probe sets in LINCS data were converted to corresponding genes using annotations from HG-U133A platform and the overlapped genes were kept for GSEA of the two datasets.

The luminal A specific signature was queried against every LINCS RNAi perturbation gene expression profile using weighted-GSEA to calculate an enrichment score estimating the correlation of gene knock-down and the phenotype. The enrichment score was normalized to a connectivity score ranging from -1 to 1. Genes with connectivity score less than -0.7 were considered as candidate gene targets. These targets are used to infer for the other three breast cancer subtypes and further validated by transcriptome analysis, methylation analysis, cancer gene analysis and drug target analysis.

2.3. Differential Expression Analayis

Differential gene expression analysis for the four breast cancer subtypes was conducted with R/Bioconductor package limma [41]. Note corresponding normal sample of a specific tumor sample has sampleID with the same prefix and ending with "11" according to TCGA barcode definition (e.g., TCGA-A7-A0D9-11 is corresponding normal sample of TCGA-A7-A0D9-01). Before differential gene expression calculation, outliers were removed based on hierarchical clustering result of cancer and corresponding normal samples for each subtype. Analysis was limited to the 12,555 overlapping genes between TCGA RNA-seq data and LINCS L1000 data.

2.4. Methylation Analysis

TCGA methylation data for breast cancer includes two different arrays, the Illumina HumanMethylation27 BeadChip (27k array) and the Illumina HumanMethylation450 BeadChip (450k array). The 450k array contains 485,577 probes covering 99% of RefSeq genes, while the 27k array only contains 27,578 probes. The 450k array is an extension of 27k array and contains most probes of the latter array. Thus, only the 450k array was used for analysis. The value range −0.5 to 0.5 provided in hgHeatmap methylation data was converted it to beta value by offsetting 0.5. Beta value was further converted to M value with beta2m function in R package wateRmelon. M value was used for differential methylation analysis with R/Bioconductor package limma. Probes with False Discovery Rate (FDR) value less than 0.05 and an average beta value difference of at least 0.2 between cancer and normal group was considered as differentially methylated CpG sites. The probes were then converted to genes with annotations contained in the methylation dataset.

2.5. Survival Analysis

Breast cancer samples were classified into "high expression" and "low expression" group using median of the gene expression level as the group's cut point. Survival analysis was performed using Kaplan-Meier method [42] with R package survival. Significance of the two groups were determined with the log-rank test. Hazard ration (HR) was calculated using Cox proportional-hazards regression model implemented in coxph function of R package survival. To reduce false negative, tertile and quartile cut-off strategies which dived the samples into three and four groups respectively based on gene expression were also used. Survival analysis was only performed for the group with "highest" and "lowest" gene expression. All differences were considered statistically significant at the level of $p < 0.05$.

2.6. Gene Functional Annotation

Gene Ontology (GO) and Kyoto Encyclopedia of Genes and Genomes (KEGG) pathway enrichment analysis were performed with DAVID functional annotation tool [43]. Gene functional classification was performed with PANTHER [44]. Protein product of each gene was assigned to a single functional class.

3. Results

3.1. Characteristic Pattern of Gene Expression among Breast Cancer Subtypes

Traditionally breast cancer classification relies on the expression of three markers, i.e., ER, PR and HER2. Gene-expression profiling has been used to dissect the complexity of breast cancer and to stratify tumors into intrinsic gene-expression subtypes, associated with distinct biology, patient outcome, and genomic alterations [45]. Characterization of gene expression patterns facilitates the identification of specific signature that distinguishes each subtype.

TCGA now contains multi-omics data for 33 different tumor types, with an average of several hundred patient samples for each cancer type. Breast cancer samples are well characterized in TCGA, including genomic, transcriptomic, proteomic data for 1098 cases at the time of our analysis. Therefore, we first collected RNA-seq derived transcriptome data of breast tissue samples of 119 normal and 800 tumors from TCGA patients and analyzed the gene expression patterns with regard to the four major breast cancer subtypes, i.e., luminal A, luminal B, HER2+ and TNBC.

Hierarchical clustering was used to display the expression patterns of 1000 most-variable genes [46] in the 919 breast tissue samples. Individual dendrogram branches are colored according to the strongest correlation of the corresponding tumor with the subtype centroid already defined in TCGA. As shown in Figure 2A, samples that clustered together were generally according to its pathological classification with a few exceptions. It is obvious that luminal A samples were divided into two groups, with one half of tumors clustering near each other on the right branch of the dendrogram and the other half clustering with a majority of luminal B samples on the left branch. It is noteworthy that almost all TNBC subtype samples (green branches) that showed the strongest correlations with each other are all contained within the middle branch of the dendrogram in individual tight cluster. The HER2+ and luminal B distinction was less clear, though a certain portion of HER2+ tumor samples (grey branches) within the middle branch of the dendrogram showed the strongest correlations with each other. We also performed hierarchical clustering for all samples using the 500 luminal A signature (Supplementary Materials Figure S1), and the result was consistent with above observation.

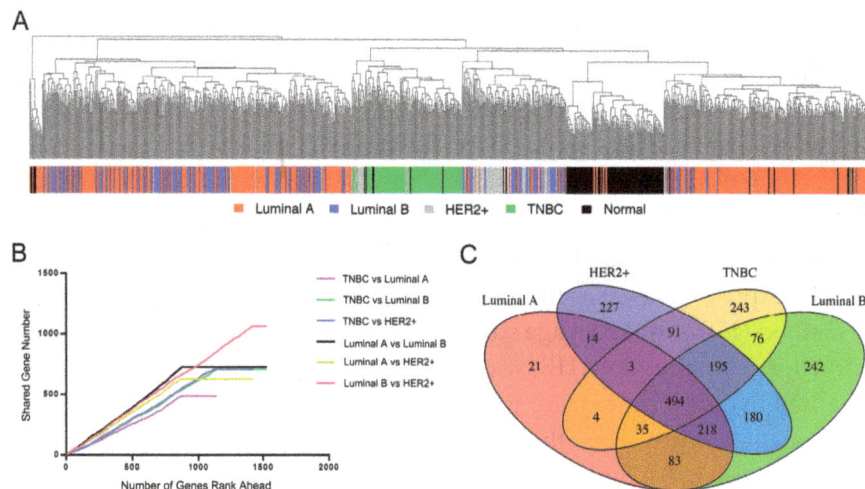

Figure 2. (**A**) Hierarchical clustering dendrogram of 919 breast samples (434 luminal A, 194 luminal B, 67 HER2+, 105 TNBC and 119 normal tissue) using 1000 most-variable genes as determined by variation; (**B**) Overlapping number of top ranking differentially expressed genes between any two cancer subtypes; (**C**) Venn diagram of differentially expressed genes for four subtypes.

After removing outliers (Supplementary Materials Figure S2–S5) for four breast cancer subtypes, we next identified the differential expression genes of each subtype using limma, calculating p value, logFC (fold change), moderated t-statistic, False Discovery Rate (FDR) value for each gene (Supplementary Materials Table S3). For simplicity, we considered genes with absolute logFC greater than 2 and FDR less than 0.05 as differentially expressed genes. As a result, the number of differentially expressed genes is 872 for luminal A, 1523 for luminal B, 1422 for HER2+ and 1141 for TNBC. We ranked the genes based on the absolute value of moderated t-statistic (larger moderated t-statistic means more differentially expressed) and calculated the shared differentially expressed genes between any two subtypes when the number of top ranking gene increases (Figure 2B). In general, any two subtypes shared approximately at least half of the differentially expressed genes (slope larger than 0.5). It was obvious that luminal A and luminal B subtype have more shared differentially expressed genes (black line with the largest slope), indicating their transcriptome similarity. Luminal B and HER2+ subtype shared the most differentially expressed genes (red line) in total, partly because their number of differentially expressed genes are the largest under our criteria. Notably, TNBC is distinctive from the other three subtypes since the curves for TNBC with other subtypes have the shallowest slopes (purple, green and blue line). These results were consistent with the gene expression patterns characterized by hierarchical clustering.

We then calculated the number of shared differentially expressed genes among four subtypes. The four subtypes shared a total of 494 differentially expressed genes, while the number of subtype-specific differentially expressed genes is 21 for luminal A, 242 for luminal B, 227 for HER2+ and 243 for TNBC (Figure 2C). This was consistent with the above analysis results and further confirmed the transcriptome similarity of luminal A with other subtypes, as well as the distinctive transcriptome of TNBC. We performed functional annotations for the TNBC-specific differentially expressed genes with DAVID. Top enriched GO BP (Biological Process) terms are cell-cell signaling, synaptic transmission, transmission of nerve impulse, neuron differentiation, etc. Top enriched KEGG pathways are glycine, serine and threonine metabolism and drug metabolism (Supplementary Materials Table S4). Collectively, the gene expression pattern analysis and further statistics characterized the extent of transcriptome similarity of four breast cancer subtypes, which provides foundation for extrapolation of predicted gene targets using genome-wide RNAi transcriptome from a single luminal A subtype.

3.2. Identifying Candidate Gene Targets for Breast Cancer Luminal A Subtype

To predict candidate targets, we followed the typical connectivity map paradigm using GSEA [40] methodology, which generally composes three steps, i.e., a reference database composed of gene-expression profiles derived from the treatment of cultured human cells with a large number of perturbagens; a list of genes as query signature, which can be obtained by differential expression analysis between disease and normal state; a pattern-matching algorithm that scores each reference profile for the direction and strength of enrichment with the query signature. Theoretically, a negative score means the genes when knocked down may reverse the cancer transcriptome to normal state, which can be used as candidate gene targets for RNAi therapy.

At the time of our analysis (Sept. 30th, 2016), LINCS L1000 contained 2922 control versus 53,763 gene expression profiles upon genome-wide RNAi treatment in a single breast cancer cell line MCF7, which were used as reference database. Pathologically, MCF7 is grouped into luminal subtype A [47]. With consideration of cell line correspondence and evaluation accuracy, we used the RNA-seq data of 434 luminal A breast cancer tissue samples from TCGA versus 61 normal samples to generate the luminal A specific signature. To compose a signature, 250 most significantly up and down-regulated genes were selected based on moderate t-statistic ranking in prior differential expression analysis (Supplementary Materials Table S5). Functional annotations of these genes found that the 500-gene luminal A specific signature was enriched in cell cycle related GO BP terms such as nuclear division (GO:0000280), mitosis (GO:0007067), M phase (GO:0000279), cell cycle phase (GO:0022403) and KEGG pathways such as cell cycle (hsa04110) (Supplementary Materials Table S4). Up-regulated genes, such as COL10A1, MMP11, NEK2, PAFAH1B3, KIF4A are highly related to breast cancer [48–52].

This signature was then used to query against the LINCS MCF7 expression profiles of 3689 genes upon RNAi treatment respectively. The pattern matching procedure resulted in a number of weighted enrichment score (ESs) ranging from −0.625 to 0.479. ES was then normalized to a connectivity score ranging from −1 to 1 [40]. Therefrom, we obtained a list of 510 genes with a connectivity score less than −0.7 as preliminary candidate targets for breast cancer luminal A subtype (Supplementary Materials Table S5). The 510 candidate genes were converted to HGNC standard gene symbols first. After removal of duplicates (SLC38A1 and SAT1), we classified the 510 genes into major functional classes with PANTHER [44] (Figure 3A, Detailed information in Supplementary Materials Table S6). In PANTHER, the 510 genes were mapped to 533 proteins. For simplicity, we assigned every gene product to a single functional class. The main functional classes represented in the candidate genes are as follows: enzyme (42%), transcription factor (17%), binding (11%), membrane receptor (8%), structural protein (5%) and transcription regulator (3%) (Figure 3A). Of all the enzymes, 56 are protein kinases, accounting for 10% of all candidate gene products. We also compared the connectivity score of genes with regard to their functional groups. Obviously, the enzyme and transcription factor group have more genes with high connectivity score (absolute value > 0.8), indicating higher potential to restore the dysregulated transcriptome at cancer state (Figure 3B).

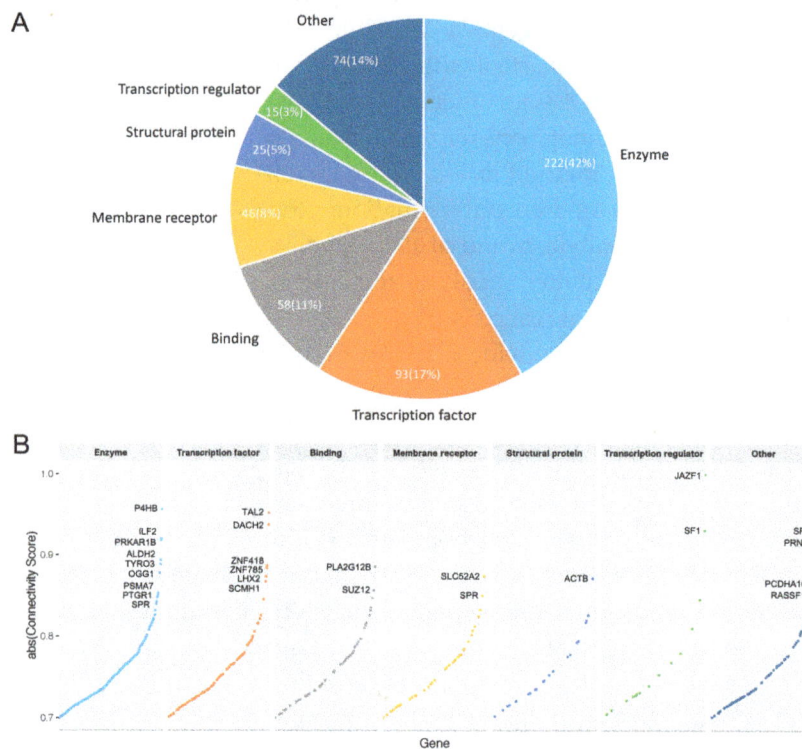

Figure 3. (**A**) Functional classes of protein products from the 510 candidate genes. Enzymes (predominantly kinases) and transcription factors constitute over half the candidate genes. A total of 14% of proteins, here labelled "Other", fall into classes including transporters, cytokines, splicing factors, kinase activator, etc.; (**B**) Absolute connectivity score of candidate genes in each functional group. Genes with absolute connectivity score above 0.85 are labeled with gene symbols.

3.3. Multifaceted Assessment for Priotizing Theraperutic Targets for Four Breast Cancer Subtypes

We hypothesized the candidate targets predicted for luminal A can be extrapolated to other breast cancer subtypes by integrating multiple filters. We then performed multifaceted assessment to analyze the biological and pathological importance of 510 candidate genes for each breast cancer subtype, from the perspectives of dysregulation in transcriptome, influence of transcriptional dysregulation on patient survival, influence of genomic abnormal methylation on gene expression, critical roles in other cancer pathogenesis, and also as novel drug targets or for repositioning purposes. In this way, we further prioritized the 510 candidate genes to screen out the targets with most potential for experimental validation.

3.3.1. Dysregulation in Transcriptome and Determinant Roles in Survival

We first evaluate the expression levels of 510 genes in transcriptomes of breast tumor samples from each breast cancer subtype. Genes with logFC above 1 and FDR value less than 0.05 were considered significantly up-regulated. We illustrate the distribution of 510 genes according to their expression levels in transcriptomes and the ratio of distribution in each subtype (Supplementary

Materials Figure S6). Notably, in the transcriptomes of tumor samples of luminal A, luminal B, HER2+ and TNBC, there were only 37, 50, 71, and 58 significantly up-regulated, and 59, 94, 85, and 74 significantly down-regulated candidate genes, respectively (Supplementary Materials Table S7). Consistently, the majority of candidate genes (69% to 81%) fall into the intermedia expression range regardless of the cancer subtype.

Since our evaluation was based on genome-wide RNAi transcriptomes, it is of great importance to unveil the impact of cancer transcriptomes on clinical prognosis. Overall survival (OS) and recurrence free survival (RFS) measured in days were included in TCGA for 800 breast cancer samples. Thus, we performed survival analysis to further explore the relationship of gene expression level of 510 candidate genes with survival time of breast cancer patients. Since gene expression is continuous, we applied three cut-off strategies to categorize the breast tumor samples into "high" and "low" expression group and used Kaplan-Meier method to analyze the censored data respectively. Genes with p value less than 0.05 and hazard ration (HR) larger than one were kept, resulting 44 genes using median cut-off strategy, 28 genes using tertile cut-off strategy and 52 genes using quartile cut-off strategy (Supplementary Materials Table S8).

Therefrom, we identified a total of 70 genes whose dysregulation in transcriptome are strongly associated with poor prognosis, especially low overall survival for breast cancer patients of all subtypes. Detailed information is provided in Supplementary Materials Table S8, and example Kaplan-Meier curves of gene MCU1 and TIAM1 are provided in Supplementary Materials Figure S7 to represent the distinctive "high" and "low" group. As shown in Supplementary Materials Figure S6 (red dots), we marked the expression level of these 70 genes for each breast cancer subtype (Supplementary Materials Table S8) and further evaluated the enrichment of these genes with high transcriptome reverse potential, i.e., absolute connectivity score above 0.8. There were only 15 poor-survival related genes with high absolute connectivity score (out of the 79 high scoring genes, hypergeometric test, $p = 0.10$). It is noteworthy that about half of the 70 poor-survival related genes were down-regulated in transcriptome (logFC < 0), which is contrary to the assumptions that patient with higher gene expression has poor prognosis. Thus, we only focused on those genes in each breast cancer subtype whose up-regulation in transcriptome are highly associated with short survival time. As a result, we found a total of 46 such genes, including 39 for luminal A, 29 for luminal B, 38 for HER2+ and 27 for TNBC, respectively. Among them, TRAPPC3, DHRS7, HYAL2, SRRT, PQBP1 and ADORA2A were of high absolute connectivity scores, and notably the latter three were valid for all subtypes (Supplementary Materials Figure S6).

3.3.2. Pathogenic Importance in Cancer and Methylation at Genomic Level

Cancer Gene Census list, which is a manually curated set of genes that have mutations or other genomic abnormalities associated with cancer and are likely to be causative, as identified from genetic studies [38]. This 602-gene data set exemplifies large gene lists that have been generated from initial experimental validation. Therefrom, we found 40 out of the 510 candidate genes playing roles in specific cancers, and among them 5 genes, i.e., ALDH2, BTG1, TNFRSF14, MUC1 and STAT6, whose high expression in cancer transcriptome are associated with poor survival. Using annotations from Cancer Gene Census, 34 genes were related to breast cancer, and 4 genes, i.e., PBRM1, TP53, AKT1, CDKN1B, were in our candidate gene list.

DNA methylation of tumor suppressor genes has been the focus of numerous studies that have aimed to identify DNA methylation biomarkers of cancer [53]. Meanwhile, it is becoming clear that hypomethylation is equally important a driving force in breast cancer metastasis [54,55]. Thus, we further investigated DNA methylation status of the 23,423 genes in tumor samples of each breast cancer subtype, with the aim to identify genes that undergo significantly differential methylation at cancer state. Using TCGA methylation data generated from Illumina HumanMethylation450 BeadChip platform, we performed differential methylation analysis for 288 luminal A, 127 luminal B, 31 HER2+, and 59 TNBC tumor samples as well as 87 normal breast tissue samples. We observed a bimodal distribution of the calculated beta values, with two peaks around 0.1 and 0.9 and a relatively flat valley around 0.2–0.8 (Supplementary Materials Figure S8). In differential methylation analysis, Beta-value has a more intuitive biological interpretation, but the M-value is more statistically valid for the differential analysis of methylation levels [56]. As recommended, we used M-value to calculate differential methylation positions (DMPs) with limma. The number of normal samples was 33 for luminal A, 18 for luminal B, 6 for HER2+ and 5 for TNBC, respectively. Probes with FDR value less than 0.05 and an average Beta-value difference of 0.2 between cancer and normal samples were considered as differentially methylated probes (Supplementary Materials Table S9).

As a result, the number of significantly methylated probes (genes) was 18,747 (6909) for luminal A, 32,558 (9314) for luminal B, 20,494 (7197) for HER2+ and 6469 (3172) for TNBC. After annotation, we identified from the 510 preliminary candidate genes a set of hypermethylated and hypomethylated genes for each subtype respectively, i.e., 104 and 70 for luminal A, 128 and 111 for luminal B, 96 and 73 for HER2+, 30 and 52 for TNBC. We then focused on the hypomethylated genes that were significantly up-regulated at cancer state and calculated the enrichment of these genes with high transcriptome reverse potential (i.e., absolute connectivity score above 0.8) and/or associated with poor survival. The number of hypomethylated genes that were significantly up-regulated at cancer state was 35 for luminal A, 48 for luminal B, 41 for HER2+, 34 for TNBC. Out of the 79 genes with high transcriptome reverse potential, the number of genes that satisfy the two conditions was 4 for luminal A (TAL2, OPN3, IL20, ARID3A, $p = 0.82$), 5 for luminal B (TAL2, UMODL1, OPN3, IL20, ARID3A, $p = 0.90$), 5 for HER2+ (TAL2, OPN3, IL20, CPT1A, ARID3A, hypergeometric test, $p = 0.79$) and 4 for TNBC (UMODL1, OPN3, LDHB, ARID3A, hypergeometric test, $p = 0.80$). Out of the 70 genes associated with poor survival, the number of genes that satisfy the two conditions was 5 for luminal A (MUC1, HLA-DRA, WNT7B, XBP1, EFCAB2, hypergeometric test, $p = 0.54$), 4 for luminal B (ATG16L2, MUC1, WNT7B, XBP1, $p = 0.92$), 5 for HER2+ (C1QTNF6, NDUFS6, MUC1, HLA-DRA, XBP1, hypergeometric test, $p = 0.69$) and 4 for TNBC (CHERP, MUC1, TIAM1, GABRP, hypergeometric test, $p = 0.71$).

3.3.3. Druggable Candidate Genes

To further evaluate the potential of candidate genes as novel targets as well as drug repositioning opportunities, we have identified proteins from the 510 candidate genes that are themselves targets of FDA-approved drugs or experimental compounds. These proteins, examples of drugs and associated therapeutic indications are detailed in Supplementary Materials Table S10. We have found that a total of 80 candidate proteins are targets of FDA-approved drugs and 47 are targets of experimentally validated active compounds (nutraceuticals excluded). When limited to targets of the 147 approved anti-neoplastic drugs (ATC code L01) in DrugBank (version 5.0), a total of 24 gene products currently have small-molecule drugs indicated for cancer therapy. We further calculated the druggable proteins from those of the 70 candidate genes that affect survival of breast cancer patients, and found 17 were targets of known drugs. For example, ICAM1 is targeted by natalizumab, which is approved for treatment of multiple sclerosis. ADORA2A is targeted by multiple drugs such as oxtriphylline for treatment of the symptoms of asthma, bronchitis, COPD, and emphysema. Among these genes, only proteins of two candidate gene FPGS and POLE2 currently have small-molecule drugs indicated for cancer therapy, i.e., raltitrexed for malignant neoplasm of colon and rectum and cladribine for lymphoproliferative diseases such as hairy cell leukemia, Non-Hodgkin's lymphoma, etc. Together, these results indicate great repositioning potential of candidate genes as drug targets for anti-neoplastic purposes in breast cancer.

3.3.4. Prioritized Therapeutic Targets for Four Breast Cancer Subtypes

Finally, combining all of the above assessment, we obtained a small list of 11 genes, i.e., MUC1, HLA-DRA, WNT7B, XBP1, EFCAB2, ATG16L2, C1QTNF6, NDUFS6, CHERP, TIAM1 and GABRP, as the most potential targets for four breast cancer subtypes (Table 1). The number of final candidate targets for each subtype is 5 for luminal A, 4 for luminal B, 5 for HER2+ and 4 for TNBC (Figure 4, Supplementary Materials Table S11). These genes show averagely high transcriptome reverse potential (absolute connectivity score > 0.7) when knocked down by RNAi reagents, indicating feasibility as targets for developing RNAi therapeutics. Meanwhile, these genes were verified by genome and transcriptome data from tumors samples of clinical breast cancer patients as significantly hypo-methylated at genomic level and transcriptionally up-regulated, despite their distinct dysregulation levels in each subtype. Notably, we found MUC1 a commonly valid target for all four subtypes (Supplementary Materials Figure S9), XBP1 for four subtypes except for TNBC, WNT7B for the two luminal subtypes, and HLA-DRA for luminal A and HER2+. Most importantly, there are several genes that can distinguish the four breast cancer subtype as specific therapeutic targets for each subtype, including EFCAB2 for luminal A, ATG16L2 for luminal B, C1QTNF6 and NDUFS6 for HER2+, as well as CHERP, TIAM1, and GABRP for TNBC (Table 1). Among them, only GABRP has known drug targeting for the treatment of diseases such as insomnia and epilepsy, and this leaves potential for drug repurposing.

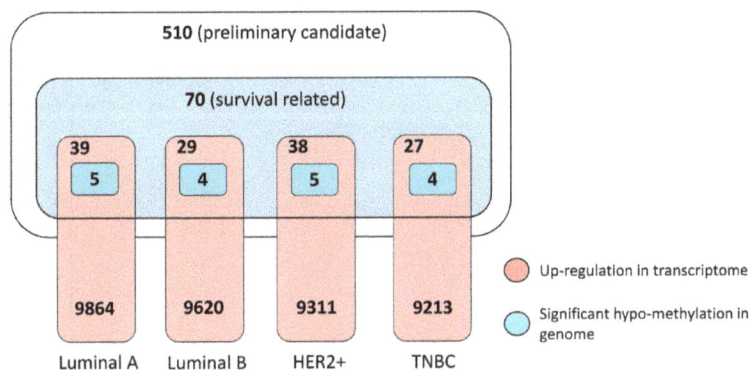

Figure 4. Statistical summarization of final candidate targets for four breast cancer subtypes from multifaceted assessment.

Obviously, most of the common targets are highly related to breast cancer, and their pathogenic importance in breast cancer has been experimentally validated. For example, MUC1 encodes glycoprotein Mucin 1 with extensive O-linked glycosylation of its extracellular domain and overexpression of MUC1 is often associated with colon, breast, ovarian, lung and pancreatic cancers [57]. It is a multifaceted oncoprotein which promotes growth, metastasis, and resistance to drugs in cancer [58]. Immune responses to MUC1 have been seen in breast and ovarian cancer patients and clinical studies have been initiated to evaluate the use of antibodies to MUC1 and of immunogens based on MUC1 for immunotherapy of breast cancer patients [59,60]. XBP1 functions as a transcription factor during endoplasmic reticulum (ER) stress by regulating the unfolded protein response (UPR). XBP1 is activated in TNBC and has a pivotal role in the tumorigenicity and progression of this human breast cancer subtype by controlling HIF1α pathway. In breast cancer cell line models, depletion of XBP1 inhibited tumor growth and tumor relapse [61]. WNT7B encoded Wnt7b is a Wnt ligand that has been demonstrated to play critical roles in several developmental processes. Myeloid WNT7b mediates the angiogenic switch and metastasis in breast cancer, and therapeutic suppression of WNT7B signaling might be advantageous due to targeting multiple aspects of tumor progression [62]. HLA-DRA encodes HLA class II histocompatibility antigen, DR alpha chain. It is part of the HLA class II molecule which is expressed in antigen presenting cells (APC) and plays a central role in the immune system by presenting peptides derived from extracellular proteins. This protein is generally invariable, yet research showed HLA-DRA was highly overexpressed in ovarian cancer, perhaps as a result of inflammatory events in the tumor microenvironment. The tumor cells may have compensatory mechanisms to reduce the production of functional MHC class II molecules, thus reducing immunogenicity and favoring tumor growth [63].

Table 1. Final candidate gene targets for four breast cancer subtypes.

Gene Symbol	Gene Name	Luminal A		Luminal B		HER2+		TNBC		Protein Name	No. of Targeted Drugs
		CS	logFC	CS	logFC	CS	logFC	CS	logFC		
MUC1	mucin 1, cell surface associated	−0.75	3.11	−0.75	1.97	−0.75	2.33	−0.75	0.43	Mucin-1	0
HLA-DRA	major histocompatibility complex, class II, DR alpha	−0.73	0.31	–	–	−0.73	0.8	–	–	HLA class II histocompatibility antigen, DR alpha chain	0
WNT7B	Wnt family member 7B	−0.73	2.14	−0.73	1.96	–	–	–	–	Protein Wnt-7b	0
XBP1	X-box binding protein 1	−0.72	1.58	−0.72	1.68	−0.72	0.94	–	–	X-box-binding protein 1	0
EFCAB2	EF-hand calcium binding domain 2	−0.72	0.08	–	–	–	–	–	–	EF-hand calcium-binding domain-containing protein 2	0
ATG16L2	autophagy related 16 like 2	–	–	−0.77	0.19	–	–	–	–	Autophagy-related protein 16-2	0
C1QTNF6	C1q and tumor necrosis factor related protein 6	–	–	–	–	−0.77	2.23	–	–	Complement C1q tumor necrosis factor-related protein 6	0
NDUFS6	NADH: ubiquinone oxidoreductase subunit S6	–	–	–	–	−0.77	1.34	–	–	NADH dehydrogenase [ubiquinone] iron-sulfur protein 6, mitochondrial	1
CHERP	calcium homeostasis endoplasmic reticulum protein	–	–	–	–	–	–	−0.77	0.46	Calcium homeostasis endoplasmic reticulum protein	0
TIAM1	T-cell lymphoma invasion and metastasis 1	–	–	–	–	–	–	−0.71	0.68	T-lymphoma invasion and metastasis-inducing protein 1	0
GABRP	gamma-aminobutyric acid type A receptor pi subunit	–	–	–	–	–	–	−0.7	1.5	Gamma-aminobutyric acid receptor subunit pi	41

Note: CS means connectivity score; FC means fold change; "–" means the corresponding values do not reach the thresholds that define statistical significance.

With regard to the specific targets predicted to TNBC, TIAM1 encodes Tiam1 (T-lymphoma invasion and metastasis 1), one of the known guanine nucleotide (GDP/GTP) exchange factors (GEFs) for Rho GTPases (e.g., Rac1) and is expressed in breast tumor cells (e.g., SP-1 cell line). Research showed ankyrin-Tiam1 interaction plays a pivotal role in regulating Rac1 signaling and cytoskeleton function required for oncogenic signaling and metastatic breast tumor cell progression [64]. GABRP encodes Gamma-aminobutyric acid receptor subunit pi, which is a component of a cell-surface receptor. Research showed GABRP stimulates basal-like breast cancer cell/ triple negative subtype migration through activation of extracellular regulated kinase 1/2 (ERK1/2). In addition, silencing GABRP in BLBC cells decreases migration, BLBC-associated cytokeratins and ERK1/2 activation [65].

4. Discussion

Ongoing large-scale loss of function studies like LINCS genome-wide RNAi screens combined with gene expression profiles are now producing tons of data that may offer limitless opportunities to discover the driving force underlying cancer pathogenesis. In this study, we systematically evaluated the transcriptome reverse potential of 3689 genes for breast cancer using LINCS profiling data from genome-wide RNAi perturbagens, resulting in a preliminary list of 510 candidate genes with absolute connectivity score above 0.7. Considering the inherent noise in LINCS L1000 data, we used this relative loose threshold of connectivity score to screen out the candidates, which is supposed to potentially reduce the false negatives. Altogether, this transcriptional bioinformatics method is more rational and straightforward since it evaluates the real cellular states when genes are knockdown without being confused by the uncertain inner interactions from simply inferring from the most dysregulated genes, thus hoping to provide more insights than existing methods.

In cancer research, cancer cell lines have long been used as experimental models because they generally carry the genomic, transcriptomic and proteomic characteristics of the primary tumor from which they were derived. Tumor is classified by stage and grade in clinic, since cancer cell lines that acquire indefinite growth tend to be high stage and poorly differentiated tumors, they are not entirely capable of representing the clinical spectrum of cancers at that site [66]. Thus, a critical issue is to match cell line to corresponding disease when computationally compare the expression profiles generated from clinical tissue samples of certain disease and those from the LINCS cell line. In an idea way, our reference signatures should also include those generated from breast cancer cell lines of different breast cancer subtypes other than luminal A. However, due to the limited breast cancer cell line used in LINCS project, we performed computational evaluation based on transcriptome similarity analysis of TCGA RNA-seq data generated from clinical breast tumor samples of different subtypes, and therefrom identification of the query signature of breast cancer luminal A subtype, which is in complete congruent with the subtype classification of all the LINCS reference signatures. As the gene expression pattern of luminal A is less distinctive and heterogeneous than other subtypes, we are able to extrapolate and prioritize the candidate targets for each other subtype by further integrating multifaceted assessment.

After preliminary identification, objective prioritization to a more manageable gene list is essential, as detailed biological validation of each individual target is a challenging, long and expensive process. Our multifaceted assessment takes into account of essential factors of both pathogenic importance and druggability. We have shown that, by integrating information from different large-scale research initiatives (comprising information on dysregulation in transcriptome that affect patient survival, in other cancer types, and in DNA methylation), we are able to effectively annotate a biologically and pathogenically important gene list containing potential targets for RNAi therapies. Further, by integrating information from protein functional class as targets of approved drugs and experimental active small molecules, we have also identified targets for repurposing known drugs or active chemical compounds that can be tested for activity in breast cancer models. Of particular interest for new drug discovery, we have identified 11 proteins in the final list (as highlighted in Table 1, with specific examples discussed above) that need to be examined carefully. In addition to having biological and pathogenic importance, they lack chemical compounds to modulate them, representing potential novel biological targets for chemical exploitation. Besides RNAi therapy, these targets can be addressed with small-molecule compounds through two strategies: by investing in medicinal chemistry approaches to expand the boundaries of druggability; or via the identification of alternative druggable targets within the pathway or subnetwork in question.

5. Conclusions

Collectively, this global, systematic transcriptome-based evaluation and objective, multidisciplinary computational assessment presented here allows for the effective, unbiased and data-driven identification and prioritization of large, biologically compelling gene lists for the purpose of RNAi therapy or drug discovery. This approach, applied here to the LINCS breast cancer genome-wide RNAi profiles an exemplar, can be adapted to any other cell line based LINCS profiles under different types of perturbagens, e.g., over expression vectors and CRISPR/Cas9 system. Meanwhile, the following multifaceted assessment can also be customized for specific purposes using annotations of interest that emerge from large-scale omics studies, functional screening initiatives or from any therapeutic area. In this way, researchers can find more reasonable driving force behind breast cancer, and significantly reduce the time of therapeutic target identification and risk in following preclinical drug development.

Supplementary Materials: Table S1: This table gives the detailed clinical information of 434 luminal A samples for TCGA breast cancer, Table S2: This table gives the detailed information of LINCS MCF7 RNAi experiments data and control untreated data, Table S3: This table gives the differential expression analysis result of limma for luminal subtype A, Table S4: KEGG pathway analysis result of luminal A signature genes with DAVID, Table S5: This sheet contains luminal A signature for GSEA calculation, i.e., top 250 up- and down-regulated genes for luminal A subtype, Table S6: Functional classification of 510 candidate genes with PANTHER, Table S7: Transcriptome analysis of 510 candidate genes for TNBC, Table S8: Information of 70 survival related genes for TNBC, Table S9: This sheet contains methylation analysis result for TNBC (gene centric), Table S10: This sheet contains DrugBank drugs which target the 70 survival related genes predicted with our pipeline, Table S11: Final candidate targets for HER2+, Figure S1: Hierarchical clustering result of 919 breast samples (434 luminal A, 194 luminal B, 67 HER2+, 105 TNBC and 119 normal tissue) using 500 luminal A specific signature, Figure S2: Hierarchical clustering result of 434 luminal A breast samples and corresponding normal samples using 1000 most-variable genes as determined by variation, Figure S3: Hierarchical clustering result of luminal B breast samples and corresponding normal samples using 1000 most-variable genes as determined by variation, Figure S4: Hierarchical clustering result of HER2+ breast samples and corresponding normal samples using 1000 most-variable genes as determined by variation, Figure S5: Hierarchical clustering result of TNBC samples and corresponding normal samples using 1000 most-variable genes as determined by variation, Figure S6: Distribution of logFC (log2-fold change) values of 510 candidate genes in transcriptome of (A) Luminal A, (B) Luminal B, (C) HER2+, (D) TNBC breast tumor samples, Figure S7: Kaplan–Meier curve of selected genes (A) MCU1 with p value 0.018 and HR 2.06. (B) TIAM1 with p value 0.022 and HR 1.57, Figure S8: Bimodal distribution of the calculated beta values, with two peaks around 0.1 and 0.9 and a relatively flat valley around 0.2–0.8, Figure S9: Venn diagram of final candidate gene targets for four breast cancer subtypes.

Acknowledgments: This work was supported by grants from Major Research Plan of National Natural Science Foundation of China (No. U1435222), National High Technology Research and Development Program of China (No. 2015AA020108), Key New Drug Discovery Project of 12th Five-Years Plan (2013ZX09102051), the Ministry of Science and Technology of the People's Republic of China and Key Science and Technology Project of Huzhou City (2013KC02).

Author Contributions: Yang Liu and Hui Bai designed the research; Xiaoyao Yin and Jing Zhong analyzed the data; Naiyang Guan, Zhigang Luo, Lishan Min and Xing Yao provided advice for the analysis; Yang Liu, Hui Bai and Xiaochen Bo wrote the manuscript; Licheng Dai and Hui Bui revised the paper and supervised the entire study.

Abbreviations

The following abbreviations are used in this manuscript:

TNBC   Triple Negative Breast Cancer
LINCS   Library of Integrated Network-based Cellular Signatures
GSEA   Gene Set Enrichment Analysis
TCGA   The Cancer Genomics Atlas

References

1. Siegel, R.L.; Miller, K.D.; Jemal, A. Cancer statistics, 2016. *CA Cancer J. Clin.* **2016**, *66*, 7–30. [CrossRef] [PubMed]
2. Torre, L.A.; Bray, F.; Siegel, R.L.; Ferlay, J.; Lortet-Tieulent, J.; Jemal, A. Global cancer statistics, 2012. *CA Cancer J. Clin.* **2015**, *65*, 87–108. [CrossRef] [PubMed]
3. Kalimutho, M.; Parsons, K.; Mittal, D.; Lopez, J.A.; Srihari, S.; Khanna, K.K. Targeted therapies for triple-negative breast cancer: Combating a stubborn disease. *Trends Pharmacol. Sci.* **2015**, *36*, 822–846. [CrossRef] [PubMed]
4. Foulkes, W.D.; Smith, I.E.; Reis-Filho, J.S. Triple-negative breast cancer. *N. Engl. J. Med.* **2010**, *363*, 1938–1948. [CrossRef] [PubMed]
5. Carey, L.; Winer, E.; Viale, G.; Cameron, D.; Gianni, L. Triple-negative breast cancer: Disease entity or title of convenience? *Nat. Rev. Clin. Oncol.* **2010**, *7*, 683–692. [CrossRef] [PubMed]
6. Perou, C.M.; Sorlie, T.; Eisen, M.B.; van de Rijn, M.; Jeffrey, S.S.; Rees, C.A.; Pollack, J.R.; Ross, D.T.; Johnsen, H.; Akslen, L.A.; et al. Molecular portraits of human breast tumours. *Nature* **2000**, *406*, 747–752. [CrossRef] [PubMed]
7. Liedtke, C.; Mazouni, C.; Hess, K.R.; Andre, F.; Tordai, A.; Mejia, J.A.; Symmans, W.F.; Gonzalez-Angulo, A.M.; Hennessy, B.; Green, M.; et al. Response to neoadjuvant therapy and long-term survival in patients with triple-negative breast cancer. *J. Clin. Oncol.* **2008**, *26*, 1275–1281. [CrossRef] [PubMed]
8. Carey, L.A.; Dees, E.C.; Sawyer, L.; Gatti, L.; Moore, D.T.; Collichio, F.; Ollila, D.W.; Sartor, C.I.; Graham, M.L.; Perou, C.M. The triple negative paradox: Primary tumor chemosensitivity of breast cancer subtypes. *Clin. Cancer Res.* **2007**, *13*, 2329–2334. [CrossRef] [PubMed]
9. von Minckwitz, G.; Untch, M.; Blohmer, J.U.; Costa, S.D.; Eidtmann, H.; Fasching, P.A.; Gerber, B.; Eiermann, W.; Hilfrich, J.; Huober, J.; et al. Definition and impact of pathologic complete response on prognosis after neoadjuvant chemotherapy in various intrinsic breast cancer subtypes. *J. Clin. Oncol.* **2012**, *30*, 1796–1804. [CrossRef] [PubMed]
10. Xu, H.; Eirew, P.; Mullally, S.C.; Aparicio, S. The omics of triple-negative breast cancers. *Clin. Chem.* **2014**, *60*, 122–133. [CrossRef] [PubMed]
11. Speers, C.; Tsimelzon, A.; Sexton, K.; Herrick, A.M.; Gutierrez, C.; Culhane, A.; Quackenbush, J.; Hilsenbeck, S.; Chang, J.; Brown, P. Identification of novel kinase targets for the treatment of estrogen receptor-negative breast cancer. *Clin. Cancer Res.* **2009**, *15*, 6327–6340. [CrossRef] [PubMed]
12. Hartman, Z.C.; Poage, G.M.; den Hollander, P.; Tsimelzon, A.; Hill, J.; Panupinthu, N.; Zhang, Y.; Mazumdar, A.; Hilsenbeck, S.G.; Mills, G.B.; et al. Growth of triple-negative breast cancer cells relies upon coordinate autocrine expression of the proinflammatory cytokines IL-6 and IL-8. *Cancer Res.* **2013**, *73*, 3470–3480. [CrossRef] [PubMed]
13. Al-Ejeh, F.; Simpson, P.T.; Sanus, J.M.; Klein, K.; Kalimutho, M.; Shi, W.; Miranda, M.; Kutasovic, J.; Raghavendra, A.; Madore, J.; et al. Meta-analysis of the global gene expression profile of triple-negative breast cancer identifies genes for the prognostication and treatment of aggressive breast cancer. *Oncogenesis* **2014**, *3*, e100. [CrossRef] [PubMed]
14. Lawrence, R.T.; Perez, E.M.; Hernandez, D.; Miller, C.P.; Haas, K.M.; Irie, H.Y.; Lee, S.I.; Blau, C.A.; Villen, J. The proteomic landscape of triple-negative breast cancer. *Cell Rep.* **2015**, *11*, 630–644. [CrossRef] [PubMed]
15. Osmanbeyoglu, H.U.; Pelossof, R.; Bromberg, J.F.; Leslie, C.S. Linking signaling pathways to transcriptional programs in breast cancer. *Genome Res.* **2014**, *24*, 1869–1880. [CrossRef] [PubMed]

16. Song, H.; Hedayati, M.; Hobbs, R.F.; Shao, C.; Bruchertseifer, F.; Morgenstern, A.; Deweese, T.L.; Sgouros, G. Targeting aberrant DNA double-strand break repair in triple-negative breast cancer with alpha-particle emitter radiolabeled anti-egfr antibody. *Mol. Cancer Ther.* **2013**, *12*, 2043–2054. [CrossRef] [PubMed]

17. Ueno, N.T.; Zhang, D. Targeting egfr in triple negative breast cancer. *J. Cancer* **2011**, *2*, 324–328. [CrossRef] [PubMed]

18. Dent, S.F. The role of vegf in triple-negative breast cancer: Where do we go from here? *Ann. Oncol.* **2009**, *20*, 1615–1617. [CrossRef] [PubMed]

19. Furth, P.A. Stat signaling in different breast cancer sub-types. *Mol. Cell. Endocrinol.* **2014**, *382*, 612–615. [CrossRef] [PubMed]

20. Gordon, V.; Banerji, S. Molecular pathways: Pi3k pathway targets in triple-negative breast cancers. *Clin. Cancer Res.* **2013**, *19*, 3738–3744. [CrossRef] [PubMed]

21. Yunokawa, M.; Koizumi, F.; Kitamura, Y.; Katanasaka, Y.; Okamoto, N.; Kodaira, M.; Yonemori, K.; Shimizu, C.; Ando, M.; Masutomi, K.; et al. Efficacy of everolimus, a novel mtor inhibitor, against basal-like triple-negative breast cancer cells. *Cancer Sci.* **2012**, *103*, 1665–1671. [CrossRef] [PubMed]

22. Chin, Y.R.; Yoshida, T.; Marusyk, A.; Beck, A.H.; Polyak, K.; Toker, A. Targeting akt3 signaling in triple-negative breast cancer. *Cancer Res.* **2014**, *74*, 964–973. [CrossRef] [PubMed]

23. Montero, J.C.; Esparis-Ogando, A.; Re-Louhau, M.F.; Seoane, S.; Abad, M.; Calero, R.; Ocana, A.; Pandiella, A. Active kinase profiling, genetic and pharmacological data define mtor as an important common target in triple-negative breast cancer. *Oncogene* **2014**, *33*, 148–156. [CrossRef] [PubMed]

24. Murai, J.; Huang, S.Y.; Das, B.B.; Renaud, A.; Zhang, Y.; Doroshow, J.H.; Ji, J.; Takeda, S.; Pommier, Y. Trapping of PARP1 and PARP2 by clinical PARP inhibitors. *Cancer Res.* **2012**, *72*, 5588–5599. [CrossRef] [PubMed]

25. Johnson, N.; Johnson, S.F.; Yao, W.; Li, Y.C.; Choi, Y.E.; Bernhardy, A.J.; Wang, Y.; Capelletti, M.; Sarosiek, K.A.; Moreau, L.A.; et al. Stabilization of mutant brca1 protein confers parp inhibitor and platinum resistance. *Proc. Natl. Acad. Sci. USA* **2013**, *110*, 17041–17046. [CrossRef] [PubMed]

26. Crown, J.; O'Shaughnessy, J.; Gullo, G. Emerging targeted therapies in triple-negative breast cancer. *Ann. Oncol.* **2012**, *23*, vi56–vi65. [CrossRef] [PubMed]

27. Jamdade, V.S.; Sethi, N.; Mundhe, N.A.; Kumar, P.; Lahkar, M.; Sinha, N. Therapeutic targets of triple-negative breast cancer: A review. *Br. J. Pharmacol.* **2015**, *172*, 4228–4237. [CrossRef] [PubMed]

28. Saha, P.; Nanda, R. Concepts and targets in triple-negative breast cancer: Recent results and clinical implications. *Ther. Adv. Med. Oncol.* **2016**, *8*, 351–359. [CrossRef] [PubMed]

29. Lin, A.; Li, C.; Xing, Z.; Hu, Q.; Liang, K.; Han, L.; Wang, C.; Hawke, D.H.; Wang, S.; Zhang, Y.; et al. The link-a lncrna activates normoxic hif1alpha signalling in triple-negative breast cancer. *Nat. Cell Biol.* **2016**, *18*, 213–224. [CrossRef] [PubMed]

30. Lehmann, B.D.; Pietenpol, J.A.; Tan, A.R. Triple-negative breast cancer: Molecular subtypes and new targets for therapy. *Am. Soc. Clin. Oncol. Educ. Book* **2015**. [CrossRef] [PubMed]

31. Wu, S.Y.; Lopez-Berestein, G.; Calin, G.A.; Sood, A.K. RNAi therapies: Drugging the undruggable. *Sci. Transl. Med.* **2014**. [CrossRef] [PubMed]

32. Mansoori, B.; Sandoghchian Shotorbani, S.; Baradaran, B. RNA interference and its role in cancer therapy. *Adv. Pharm. Bull.* **2014**, *4*, 313–321. [PubMed]

33. Abdelrahim, M.; Safe, S.; Baker, C.; Abudayyeh, A. RNAi and cancer: Implications and applications. *J. RNAi Gene Silencing* **2006**, *2*, 136–145. [PubMed]

34. Trevino, V.; Falciani, F.; Barrera-Saldana, H.A. DNA microarrays: A powerful genomic tool for biomedical and clinical research. *Mol. Med.* **2007**, *13*, 527–541. [CrossRef] [PubMed]

35. Peck, D.; Crawford, E.D.; Ross, K.N.; Stegmaier, K.; Golub, T.R.; Lamb, J. A method for high-throughput gene expression signature analysis. *Genome Biol.* **2006**. [CrossRef] [PubMed]

36. Zhu, J.; Sanborn, J.Z.; Benz, S.; Szeto, C.; Hsu, F.; Kuhn, R.M.; Karolchik, D.; Archie, J.; Lenburg, M.E.; Esserman, L.J.; et al. The ucsc cancer genomics browser. *Nat. Methods* **2009**, *6*, 239–240. [CrossRef] [PubMed]

37. Parker, J.S.; Mullins, M.; Cheang, M.C.; Leung, S.; Voduc, D.; Vickery, T.; Davies, S.; Fauron, C.; He, X.; Hu, Z.; et al. Supervised risk predictor of breast cancer based on intrinsic subtypes. *J. Clin. Oncol.* **2009**, *27*, 1160–1167. [CrossRef] [PubMed]

38. Futreal, P.A.; Coin, L.; Marshall, M.; Down, T.; Hubbard, T.; Wooster, R.; Rahman, N.; Stratton, M.R. A census of human cancer genes. *Nat. Rev. Cancer* **2004**, *4*, 177–183. [CrossRef] [PubMed]

39. Wishart, D.S.; Knox, C.; Guo, A.C.; Shrivastava, S.; Hassanali, M.; Stothard, P.; Chang, Z.; Woolsey, J. Drugbank: A comprehensive resource for in silico drug discovery and exploration. *Nucleic Acids Res.* **2006**, *34*, D668–D672. [CrossRef] [PubMed]

40. Subramanian, A.; Tamayo, P.; Mootha, V.K.; Mukherjee, S.; Ebert, B.L.; Gillette, M.A.; Paulovich, A.; Pomeroy, S.L.; Golub, T.R.; Lander, E.S.; et al. Gene set enrichment analysis: A knowledge-based approach for interpreting genome-wide expression profiles. *Proc. Natl. Acad. Sci. USA* **2005**, *102*, 15545–15550. [CrossRef] [PubMed]

41. Ritchie, M.E.; Phipson, B.; Wu, D.; Hu, Y.; Law, C.W.; Shi, W.; Smyth, G.K. Limma powers differential expression analyses for rna-sequencing and microarray studies. *Nucleic Acids Res.* **2015**. [CrossRef] [PubMed]

42. Kaplan, E.L.; Meier, P. Nonparametric estimation from incomplete observations. *J. Am. Stat. Assoc.* **1958**, *53*, 457–481. [CrossRef]

43. Huang da, W.; Sherman, B.T.; Lempicki, R.A. Systematic and integrative analysis of large gene lists using david bioinformatics resources. *Nat. Protoc.* **2009**, *4*, 44–57. [CrossRef] [PubMed]

44. Mi, H.; Poudel, S.; Muruganujan, A.; Casagrande, J.T.; Thomas, P.D. Panther version 10: Expanded protein families and functions, and analysis tools. *Nucleic Acids Res.* **2016**, *44*, D336–D342. [CrossRef] [PubMed]

45. Sorlie, T.; Tibshirani, R.; Parker, J.; Hastie, T.; Marron, J.S.; Nobel, A.; Deng, S.; Johnsen, H.; Pesich, R.; Geisler, S.; et al. Repeated observation of breast tumor subtypes in independent gene expression data sets. *Proc. Natl. Acad. Sci. USA* **2003**, *100*, 8418–8423. [CrossRef] [PubMed]

46. Klijn, C.; Durinck, S.; Stawiski, E.W.; Haverty, P.M.; Jiang, Z.; Liu, H.; Degenhardt, J.; Mayba, O.; Gnad, F.; Liu, J.; et al. A comprehensive transcriptional portrait of human cancer cell lines. *Nat. Biotechnol.* **2015**, *33*, 306–312. [CrossRef] [PubMed]

47. Holliday, D.L.; Speirs, V. Choosing the right cell line for breast cancer research. *Breast Cancer Res.* **2011**, *13*, 215. [CrossRef] [PubMed]

48. Chapman, K.B.; Prendes, M.J.; Sternberg, H.; Kidd, J.L.; Funk, W.D.; Wagner, J.; West, M.D. COL10A1 expression is elevated in diverse solid tumor types and is associated with tumor vasculature. *Future Oncol.* **2012**, *8*, 1031–1040. [CrossRef] [PubMed]

49. Cheng, C.W.; Yu, J.C.; Wang, H.W.; Huang, C.S.; Shieh, J.C.; Fu, Y.P.; Chang, C.W.; Wu, P.E.; Shen, C.Y. The clinical implications of MMP-11 and CK-20 expression in human breast cancer. *Clin. Chim. Acta* **2010**, *411*, 234–241. [CrossRef] [PubMed]

50. Cappello, P.; Blaser, H.; Gorrini, C.; Lin, D.C.; Elia, A.J.; Wakeham, A.; Haider, S.; Boutros, P.C.; Mason, J.M.; Miller, N.A.; et al. Role of Nek2 on centrosome duplication and aneuploidy in breast cancer cells. *Oncogene* **2014**, *33*, 2375–2384. [CrossRef] [PubMed]

51. Mulvihill, M.M.; Benjamin, D.I.; Ji, X.; Le Scolan, E.; Louie, S.M.; Shieh, A.; Green, M.; Narasimhalu, T.; Morris, P.J.; Luo, K.; et al. Metabolic profiling reveals PAFAH1B3 as a critical driver of breast cancer pathogenicity. *Chem. Biol.* **2014**, *21*, 831–840. [CrossRef] [PubMed]

52. Wang, H.; Lu, C.; Li, Q.; Xie, J.; Chen, T.; Tan, Y.; Wu, C.; Jiang, J. The role of kif4a in doxorubicin-induced apoptosis in breast cancer cells. *Mol. Cells* **2014**, *37*, 812–818. [CrossRef] [PubMed]

53. Mikeska, T.; Craig, J.M. DNA methylation biomarkers: Cancer and beyond. *Genes* **2014**, *5*, 821–864. [CrossRef] [PubMed]

54. Widschwendter, M.; Jones, P.A. DNA methylation and breast carcinogenesis. *Oncogene* **2002**, *21*, 5462–5482. [CrossRef] [PubMed]

55. Pakneshan, P.; Szyf, M.; Farias-Eisner, R.; Rabbani, S.A. Reversal of the hypomethylation status of urokinase (uPA) promoter blocks breast cancer growth and metastasis. *J. Biol. Chem.* **2004**, *279*, 31735–31744. [CrossRef] [PubMed]

56. Du, P.; Zhang, X.; Huang, C.C.; Jafari, N.; Kibbe, W.A.; Hou, L.; Lin, S.M. Comparison of beta-value and m-value methods for quantifying methylation levels by microarray analysis. *BMC Bioinformatics* **2010**, *11*, 587. [CrossRef] [PubMed]

57. Gendler, S.J. MUC1, the renaissance molecule. *J. Mammary Gland Biol. Neoplasia* **2001**, *6*, 339–353. [CrossRef] [PubMed]

58. Nath, S.; Mukherjee, P. Muc1: A multifaceted oncoprotein with a key role in cancer progression. *Trends Mol. Med.* **2014**, *20*, 332–342. [CrossRef] [PubMed]

59. Apostolopoulos, V.; Pietersz, G.A.; McKenzie, I.F. Muc1 and breast cancer. *Curr. Opin. Mol. Ther.* **1999**, *1*, 98–103. [PubMed]

60. Kufe, D.W. MUC1-C oncoprotein as a target in breast cancer: Activation of signaling pathways and therapeutic approaches. *Oncogene* **2013**, *32*, 1073–1081. [CrossRef] [PubMed]

61. Chen, X.; Iliopoulos, D.; Zhang, Q.; Tang, Q.; Greenblatt, M.B.; Hatziapostolou, M.; Lim, E.; Tam, W.L.; Ni, M.; Chen, Y.; et al. Xbp1 promotes triple-negative breast cancer by controlling the hif1alpha pathway. *Nature* **2014**, *508*, 103–107. [CrossRef] [PubMed]

62. Yeo, E.J.; Cassetta, L.; Qian, B.Z.; Lewkowich, I.; Li, J.F.; Stefater, J.A.; Smith, A.N.; Wiechmann, L.S.; Wang, Y.; Pollard, J.W.; et al. Myeloid WNT7B mediates the angiogenic switch and metastasis in breast cancer. *Cancer Res.* **2014**, *74*, 2962–2973. [CrossRef] [PubMed]

63. Rangel, L.B.; Agarwal, R.; Sherman-Baust, C.A.; Mello-Coelho, V.; Pizer, E.S.; Ji, H.; Taub, D.D.; Morin, P.J. Anomalous expression of the HLA-DR alpha and beta chains in ovarian and other cancers. *Cancer Biol. Ther.* **2004**, *3*, 1021–1027. [CrossRef] [PubMed]

64. Bourguignon, L.Y.; Zhu, H.; Shao, L.; Chen, Y.W. Ankyrin-Tiam1 interaction promotes Rac1 signaling and metastatic breast tumor cell invasion and migration. *J. Cell. Biol.* **2000**, *150*, 177–191. [CrossRef] [PubMed]

65. Sizemore, G.M.; Sizemore, S.T.; Seachrist, D.D.; Keri, R.A. GABA(A) receptor pi (GABRP) stimulates basal-like breast cancer cell migration through activation of extracellular-regulated kinase 1/2 (ERK1/2). *J. Biol. Chem.* **2014**, *289*, 24102–24113. [CrossRef] [PubMed]

66. Masters, J.R. Human cancer cell lines: Fact and fantasy. *Nature Rev. Mol. Cell Biol.* **2000**, *1*, 233–236. [CrossRef] [PubMed]

13

Applying Human ADAR1p110 and ADAR1p150 for Site-Directed RNA Editing—G/C Substitution Stabilizes GuideRNAs against Editing

Madeleine Heep, Pia Mach, Philipp Reautschnig, Jacqueline Wettengel and Thorsten Stafforst *

Interfaculty Institute of Biochemistry, University of Tübingen, Auf der Morgenstelle 15, 72076 Tübingen, Germany; mad.heep@web.de (M.H.); pia.mach@student.uni-tuebingen.de (P.M.); philipp.reautschnig@uni-tuebingen.de (P.R.); jacqueline.wettengel@uni-tuebingen.de (J.W.)
* Correspondence: thorsten.stafforst@uni-tuebingen.de

Academic Editor: H. Ulrich Göringer

Abstract: Site-directed RNA editing is an approach to reprogram genetic information at the RNA level. We recently introduced a novel guideRNA that allows for the recruitment of human ADAR2 to manipulate genetic information. Here, we show that the current guideRNA design is already able to recruit another human deaminase, ADAR1, in both isoforms, p110 and p150. However, further optimization seems necessary as the current design is less efficient for ADAR1 isoforms. Furthermore, we describe hotspots at which the guideRNA itself is edited and show a way to circumvent this auto-editing without losing editing efficiency at the target. Both findings are important for the advancement of site-directed RNA editing as a tool in basic biology or as a platform for therapeutic editing.

Keywords: site-directed RNA editing; ADAR; guideRNA; genetic disease; RNA repair

1. Introduction

Site-directed RNA editing is a method to recode genetic information at the RNA level [1]. The approach is based on the enzymatic conversion of adenosine (A) to inosine (I). Inosine is interpreted as guanosine in many biochemical processes including translation, thus A-to-I RNA editing allows the recoding of amino acids, splice elements, miRNAs, and miRNA binding sites among others [2]. We and others have recently developed methods, called site-directed RNA editing, that employ engineered deaminases in combination with short guideRNAs to recode single adenosine bases at specific sites in any user-defined transcript [3,4]. Due to the usage of guideRNAs, the target selection and specificity is easily and rationally programmed based on simple Watson–Crick base pairing rules. We and others have shown the functioning of site-directed editing in the PCR tube, in human cell lines and even in simple organisms for the repair of reporter genes, but also disease-relevant genes like CFTR [4–6]. However, with respect to the ease of the procedure and in particular with respect to application in medicine, the requirement for the expression of an engineered deaminase is a limiting factor. Hence, we have recently developed a guideRNA architecture derived from the natural R/G-site of the GluR2 transcript that enables the recruitment of human ADAR2 for site-directed RNA editing (Figure 1B) [7]. Applying such guideRNAs allows for the editing of endogenous housekeeping genes by expression of a guideRNA only. Furthermore, we demonstrated the recoding of a nonsense mutation in PINK1 to a level sufficient to rescue the mitophagy phenotype that links dysfunctional PINK1 to the etiology of Parkinson's disease.

Human cells express three different forms of ADAR (adenosine deaminase acting on RNA), called ADAR1, 2, and 3, which are potentially re-addressable for site-directed RNA editing (Figure 1A) [2,8].

Besides ADAR2, the recruitment of ADAR1 is particularly interesting. On one hand ADAR1 is highly active for the deamination reaction and has shown to exhibit a slightly different substrate preference [9]. On the other hand, ADAR1 is more ubiquitously and to a higher level expressed as compared to ADAR2 [10], which is mainly expressed in neurons and which is believed to be exclusively localized to the nucleus [11]. ADAR1 is expressed in two isoforms, an interferon-inducible long form, called p150, and a constitutive, short form, called p110 (Figure 1A). The long form (p150) comprises an additional N-terminal stretch containing the Z-DNA/Z-RNA binding domain Zα and a nuclear export signal (NES) [12]. Due to the latter, ADAR1p150 is mainly found in the cytosol. ADAR1 and ADAR2 have distinct but also overlapping substrates. Both are essential as we know from knockout studies in mice [13–15]. Alteration of RNA editing is linked to various neurological diseases including behavioral disorders, epilepsy, and the Prader-Willi syndrome [16–19]. Mutations in ADAR1 are linked to the Aicardi-Goutieres syndrome [20], an autoimmune disorder, and others [21]. Both, hyper- [22] and hypoediting [23] have been associated with cancer [24–26].

Here, we demonstrate that a trans-acting guideRNA that has recently been developed for the recruitment of ADAR2 is also capable of recruiting both ADAR1 isoforms. Furthermore, we studied the auto-editing of the guideRNA itself, and present novel guideRNA sequences that are less prone to auto-editing but still allow for the recruitment of ADAR2.

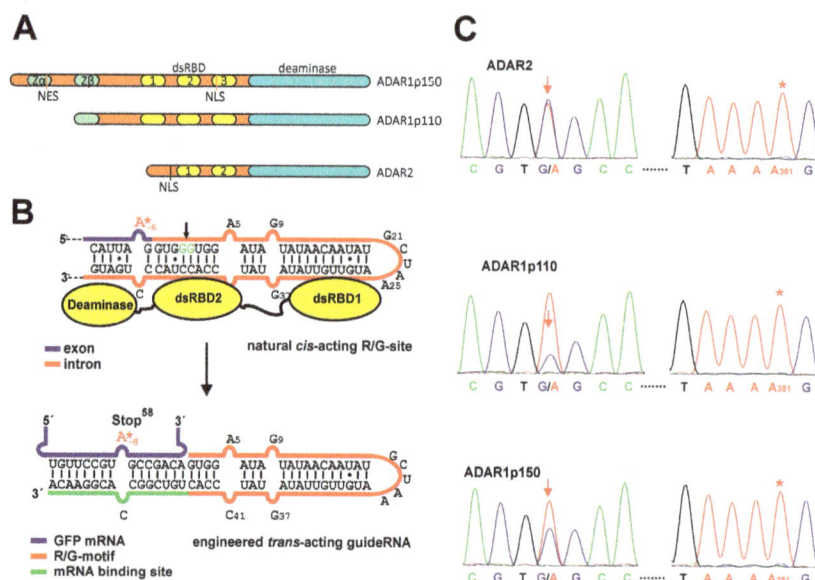

Figure 1. Site-directed RNA editing with R/G-guideRNAs. (**A**) Scheme of the three human ADARs used in this study; (**B**) Design principle: the R/G-guideRNAs have been developed as trans-acting guideRNA from the natural cistronic R/G-motif of the GluR2 transcript. The binding sites of the dsRBDs (dsRNA-binding domains) of ADAR2 are indicated; (**C**) Sanger sequencing of the editing experiments when applying the R/G-guideRNA with ADAR1p110 or p150 compared to ADAR2 in the repair the W58x codon in eGFP. Shown is the sequence around the editing site (arrow) and around a typical off-target site, A381 (*). For full Sanger sequences see Figures S2 and S3.

2. Materials and Methods

2.1. ADAR1-Expressing Cells

The 293 Flp-In T-REx system (Life Technologies) was used for stable integration of a single copy of the cDNA of ADAR1p110 or ADAR1p150 at a genomic FRT-site in engineered 293 cells. Briefly, 4×10^6 cells were seeded on a 10 cm dish. After one day, 1 µg of the respective ADAR1-form in a pcDNA5 vector under control of the tet-on CMV promoter, and 9 µg of pOG44 expressing the Flp recombinase were transfected with lipofectamine 2000 (30 µL). One day later, the medium was changed for at least two weeks to selection medium (DMEM, 10% FBS, 100 μg/mL hygromycin B, 15 μg/mL blasticidin S). Cells were kept in selection medium prior to the editing experiment, which was then performed in the absence of antibiotics.

2.2. Editing of W58amber eGFP

Protocol for experiments shown in Figures 1 and 3B: The R/G-guideRNAs were subcloned into the pSilencer2.1-U6hygro vector under control of U6 promoter and terminator, the W58x eGFP gene was delivered on a pcDNA3.1 vector as described before [7]. ADAR-expressing 293T cells were cultivated with DMEM, 10% FBS, 1% H/B, 37 °C, 5% CO_2. Cells (3×10^5/well) were seeded into poly-D-lysin-coated 24-well plates and induced with doxycycline (10 ng/mL). Transfection was carried out 24 h later with lipofectamine 2000, applying 300 ng of W58x GFP and 1300 ng of guideRNA per well. Editing was evaluated 72 h after transfection by isolation and sequencing analysis of the target mRNA. For the latter, total RNA from the cells (NucleoSpin RNA Plus kit, Macherey Nagel, Düren, Germany) was DNaseI-digested, followed by reverse transcription, Taq-PCR amplification, and Sanger sequencing.

2.3. Defining Hotspots of Auto-Editing

Protocol for the experiment shown in Figure 2: The respective eCFP W66x mRNA containing the guideRNA in cis was in-vitro-transcribed with T7 RNA pol as described earlier [7]. ADAR2 was produced and purified as described earlier [7]. Editing was carried out as described earlier [7] with (mRNA) = 25 nM, (ADAR2) = 350 nM, (Mg^{2+}) = 3 mM, no heparin, no spermidine, and incubation in editing buffer (12.5 mM Tris, 12.5 mM Tris-HCl, 75 mM KCl, 2 mM DTT) for 180 min while cycling between 30 °C and 37 °C.

2.4. Testing Auto-Editing Inside the R/G-Motif for Variants 1–4

Protocol for experiments shown in Figure 3A: The R/G-guideRNAs were subcloned into the 3'-UTR of wild-type eGFP in a pcDNA3.1 vector. GFP expression served as a transfection control. Editing was carried out in 293T cells under transient expression of one of the respective three ADAR forms from plasmid vectors (CMV promotor). For this, 293T cells (2×10^5/well) were seeded into 24-well plates. Transfection was carried out 24 h later with lipofectamine 2000, applying 300 ng of wt eGFP-guideRNA and 600 ng of the respective ADAR per well. RNA was isolated for sequencing 48 h after co-transfection as described above.

3. Results

3.1. Site-Directed RNA Editing with ADAR1

In this study, we started from a recently developed guideRNA architecture [7] that is based on the cistronic R/G-motif of the GluR2 transcript (Figure 1B). Compared to the natural R/G-site, the targeted adenosine in the mRNA is put to position −8, thus 2 nt further upstream of the R/G-motif (Figure 1B). We had shown the functioning of the guideRNA by co-transfection of the plasmid-borne guideRNA together with ADAR2 on a plasmid or by transfection of the guideRNA-plasmid into engineered ADAR2-expressing 293 Flp-In cells [7]. To test if the original guideRNA design is able to recruit ADAR1 isoforms, we have now created 293 Flp-In cells that express ADAR1p110 or ADAR1p150 from a single genomic copy under control of a CMV-tet on promoter. In such cell lines, the maximal induction-level of all three ADAR forms was similar and comparable to the level of β-actin mRNA (Figure S1). After induction (24 h) of the respective ADAR with doxycycline, the respective guideRNA construct was co-transfected together with a GFP reporter construct containing a single W58amber missense codon for site-directed repair. 72 h after co-transfection, the editing yield was determined by Sanger sequencing of the RNA (Figure 1), as described earlier [7]. In ADAR2-expressing cells, editing yields around 50% are typically obtained, with no detectable off-target editing in the ORF of the reporter gene [7]. This was again confirmed here (Figure 1C). The editing reaction in ADAR1-expressing cells (Figure 1C) gave slightly reduced yields for the p150 form (35%–40%), and markedly reduced yields (20%) for the p110 form. Again, no off-target editing was detectable for either ADAR1 form in the ORF of the reporter gene. One site, adenosine 381, which was prone to off-target editing under transient overexpression of ADAR2 before [7], is shown in Figure 1C to exemplify the lack of off-target editing in the ORF of eGFP under genomic expression of all three ADARs (see Figures S2 and S3 for full Sanger sequencing traces).

3.2. Auto-Editing inside the GuideRNA

The original guideRNA design is relatively AU-rich. Editing could suffer from the auto-editing of the guideRNA itself thus destabilizing the hairpin. We tested this in an initial editing experiment on an in-vitro transcribed RNA substrate that contained the eCFP W66amber sequence in frame with a cis-acting guideRNA inside the PCR tube (Figure 2). Under these conditions, we found five sites in the R/G-motif prone to auto-editing: A8, A14, A34, A36, and A39. All five editing events can be assumed to destabilize the helix. Notably, the editing yields at the sites in the 3′-half of the R/G hairpin were higher compared to those in the 5′-half of the hairpin. We further tested the auto-editing by co-transfection of the same cis-acting CFP-guideRNA construct with ADAR2 in 293T cells (Figure S4). All auto-editing sites were confirmed, the editing yields stayed mostly unchanged, only at A-16 and A34 was the editing yield increased, and, at one new site, A11, auto-editing occurred additionally.

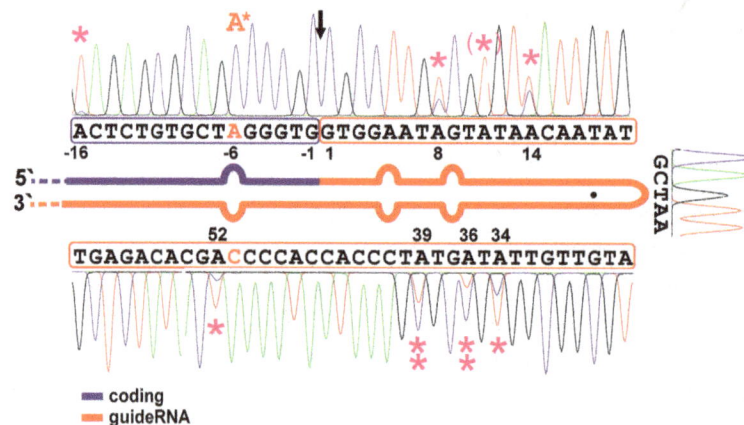

Figure 2. Identification of auto-editing hotspots. An in-vitro transcribed RNA substrate containing a part of the eCFP ORF around the W66x site (blue) in *cis* with an R/G-guideRNA (red) was edited with purified ADAR2 enzyme in a PCR tube. Hotspots for auto-editing have been marked by magenta asterisks; double asterisks mark strongly edited sites; asterisk in brackets mark an editing site only found in a similar experiment inside the cell (see Figure S4). The red A* marks the targeted editing site, full conversion was achieved. The black arrow shows the site where the cistronic motif is cut when the guideRNA is applied in *trans*.

3.3. G/C-Substitution inside the R/G-Motif Reduces Auto-Editing

To reduce auto-editing of the guideRNA, 3, 6, or 13 A/U base pairs in the original R/G-motif have been substituted by G/C base pairs, creating variants 2, 3, and 4 (Figure 3A). To assess auto-editing inside these motif variants, we constructed for every guideRNA an eGFP reporter transcript that contains the complete guideRNA in the 3′-UTR of the transcript, thus ca. 550 nt downstream the targeted editing site. This allows for testing the guideRNAs in a situation that resembles the trans-acting guideRNA under concurrent possibility to access auto-editing by Sanger sequencing. Editing was tested by co-transfection of these constructs (300 ng) with each of the three forms of ADAR (600 ng) into 293T cells. Editing in the R/G-motif was analyzed 48 h after co-transfection by RNA isolation and Sanger sequencing. The editing yields inside the R/G-motif are given in color-coded circles in Figure 3A.

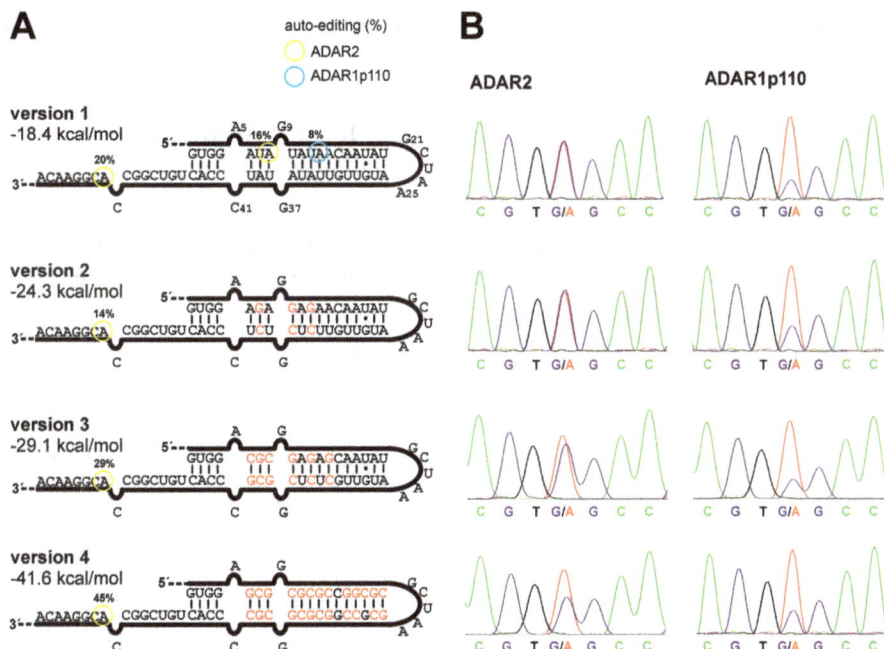

Figure 3. GuideRNA variants that avoid auto-editing. (**A**) Three guideRNA versions with differing degree of G/C substitution were tested for auto-editing. The shown sequence was introduced into the 3′-UTR of the eGFP transcript for easier Sanger sequencing. The folding energies have been estimated using mfold. For ADAR1p150, no auto-editing site was detectable; (**B**) Performance of the three new guideRNA versions for the editing of W58x with ADAR2 or ADAR1p110, respectively, when applied in *trans* in the respective ADAR-expressing cell line. For fluorescence imaging, see Figure S5.

Under these assay conditions, the starting guideRNA architecture, version 1, gives only subtle auto-editing, and only at positions A8 by ADAR2 and A13 by ADAR1p110. With ADAR2, we found an additional off-target editing at A52, which is part of the mRNA binding sequence. For ADAR1p150, no auto-editing was detectable. It is reasonable to infer that the 36 bp RNA substrate, given in Figure 2, is much more heavily edited in the R/G-motif compared to the lone-standing, 20 bp R/G-motif, given in Figure 3, as the ADAR enzymes are known to accept dsRNA with ≥30 bp much better than those ≤30 bp [2,11]. When introducing G/C-substitutions into the R/G-motif, no auto-editing was detectable with any of the three enzymes. This was already achieved with only 3 G/C substitutions, which have been chosen in a way to target the main auto-editing sites (A8, A14, A34, A36, and A39) simultaneously with a minimal number of substitutions.

The G/C substitutions do not only remove editable adenosines and put adenosines into contexts non-preferred for editing (e.g., 5′-GA), but they rather increase the free energy gain of folding from −18 kcal/mol up to −42 kcal/mol as estimated by the software mfold [27] (Figure 3A). We tested how the new guideRNA variants 2 to 4 behave compared to the original version 1 for the editing of the reporter W58x eGFP in ADAR2-expressing cells. For versions 2 and 3, virtually no change in the editing yield was observed. Only the highly substituted version 4 showed a reduction in editing yield (Figure 3B, Figure S5). For ADAR1p110, we found a virtually unchanged editing yield with all versions but at comparably low editing efficiency (Figure 3B).

4. Discussion

This is the first report about site-directed RNA editing with human ADAR1 isoforms, the constitutive (p110) and the inducible one (p150). As the R/G-guideRNA was developed from a classical

ADAR2 substrate for the recruitment of ADAR2, it could be expected that the recruitment of ADAR1 isoforms is less efficient. However, we achieved site-directed RNA editing with the R/G-guideRNAs with both ADAR1 isoforms, in particular with the p150 form; thus, the general strategy is feasible with ADAR1, and our R/G-guideRNA provides a starting point for the development of more appropriate guideRNAs for ADAR1.

Furthermore, we show the first results on auto-editing of the guideRNA itself. We have defined the hotspots for auto-editing and show a simple strategy to circumvent it by minimal substitution in the hairpin secondary structure. The respective optimized guideRNAs are much less auto-edited; however, our results show that a high substitution degree may interfere with editing efficiency at the target. Even though the confinement of auto-editing was not improving the overall editing yield under the given continuous expression of the guideRNA, it could improve editing when drug-like, chemically stabilized R/G-guideRNAs are explored in the future. Notably, when defining the hotspots of auto-editing by applying an artificial eCFP mRNA containing the guideRNA in *cis*, we found full conversion at the targeted adenosine even in cell culture (Figure S4). In contrast, when applying the same guideRNA in *trans*, editing levels above 60% are difficult to obtain. It is likely that there is still space for further optimization of the guideRNA's design to obtain better editing results with all three enzymes. In summary, both findings are important for the advancement of site-directed RNA editing as a tool in basic biology or as a platform for therapeutic editing [7,28].

Acknowledgments: We gratefully acknowledge support from the University of Tübingen and the Deutsche Forschungsgemeinschaft (STA 1053/3-2). This work has received funding from the European Research Council (ERC) under the European Union's Horizon 2020 research and innovation program (grant agreement No 647328). Open access fees have been paid from University and DFG funds.

Author Contributions: P.R., J.W., and T.S. conceived and designed the experiments; P.R., M.H., P.M., and J.W. performed cloning; M.H. and J.W. performed editing experiments; P.M. and M.H. performed RT-qPCR; all authors analyzed the data; P.M. engineered the ADAR1-expressing cells; T.S. wrote the paper.

References

1. Vogel, P.; Stafforst, T. Site-directed RNA editing with antagomir deaminases—A tool to study protein and RNA function. *ChemMedChem* **2014**, *9*, 2021–2025. [CrossRef] [PubMed]
2. Nishikura, K. Functions and regulation of RNA editing by ADAR deaminases. *Annu. Rev. Biochem.* **2010**, *79*, 321–349. [CrossRef] [PubMed]
3. Stafforst, T.; Schneider, M.F. An RNA–Deaminase conjugate selectively repairs point mutations. *Angew. Chem. Int. Ed.* **2012**, *51*, 11166–11169. [CrossRef] [PubMed]
4. Montiel-Gonzalez, M.F.; Guillermo, I.; Yudowski, A.; Rosenthal, J.J.C. Correction of mutations within the cystic fibrosis transmembrane conductance regulator by site-directed RNA editing. *Proc. Natl. Acad. Sci. USA* **2013**, *110*, 18285–18290. [CrossRef] [PubMed]
5. Vogel, P.; Schneider, M.F.; Wettengel, J.; Stafforst, T. Improving site-directed RNA editing in vitro and in cell culture by chemical modification of the guideRNA. *Angew. Chem. Int. Ed.* **2014**, *53*, 6267–6271. [CrossRef] [PubMed]
6. Hanswillemenke, A.; Kuzdere, T.; Vogel, P.; Jékely, G.; Stafforst, T. Site-directed RNA editing in vivo can be triggered by the light-driven assembly of an artificial riboprotein. *J. Am. Chem. Soc.* **2015**, *137*, 15875–15881. [CrossRef] [PubMed]
7. Wettengel, J.; Reautschnig, J.; Geisler, S.; Kahle, P. J.; Stafforst, T. Harnessing human ADAR2 for RNA repair—Recoding a PINK1 mutation rescues mitophagy. *Nucl. Acids Res.* **2016**. [CrossRef] [PubMed]
8. Bass, B.L. RNA editing by adenosine deaminases that act on RNA. *Annu. Rev. Biochem.* **2002**, *71*, 817–846. [CrossRef] [PubMed]

9. Schneider, M.F.; Wettengel, J.; Hoffmann, P.C.; Stafforst, T. Optimal guideRNAs for re-directing deaminase activity of hADAR1 and hADAR2 in trans. *Nucl. Acids Res.* **2004**. [CrossRef] [PubMed]

10. Picardi, E.; Manzari, C.; Mastropasqua, F.; Aiello, I.; D'Erchia, A.M.; Pesole, G. Profiling RNA editing in human tissues: towards the inosinome Atlas. *Sci. Rep.* **2015**. [CrossRef] [PubMed]

11. Nishikura, K. A-to-I editing of coding and non-coding RNAs by ADARs. *Nat. Rev. Mol. Cell Biol.* **2016**, *17*, 83–96. [CrossRef] [PubMed]

12. Barraud, P.; Allain, F.H.-T. ADAR Proteins: Double-stranded RNA and Z-DNA binding domains. *Curr. Topics Microbiol. Immun.* **2012**, *353*, 35–60.

13. Higuchi, M.; Maas, S.; Single, F.N.; Hartner, J.; Rozov, A.; Burnashev, N.; Feldmeyer, D.; Sprengel, R.; Seeburg, P.H. Point mutation in an AMPA receptor gene rescues lethality in mice deficient in the RNA-editing enzyme ADAR2. *Nature* **2000**, *406*, 78–81. [PubMed]

14. Hartner, J.C.; Schmittwolf, C.; Kispert, A.; Müller, A.M.; Higuchi, M.; Seeburg, P.H. Liver disintegration in the mouse embryo caused by deficiency in the RNA-editing enzyme ADAR1. *J. Biol. Chem.* **2004**, *279*, 4894–4902. [CrossRef] [PubMed]

15. Wang, Q.; Miyakoda, M.; Yang, W.; Khillan, J.; Stachura, D.L.; Weiss, M.J.; Nishikura, K. Stress-induced apoptosis associated with null mutation of ADAR1 RNA editing deaminase gene. *J. Biol. Chem.* **2004**, *279*, 4952–4961. [CrossRef] [PubMed]

16. Morabito, M.V.; Abbas, A.I.; Hood, J.L.; Kesterson, R.A.; Jacobs, M.M.; Kump, D.S.; Hachey, D.L.; Roth, B.L.; Emeson, R.B. Mice with altered serotonin 2C receptor RNA editing display characteristics of Prader-Willi syndrome. *Neurobiol. Dis.* **2010**, *39*, 169–180. [CrossRef] [PubMed]

17. Maas, S.; Kawahara, Y.; Tamburro, K.M.; Nishikura, K. A-to-I RNA editing and human disease. *RNA Biol.* **2006**, *3*, 1–9. [CrossRef] [PubMed]

18. Slotkin, W.; Nishikura, K. Adenosine-to-inosine RNA editing and human disease. *Genome Med.* **2013**. [CrossRef] [PubMed]

19. Silberberg, G.; Lundin, D.; Navon, R.; Öhman, M. Deregulation of the A-to-I RNA editing mechanism in psychiatric disorders. *Hum. Mol. Genet.* **2012**, *21*, 311–321. [CrossRef] [PubMed]

20. Rice, G.I.; Kasher, P.R.; Forte, G.M.; Mannion, N.M.; Greenwood, S.M.; Szynkiewicz, M.; Dickerson, J.E.; Bhaskar, S.S.; Zampini, M.; Briggs, T.A.; et al. Mutations in ADAR1 cause Aicardi-Goutières syndrome associated with a type I interferon signature. *Nat. Genet.* **2012**, *44*, 1243–1248. [CrossRef] [PubMed]

21. Zhang, X.J.; He, P.P.; Li, M.; He, C.D.; Yan, K.L.; Cui, Y.; Yang, S.; Zhang, K.Y.; Gao, M.; Chen, J.J.; et al. Seven novel mutations of the ADAR gene in Chinese families and sporadic patients with dyschromatosis symmetrica hereditaria (DSH). *Hum. Mutat.* **2004**, *23*, 629–630. [CrossRef] [PubMed]

22. Chen, L.; Li, Y.; Lin, C.H.; Chan, T.H.M.; Chow, R.K.K.; Song, Y.; Liu, M.; Yuan, Y.F.; Fu, L.; Kong, K.L.; et al. Recoding RNA editing of antizyme inhibitor 1 predisposes to hepatocellular carcinoma. *Nat. Med.* **2013**, *19*, 209–216. [CrossRef] [PubMed]

23. Shimokawa, T.; Rahman, M.F.; Tostar, U.; Sonkoly, E.; Ståhle, M.; Pivarcsi, A.; Palaniswamy, R.; Zaphiropoulos, P.G. RNA editing of the GLI1 transcription factor modulates the output of Hedgehog signaling. *RNA Biol.* **2013**, *10*, 321–333. [CrossRef] [PubMed]

24. Gallo, A. RNA editing enters the limelight in cancer. *Nat. Med.* **2013**, *19*, 130–131. [CrossRef] [PubMed]

25. Paz-Yaacov, N.; Bazak, L.; Buchumenski, I.; Porath, H.T.; Danan-Gotthold, M.; Knisbacher, B.A.; Eisenberg, E.; Levanon, E.Y. Elevated RNA editing activity is a major contributor to transcriptomic diversity in tumors. *Cell Rep.* **2015**, *13*, 267–276. [CrossRef] [PubMed]

26. Han, L.; Diao, L.; Yu, S.; Xu, X.; Li, J.; Zhang, R.; Yang, Y.; Werner, H.M.; Eterovic, A.K.; Yuan, Y.; et al. The genomic landscape and clinical relevance of A-to-I RNA editing in human cancers. *Cancer Cell* **2015**, *28*, 515–528. [CrossRef] [PubMed]

27. Zuker, M. Mfold web server for nucleic acid folding and hybridization prediction. *Nucl. Acids Res.* **2003**, *31*, 3406–3415.

28. Reautschnig, P.; Vogel, P.; Stafforst, T. The notorious RNA in the spotlight—Drug or target for the treatment of disease. *RNA Biol.* **2016**. [CrossRef]

14

Type 2 Diabetes Susceptibility in the Greek-Cypriot Population: Replication of Associations with *TCF7L2*, *FTO*, *HHEX*, *SLC30A8* and *IGF2BP2* Polymorphisms

Christina Votsi [1], **Costas Toufexis** [2], **Kyriaki Michailidou** [3], **Athos Antoniades** [4], **Nicos Skordis** [5], **Minas Karaolis** [4], **Constantinos S. Pattichis** [4] and **Kyproula Christodoulou** [1,*]

[1]   Department of Neurogenetics, The Cyprus Institute of Neurology and Genetics and the Cyprus School of Molecular Medicine, Ayios Dhometios, 2370 Nicosia, Cyprus; votsi@cing.ac.cy

[2]   Department of Endocrinology and Diabetes, Hippocrateon Private Hospital, Engomi, 2408 Nicosia, Cyprus; c.toufexis@hippocrateon.com

[3]   Department of Electron Microscopy/Molecular Pathology, The Cyprus Institute of Neurology and Genetics, Ayios Dhometios, 2370 Nicosia, Cyprus; kyriakimi@cing.ac.cy

[4]   Department of Computer Science, University of Cyprus, 1678 Nicosia, Cyprus; athos.antoniades@stremble.com (A.A.); karaolis@spidernet.com.cy (M.K.); pattichi@ucy.ac.cy (C.S.P.)

[5]   St. George's University Medical School at the University of Nicosia, Engomi, 2408 Nicosia, Cyprus and Department of Pediatric Endocrinology, Paedi Center for Specialized Pediatrics, Strovolos, 2025 Nicosia, Cyprus; nskordis@cytanet.com.cy

*   Correspondence: roula@cing.ac.cy

Academic Editor: Bernhard O. Boehm

Abstract: Type 2 diabetes (T2D) has been the subject of numerous genetic studies in recent years which revealed associations of the disease with a large number of susceptibility loci. We hereby initiate the evaluation of T2D susceptibility loci in the Greek-Cypriot population by performing a replication case-control study. One thousand and eighteen individuals (528 T2D patients, 490 controls) were genotyped at 21 T2D susceptibility loci, using the allelic discrimination method. Statistically significant associations of T2D with five of the tested single nucleotide polymorphisms (SNPs) (*TCF7L2* rs7901695, *FTO* rs8050136, *HHEX* rs5015480, *SLC30A8* rs13266634 and *IGF2BP2* rs4402960) were observed in this study population. Furthermore, 14 of the tested SNPs had odds ratios (ORs) in the same direction as the previously published studies, suggesting that these variants can potentially be used in the Greek-Cypriot population for predictive testing of T2D. In conclusion, our findings expand the genetic assessment of T2D susceptibility loci and reconfirm five of the worldwide established loci in a distinct, relatively small, newly investigated population.

Keywords: type 2 diabetes; susceptibility loci; association studies; population studies; Greek-Cypriot population

1. Introduction

Type 2 diabetes (T2D) is a chronic complex heterogeneous disease of glucose metabolism caused by multiple genetic, epigenetic and environmental factors [1]. It is characterised by high blood glucose levels caused by the combination of insulin resistance and impaired insulin secretion [2,3]. It is a serious worldwide public health burden which has reached epidemic proportions with an increasing prevalence and substantial familial clustering [4]. The disease leads to morbidity with the life expectancy being reduced, while additional implications include premature coronary heart disease, peripheral vascular disease, renal failure, stroke and amputation [2].

The heritability of the disease has been well established through twin and family studies [5] that also revealed an estimated lifetime risk of 38% by the age of 80 if one parent is affected and by the age of 60 if both parents are affected [6]. T2D has been the subject of numerous genetic studies aiming at the elucidation of the genetic mechanisms involved in the development of the disease. Linkage and candidate gene association studies initially reported many T2D linked chromosomal regions. However, only *PPARG*, *KCNJ11* and *TCF7L2* were replicated in most populations, thus being established as T2D associated genes [2,4]. Development of the high-throughput single nucleotide polymorphism (SNP) genotyping technology and completion of the HapMap project enabled the implementation of genome-wide association studies (GWAS) [7] and since 2007 they have become the leading tool for the identification of several T2D susceptibility loci [2,8]. Recent GWAS, initially performed in Caucasians and then in non-European populations, and meta-analyses of these studies, have increased the total number of the identified variants with possible association to over 88 variants, including and further confirming some variants that were previously identified by the linkage and candidate gene studies [9,10]. However, additional variants are likely to be discovered since the identified associations explain only about 10% of the heritability of the disease [2,4,8].

Further to GWAS, replication studies have been performed in various populations which aimed at investigating whether identified T2D susceptibility variants confer risk across different genetic backgrounds [4,11–16]. In this concept, we performed the first genetic study on T2D in the Greek-Cypriot population, which represents the great majority (74%) of the Cypriot population (694,700 indigenous individuals at the 2014 census) [17] and has proved to be genetically distinct through other genetic studies, such as on Thalassaemia [18] and breast cancer susceptibility [19]. The prevalence of diabetes in the adult Greek-Cypriot population was estimated to be 10.3% [20] in 2006, which is close to the global prevalence and more similar to that of Middle East and North African populations rather than European populations (International Diabetes Atlas, [21]). This pilot replication study initiates evaluation of T2D susceptibility loci in the Greek-Cypriot population. Results for 21 SNPs that were associated with T2D in other populations until the year 2010 are presented.

2. Materials and Methods

2.1. Study Participants

A total of 1018 subjects, including 528 unrelated well characterised Greek-Cypriot T2D patients and 490 controls without diabetes, were recruited for this study. The established World Health Organization (WHO) diagnostic criteria were used for the diagnosis of T2D. Control participants' selection criteria included fasting plasma glucose levels of ≤ 108 mg/dL and absence of a T2D family history. T2D is a relatively late onset disease. In order to minimize the possibility of including individuals that will later develop T2D, we mostly selected individuals above the age of 50 for the study control group.

Glucose levels were recorded for all study participants. Clinical, biochemical data (Table 1) and other information such as age of T2D first diagnosis, specific diet, medication or complications of the disease, including cardiovascular disease (CVD), ophthalmological symptoms, hypertension and nephropathy were also recorded for the majority of the study participants. The study was conducted in accordance with the Declaration of Helsinki, and the ethical approval was granted by the National Bioethics Committee of Cyprus (EEBK/EΠ/2010/19, 8 March 2011). Written informed consent was obtained from all study participants.

Table 1. Summary of the main phenotypic characteristics of the study participants.

Trait	T2D Patients	Controls	t-Test p-Value
Number	528	490	n/a
Sex (male/female)	321/207	263/227	n/a
Age at interview (years, mean $\pm$ SD)	63.73 $\pm$ 10.50	59.14 $\pm$ 11.91	1.1×10^{-10}
Age at diagnosis (years, mean $\pm$ SD)	52.08 $\pm$ 11.01	n/a	n/a
BMI (kg/m^2 $\pm$ SD)	30.02 $\pm$ 4.95	26.75 $\pm$ 4.03	1.3×10^{-25}
Glucose (mg/dL $\pm$ SD)	149.63 $\pm$ 48.37	89.11 $\pm$ 9.11	1.1×10^{-120}
HbA1c (%) (DCCT $\pm$ SD)	0.17 $\pm$ 0.90	n/a	n/a
HDL-C (mg/dL $\pm$ SD)	43.35 $\pm$ 12.23	51.52 $\pm$ 14.91	2.0×10^{-19}
LDL-C (mg/dL $\pm$ SD)	102.39 $\pm$ 31.18	136 $\pm$ 34.29	7.3×10^{-50}
TC (mg/dL $\pm$ SD)	176.11 $\pm$ 40.42	208.68 $\pm$ 40.47	2.9×10^{-33}
TG (mg/dL $\pm$ SD)	149.66 $\pm$ 86.11	118.73 $\pm$ 64.46	5.8×10^{-10}

T2D: Type 2 diabetes; BMI: body mass index; HbA1c: Haemoglobin A1c; HDL-C: high-density lipoprotein cholesterol; LDL-C: low-density lipoprotein cholesterol; TC: total cholesterol; TG: triglycerides.

2.2. DNA Extraction

DNA was extracted from whole blood using standard salting out procedures. DNA samples were diluted and aliquots were plated in duplicates in 384-deep-well storage plates at a uniform concentration of 10 ng/μL.

2.3. SNP Selection and Genotyping

Twenty-one SNPs associated with T2D in other populations until year 2010 have been investigated in this study (Table 2). SNPs were selected through a literature search of T2D GWAS and from candidate gene studies [22–30]. Samples were genotyped using pre-designed TaqMan SNP Genotyping Assays following the standard protocol provided by the manufacturer (Applied Biosystems, Foster City, CA, USA). The genotyping success rate was >99%.

2.4. Statistical Analyses

Student's t-test was used to evaluate the differences of the continuous variables (presented as mean $\pm$ standard deviation) between cases and controls. Quality control (QC) checks were performed, for samples and SNPs. We evaluated the genotype distributions of all SNPs in the control samples for Hardy–Weinberg equilibrium (HWE) using an exact test [31] with a threshold of deviation $p \leq 0.05$. Minor allele frequencies (MAF) were calculated and used for quality control (variants with MAF < 0.01 were excluded). Linear (for T2D age of onset) and logistic regression (for T2D) analyses were performed for samples and SNPs passing the QC filters. Analyses were performed in R (R Development Core Team, Vienna, Austria) [32] and PLINK (Center for Human Genetic Research, Boston, MA, USA) [33,34]. Logistic regression analyses were performed at two levels: (a) with an adjustment for age and gender; and (b) with an adjustment for age, gender and Body Mass Index (BMI). Statistical significance for T2D association with the tested variants was defined using a p-value threshold of 0.05. Linear regression analysis was performed with an adjustment for gender and BMI. Bonferroni's method was applied for multiple testing correction to determine the significance of potential novel associations with T2D age of diagnosis, using a p-value threshold of 0.0026 (0.05/19).

Table 2. Summary of the single nucleotide polymorphisms (SNPs) included in the study.

SNP	Nearest Gene(s)	Chromosome	Reference	Non Risk/Risk Allele [a]	RAF [b]	OR [b] (95% CI)	Frequency [c]	OR [c] (95% CI)	p-Value [d]
				Published Results				Current Analysis	
rs10923931	NOTCH2	1p12	[23]	G/T	0.11	1.13 (1.08–1.17)	0.06	1.09 (0.73–1.64)	0.68
rs7578597	THADA	2p21	[23]	C/T	0.90	1.15 (1.10–1.20)	0.94	0.75 (0.49–1.15)	0.19
rs4607103	ADAMTS9	3p14.1	[23]	T/C	0.76	1.09 (1.06–1.12)	0.59	0.89 (0.73–1.1)	0.29
rs4402960	IGF2BP2	3q27.2	[24]	G/T	0.30	1.14 (1.11–1.18)	0.27	1.24 (1.01–1.53)	**0.04**
rs1801282	PPARG	3p25.2	[24]	G/C	0.82	1.14 (1.08–1.20)	0.95	1.33 (0.82–2.16)	0.25
rs10010131[e]	WFS1	4p16.1	[25]	A/G	0.60	1.16 (1.05–1.28)	0.66	1.2 (0.97–1.49)	0.09
rs4457053	ZBED3	5q13.3	[26]	A/G	0.26	1.08 (1.06–1.11)	0.31	0.96 (0.77–1.19)	0.69
rs10946398	CDKAL1	6p22	[22]	A/C	0.32	1.16 (1.10–1.22)	0.32	1.21 (0.99–1.49)	0.07
rs864745	JAZF1	7p15.1	[23]	G/A	0.50	1.10 (1.07–1.13)	0.55	0.82 (0.67–1)	0.05
rs13266634	SLC30A8	8q24.11	[24]	T/C	0.61	1.12 (1.07–1.16)	0.69	1.31 (1.05–1.63)	**0.02**
rs10811661	CDKN2A	9p21	[27]	C/T	0.83	1.20 (1.12–1.28)	0.79	1.13 (0.87–1.46)	0.35
rs12779790[f]	CDC123, CAMK1D	10p13	[23]	A/G	0.18	1.11 (1.07–1.14)	-	-	-
rs5015480	HHEX	10q23.33	[22]	T/C	0.57	1.13 (1.07–1.19)	0.53	1.38 (1.13–1.69)	**0.002**
rs7901695	TCF7L2	10q25.2	[22]	T/C	0.27	1.37 (1.25–1.49)	0.41	1.35 (1.1–1.64)	**0.003**
rs10830963	MTNR1B	11q14.3	[28]	C/G	0.27	1.09 (1.05–1.12)	0.25	1.13 (0.9–1.41)	0.29
rs5219	KCNJ11	11p15.1	[24]	C/T	0.46	1.14 (1.10–1.19)	0.33	0.95 (0.77–1.19)	0.67
rs2237892	KCNQ1	11p15.5	[29]	T/C	0.61	1.45 (1.34–1.47)	0.96	1.25 (0.7–2.23)	0.44
rs7961581	TSPAN8, LGR5	12q21.1	[23]	T/C	0.27	1.09 (1.06–1.12)	0.41	1.02 (0.83–1.26)	0.84
rs8042680	PRC1	15q26.1	[26]	C/A	0.33	1.07 (1.05–1.09)	0.45	1.09 (0.88–1.33)	0.43
rs8050136	FTO	16q12.2	[22]	C/A	0.41	1.23 (1.18–1.32)	0.41	1.33 (1.08–1.63)	**0.006**
rs757210[f]	HNF1B	17q12	[30]	G + C/A	0.38	1.12 (1.07–1.18)	-	-	-

[a] Non risk/ Risk allele based on the published study; [b] Published risk allele frequency (RAF) odds ratio (OR) and 95% confidence interval (CI); [c] RAF, OR and 95% CI obtained in the current study population; [d] p-value from logistic regression adjusted for age, gender and BMI. A p-value threshold of 0.05 was used and the identified significant associations are shown in bold; [e] OR converted to be in respect of the risk allele; [f] Failed quality control in this study.

3. Results

3.1. Study Participants

The main phenotypic characteristics of the study participants are presented in Table 1. The mean age of patients and controls at the time of interviewing and sampling was slightly different (63.73 and 59.14 respectively). However, the great majority was over 50 which is close to the mean age of T2D diagnosis (52.08). Overall, the mean values of the remaining characteristics—BMI, glucose, High-density lipoprotein cholesterol (HDL-C), Low-density lipoprotein cholesterol (LDL-C), Total Cholesterol (TC), and Triglycerides (TG)—were all significantly different between the two groups with the most significant difference observed at the glucose levels ($p = 1.1 \times 10^{-120}$). Patients had significantly higher BMI, glucose and TG and lower HDL, LDL and TC levels compared to the controls. The observed differences in the lipid patterns between the two groups may be attributed to targeted LDL lowering medication taken by the patients.

3.2. Association Studies—Statistical Analyses

Twenty-one SNPs were analysed (Table 2). All SNPs, except rs12779790 and rs757210, passed QC. SNP rs12779790 failed HWE in controls ($p < 0.05$). SNP rs757210 was found to be tri-allelic in our study, a finding that was confirmed by other studies [35,36], and was thus not used in the analyses. Linkage disequilibrium (LD) testing confirmed that none of the study SNPs was in LD.

The 19 SNPs were tested for association with T2D and the results are presented in Table 2 and Supplementary Table S1. Effects and confidence intervals reported in previous studies are also presented in Table 2. The current study results were aligned in order to correspond to the published risk allele. Initial analysis was performed with an adjustment for age and gender, which resulted in four loci showing statistically significant associations with T2D (Supplementary Table S1) at nominal significance levels ($p < 0.05$); *TCF7L2* rs7901695 [odds ratio (OR) (95% confidence interval (CI)) 1.3 (1.08–1.55) p-value = 0.005], *FTO* rs8050136 [OR (95% CI) 1.34 (1.11–1.61) p-value = 0.002], *HHEX* rs5015480 [OR (95% CI) 1.36 (1.13–1.62) p-value = 0.001] and *SLC30A8* rs13266634 [OR (95% CI) 1.31 (1.08–1.6) p-value = 0.007]. Further analysis with adjustment for age, gender and BMI showed statistically significant associations with five variants; in addition to the previous four, association with locus *IGF2BP2* rs4402960 [OR (95% CI) 1.24 (1.01–1.53) p-value = 0.04] was also observed (Table 2). Furthermore, adjustment for BMI slightly strengthened the T2D association with *TCF7L2* [OR (95%CI) 1.35 (1.1–1.64) p-value = 0.003], slightly weakened the association with *HHEX* [OR (95%CI) 1.38 (1.13–1.69) p-value = 0.002] and *FTO* [OR (95% CI) 1.33 (1.08–1.63) p-value = 0.006] and significantly weakened the association with *SLC30A8* [OR (95% CI) 1.31 (1.05–1.63) p-value = 0.02].

Association analyses were also performed between the 19 SNPs and age of T2D diagnosis using linear regression and adjusting for gender and BMI (Table 3). Although two variants (*THADA* rs 7578597 and *TCFL2* rs7901695) were associated with earlier age of diagnosis at nominal significance levels ($p < 0.05$), these did not survive Bonferroni correction [$p < 0.0026$ (0.05/19)]. For each extra risk allele of SNP rs7578597, the mean age of diagnosis was reduced by 3.81 years ($p = 0.006$) and for each extra risk allele of rs7901695, the mean age of diagnosis was reduced by 1.91 years ($p = 0.005$).

Table 3. Results for single SNP association analyses with the age at T2D first diagnosis.

SNP	Gene	Beta	Standard Error	p-Value
rs10923931	NOTCH2	0.48	1.38	0.729
rs7578597	THADA	−3.81	1.38	**0.006**
rs4607103	ADAMTS9	−0.98	0.71	0.172
rs4402960	IGF2BP2	−0.64	0.73	0.376
rs1801282	PPARG	−1.97	1.81	0.277
rs10010131	WFS1	−0.54	0.75	0.473
rs4457053	ZBED3	−0.41	0.77	0.595
rs10946398	CDKAL1	−0.31	0.72	0.671
rs864745	JAZF1	−0.34	0.70	0.632
rs13266634	SLC30A8	0.25	0.80	0.753
rs10811661	CDKN2A	0.47	0.95	0.620
rs5015480	HHEX	0.26	0.71	0.713
rs7901695	TCF7L2	−1.91	0.67	**0.005**
rs10830963	MTNR1B	−0.22	0.77	0.774
rs5219	KCNJ11	−0.25	0.77	0.743
rs2237892	KCNQ1	−0.55	2.08	0.792
rs7961581	TSPAN8, LGR5	0.33	0.73	0.655
rs8042680	PRC1	−0.66	0.74	0.376
rs8050136	FTO	−0.60	0.70	0.393

Regression estimates and p-values based on linear regression adjusted for gender and BMI. Nominally significant associations ($p < 0.05$) are shown in bold, however they did not survive Bonferroni correction ($p < 0.0026$).

4. Discussion

Type 2 Diabetes is a chronic complex disease of glucose metabolism with a strong genetic contribution and a worldwide prevalence that reaches epidemic proportions [2,4]. Many GWAS [24,27,37], meta-analyses [2,5,23,26] and replication studies have been performed that focused on common variants in different ethnic groups [4,12,38] and identified a large number of SNPs that are associated with the disease. Substantial genetic heterogeneity has been observed between the different populations suggesting that exploration of T2D susceptibility in additional populations might provide further insight into the disease aetiology.

We performed the first genetic study of T2D in the Greek-Cypriot population, a population of Caucasian origin. Compared to other Caucasian populations the Cypriot population is genetically distinct with significant differences from the Northern European populations [39,40], and likely has genetic similarities with the current populations of Levant [41]. Previous studies revealed unique genetic features in this population [18,19] and in one of them, the genetic characteristics of Greek- and Turkish-Cypriots were compared with the mainland Greek and Turkish populations. This study revealed a close genetic similarity between the two Cypriot communities and considerable differences with the Greek and Turkish populations [18]. The island, inhabited by the Greeks during the Bronze age, is located at the crossroads of three continents (Africa, Asia, and Europe) and has come under the domination of successive foreign invaders in its long history; including the Phoenicians, the Assyrians, the Egyptians, the Persians, the Romans, the Arabs, the Franks, the Venetians, the Ottoman Turks and the British. This pool of interactions with other populations has probably led to the assimilation of not only cultural influences but also genetic influences, thus contributing to the genetic background of the Cypriot population.

This study focused on an initial evaluation of 21 known T2D common susceptibility loci, most of which were identified in Caucasian populations. Five of the tested loci (rs7901695 in TCF7L2, rs8050136 in FTO, rs5015480 near HHEX, rs13266634 in SLC30A8 and rs4402960 in IGF2BP2) showed statistically significant associations with T2D in this population after adjustment for age, gender and BMI. Previous studies on SNP rs8050136 (FTO) report that the association with T2D is abolished after adjustment for BMI, thus indicating that the association is probably mediated through a primary effect on BMI, a finding that necessitates the inclusion of BMI as a covariate in similar analyses [22,23,38,42,43]. In this study, adjustment for BMI weakened, but did not abolish the association of rs8050136 (*FTO*) with T2D. This result indicates that the association is likely not mediated through BMI in the Greek-Cypriot population, also in agreement with other published studies [44,45].

Nine of the non-statistically significant variants are affecting the disease in the same direction as previously reported in other populations, suggesting that these associations might not have been identified in this population due to reduced power. However, lack of association may also be attributed to true genetic diversity. At least for the significant loci and for the majority of the loci affecting the disease in the same direction, the obtained risk allele frequencies (RAF) and effect sizes in the Greek-Cypriot population, are similar, with overlapping confidence intervals with the published studies. The highest difference in the obtained RAF compared to the published study data was observed for SNP rs2237892 (*KCNQ1*), which was initially identified in the Japanese population [29] (Japanese RAF = 0.61, Greek-Cypriot RAF = 0.96). However, the equivalent reported European RAF is 0.93 [29], which is close to the RAF obtained in the present study. The trend of 14/19 SNPs to affect the disease in the same direction suggests that these variants can have a potential predictive value in the Greek-Cypriot population. Future investigation targeting a larger sample size and a larger selection of SNPs is expected to overcome the current limitations and enable the genetic characterization of T2D susceptibility in this population. Furthermore, a genome-wide study might reveal novel variants, thus expanding the knowledge on genetic susceptibility of T2D.

The association of age at T2D diagnosis with the 19 SNPs was further tested. Previous studies in other populations revealed associations of the age at T2D first diagnosis with specific loci, including variants in the *TCF7L2*, *FTO* and *TMEM* genes [22,44,46,47]. In this study, significant associations of age at T2D diagnosis with SNPs rs7578597 (*THADA*) and rs7901695 (*TCF7L2*) were detected at nominal significance, however both associations did not survive Bonferroni correction.

To our knowledge, similar replication studies in neighbouring populations have only been performed in the Lebanese population [12,48,49] and more recently, a T2D GWAS has also been reported [50]. In the Lebanese population two of the replicated associations (rs7901695 in *TCF7L2* and rs4402960 in *IGF2BP2*) are common with replicated associations in the Greek-Cypriot population. Additional SNPs (rs13266634 in *SLC30A8* and rs8050136 in *FTO*) replicated with significant association in this study, were not statistically associated with T2D in the Lebanese population [12]. The Lebanese replication studies did not include an analysis of the fifth replicated association (rs5015480 in *HHEX*) in the Cypriot population. Furthermore, a lack of association in the Lebanese population is reported for five of this study's non-significant SNPs (rs864745, rs7578597, rs10923931, rs4607103, rs10010131) [12], for three of which (rs864745, rs7578597 and rs4607103) the currently obtained effect sizes are in the opposite direction compared to the European GWAS. In the recent Lebanese GWAS study, leading variants in two loci (rs7766070 in *CDKAL1* and rs34872471 in *TCF7L2*) were reported with genome-wide significant association [50].

A few small studies relevant to T2D susceptibility or to disease complications have been reported in the Greek population focused on loci that were not identified through GWAS [51,52]. Association of T2D with *C1q* SNP rs2920001 [51] and diabetic nephropathy with *IL-6* SNP rs1800795 were identified in the Greek population [52]. In a third study, five of the established T2D susceptibility loci were tested to examine if Gestational Diabetes Mellitus (GDM) exhibits a genetic predisposition similar to that of T2D [53]. An association of GDM with two of these loci, including a *TCF7L2* gene variant (rs7903146) was obtained through the study. Overall, published studies do not allow for a comparison between the Greek and the Greek-Cypriot populations due to the absence of any common study SNPs or diabetes study type.

A replication study has not been performed in any of the remaining neighbouring populations. The majority of studies in the Turkish population were focused on variants in single genes, which identified the association of T2D with *CAPN10* variants [54,55], an *IRS1* variant [56], some *ABCC8* variants [57] and some *ADIPOQ* variants [58]. In one study, a lack of association was reported for the

KCNJ11 variant rs5219 [57], in agreement with the lack of association also reported in the Greek-Cypriot population. In the Egyptian population, T2D susceptibility was associated with *MTHFR* [59], *IL-4*, *IL-13* [60] and *GSTP1* [61] variants. Overall, published studies in both the Turkish and the Egyptian populations do not allow for a comparison with the Greek-Cypriot population because different SNPs were investigated in each study.

The majority of T2D susceptibility loci including four of the current study replicated loci (*TCF7L2*, *HHEX*, *SLC30A8* and *IGF2BP2*), have been associated with impairments in insulin secretion or sensitivity [4,62]. In addition, some loci, including two of the current study replicated loci (*FTO* and IGF2BP2), have been associated with regulation of adipogenesis [45,63]. However, less information exists on the molecular mechanisms through which the associated SNPs can alter the function of each gene. Previous studies on candidate genes and the DNA methylation profiling in pancreatic islets from patients and controls, demonstrated a key role for epigenetic modifications in T2D pathogenesis [64–67]. Epigenetic modifications have more recently been proposed as a potential mechanism through which the associated SNPs can alter the normal function of a T2D candidate gene as well. Therefore, an interaction of genetic and epigenetic mechanisms affecting T2D susceptibility may be implicated in the molecular mechanisms through which susceptibility SNPs alter the function of each gene [1,68]. A recent study was focused on 19 of the T2D associated SNPs that introduce or delete possible Cytosine-phosphate-Guanine (CpG) methylation sites, including three of the statistically significant SNPs associated with T2D in our study (rs7901695, rs5015480 and rs13266634). All tested CpG SNPs have been associated with differential methylation of these sites in human non-diabetic pancreatic islets and some of them, including rs7901695, with differential methylation of surrounding CpG sites as well. In addition, some of these SNPs exhibiting differential DNA methylation, were associated with altered gene expression, alternative splicing events (including rs13266634 and rs7901695) and hormone secretion (including rs5015480) in the human islets [1]. A more recent study on genome-wide DNA methylation of pancreatic islets showed that DNA methylation patterns of T2D candidate genes, including the *TCF7L2*, *FTO* and *HHEX*, were altered in human islets from patients with T2D compared to controls without diabetes. New target genes with altered DNA methylation and expression in human T2D islets contributing to impaired insulin and glucagon secretion, have also been identified [68].

5. Conclusions

This study initiates the genetic assessment of T2D in a distinct newly investigated population. Statistically significant associations of T2D with five established loci (*TCF7L2*, *FTO*, *HHEX*, *SLC30A8* and *IGF2BP2*) are replicated in the Greek-Cypriot population. Fourteen of the nineteen loci tested had ORs in the same direction as previously published, suggesting that these variants can potentially be used in the Greek-Cypriot population for predictive testing of T2D.

Acknowledgments: This work was supported from the European Regional Structural Funds and the Republic of Cyprus through a grant from the Cyprus Research Promotion Foundation (ΥΓΕΙΑ/ΔΥΓΕΙΑ/0609(ΒΙΕ)/03). The authors thank: (1) all the patients and controls without diabetes and the Cyprus Diabetic Association for their participation in the study; (2) the management staff of the blood donation station of Engomi, the management and nursing staff of Kyperounta hospital, the nursing homes of Solea, Agia Marina Strovolos, the Agia Varvara Kaimakli community nursing home and the Cyprus Institute of Neurology and Genetics clinics that significantly contributed to the recruitment of controls without diabetes; (3) the volunteers Anna Minaidou and Evie Votsi for their contribution in the recruitment of control samples.

Author Contributions: All authors conceived and designed the experiments; C.V. performed the experiments; K.M. and A.A. analysed the data; all authors contributed reagents/materials/analysis tools; C.V., K.M. and K.C. wrote the paper.

References

1.   Dayeh, T.A.; Olsson, A.H.; Volkov, P.; Almgren, P.; Ronn, T.; Ling, C. Identification of CpG-SNPs associated with type 2 diabetes and differential DNA methylation in human pancreatic islets. *Diabetologia* **2013**, *56*, 1036–1046. [CrossRef] [PubMed]

2.   Sanghera, D.K.; Blackett, P.R. Type 2 Diabetes Genetics: Beyond GWAS. *J. Diabetes Metab.* **2012**, *3*. [CrossRef] [PubMed]

3.   Gupta, V.; Vinay, D.G.; Rafiq, S.; Kranthikumar, M.V.; Janipalli, C.S.; Giambartolomei, C.; Evans, D.M.; Mani, K.R.; Sandeep, M.N.; Taylor, A.E.; et al. Association analysis of 31 common polymorphisms with type 2 diabetes and its related traits in indian sib pairs. *Diabetologia* **2012**, *55*, 349–357. [CrossRef] [PubMed]

4.   Ali, S.; Chopra, R.; Manvati, S.; Singh, Y.P.; Kaul, N.; Behura, A.; Mahajan, A.; Sehajpal, P.; Gupta, S.; Dhar, M.K.; et al. Replication of Type 2 Diabetes Candidate Genes Variations in Three Geographically Unrelated Indian Population Groups. *PLoS ONE* **2013**, *8*, e58881. [CrossRef] [PubMed]

5.   Saxena, R.; Elbers, C.C.; Guo, Y.; Peter, I.; Gaunt, T.R.; Mega, J.L.; Lanktree, M.B.; Tare, A.; Castillo, B.A.; Li, Y.R.; et al. Large-Scale Gene-Centric Meta-Analysis across 39 Studies Identifies Type 2 Diabetes Loci. *Am. J. Hum. Genet.* **2012**, *90*, 410–425. [CrossRef] [PubMed]

6.   Ayub, Q.; Moutsianas, L.; Chen, Y.; Panoutsopoulou, K.; Colonna, V.; Pagani, L.; Prokopenko, I.; Ritchie, G.R.; Tyler-Smith, C.; McCarthy, M.I.; et al. Revisiting the Thrifty Gene Hypothesis via 65 Loci Associated with Susceptibility to Type 2 Diabetes. *Am. J. Hum. Genet.* **2014**, *94*, 176–185. [CrossRef] [PubMed]

7.   Ali, O. Genetics of type 2 diabetes. *World J. Diabetes* **2013**, *4*, 114–123. [CrossRef] [PubMed]

8.   Basile, K.J.; Johnson, M.E.; Xia, Q.; Grant, S.F. Genetic Susceptibility to Type 2 Diabetes and Obesity: Follow-Up of Findings from Genome-Wide Association Studies. *Int. J. Endocrinol.* **2014**, *2014*, 769671. [CrossRef] [PubMed]

9.   Zhang, Y.; Liu, Y.; Su, Z. Genetic Variants of Retinoic Acid Receptor-Related Orphan Receptor Alpha Determine Susceptibility to Type 2 Diabetes Mellitus in Han Chinese. *Genes* **2016**, *7*, 54. [CrossRef] [PubMed]

10. Mohlke, K.L.; Boehnke, M. Recent advances in understanding the genetic architecture of type 2 diabetes. *Hum. Mol. Genet.* **2015**, *24*, R85–R92. [CrossRef] [PubMed]

11. Phani, N.M.; Adhikari, P.; Nagri, S.K.; D'Souza, S.C.; Satyamoorthy, K.; Rai, P.S. Replication and Relevance of Multiple Susceptibility Loci Discovered from Genome Wide Association Studies for Type 2 Diabetes in an Indian Population. *PLoS ONE* **2016**, *11*, e0157364. [CrossRef] [PubMed]

12. Almawi, W.Y.; Nemr, R.; Keleshian, S.H.; Echtay, A.; Saldanha, F.L.; AlDoseri, F.A.; Racoubian, E. A replication study of 19 GWAS-validated type 2 diabetes at-risk variants in the Lebanese population. *Diabetes Res. Clin. Pract.* **2013**, *102*, 117–122. [CrossRef] [PubMed]

13. Hu, C.; Zhang, R.; Wang, C.; Wang, J.; Ma, X.; Lu, J.; Qin, W.; Hou, X.; Bao, Y.; Xiang, K.; et al. *PPARG, KCNJ11, CDKAL1, CDKN2A-CDKN2B, IDE-KIF11-HHEX, IGF2BP2* and *SLC30A8* Are Associated with Type 2 Diabetes in a Chinese Population. *PLoS ONE* **2009**, *4*, e7643. [CrossRef] [PubMed]

14. Matsuba, R.; Sakai, K.; Imamura, M.; Tanaka, Y.; Iwata, M.; Hirose, H.; Kaku, K.; Maegawa, H.; Watada, H.; Tobe, K.; et al. Replication Study in a Japanese Population to Evaluate the Association between 10 SNP Loci, Identified in European Genome-Wide Association Studies, and Type 2 Diabetes. *PLoS ONE* **2015**, *10*, e0126363. [CrossRef] [PubMed]

15. Ng, M.C. Genetics of Type 2 Diabetes in African Americans. *Curr. Diab. Rep.* **2015**, *15*, 74. [CrossRef] [PubMed]

16. Kato, N. Insights into the genetic basis of type 2 diabetes. *J. Diabetes Investig.* **2013**, *4*, 233–244. [CrossRef] [PubMed]

17. Demographic report 2014, statistical service of Cyprus. Available online: http://www.mof.gov.cy/mof/cystat/statistics.nsf/all/2b7ccdc7c637c864c225807b0032830d/\protect\T1\textdollarfile/demographic_report-2014-271115.pdf?openelement (accessed on 1 March 2016).

18. Baysal, E.; Indrak, K.; Bozkurt, G.; Berkalp, A.; Aritkan, E.; Old, J.M.; Ioannou, P.; Angastiniotis, M.; Droushiotou, A.; Yuregir, G.T.; et al. The β-thalassaemia mutations in the population of Cyprus. *Br. J. Haematol.* **1992**, *81*, 607–609. [CrossRef] [PubMed]

19. Hadjisavvas, A.; Loizidou, M.A.; Middleton, N.; Michael, T.; Papachristoforou, R.; Kakouri, E.; Daniel, M.; Papadopoulos, P.; Malas, S.; Marcou, Y.; et al. An investigation of breast cancer risk factors in Cyprus: A case control study. *BMC Cancer* **2010**, *10*, 447. [CrossRef] [PubMed]

20. Loizou, T.; Pouloukas, S.; Tountas, C.; Thanopoulou, A.; Karamanos, V. An epidemiologic study on the prevalence of diabetes, glucose intolerance, and metabolic syndrome in the adult population of the Republic of Cyprus. *Diabetes Care* **2006**, *29*, 1714–1715. [CrossRef] [PubMed]

21. International diabetes atlas, the sixth edition, international diabetes federation 2013. Available online: www.idf.org/diabetesatlas (accessed on 15 September 2014).

22. Zeggini, E.; Weedon, M.N.; Lindgren, C.M.; Frayling, T.M.; Elliott, K.S.; Lango, H.; Timpson, N.J.; Perry, J.R.; Rayner, N.W.; Freathy, R.M.; et al. Replication of genome-wide association signals in UK samples reveals risk loci for type 2 diabetes. *Science* **2007**, *316*, 1336–1341. [CrossRef] [PubMed]

23. Zeggini, E.; Scott, L.J.; Saxena, R.; Voight, B.F.; Marchini, J.L.; Hu, T.; de Bakker, P.I.; Abecasis, G.R.; Almgren, P.; Andersen, G.; et al. Meta-analysis of genome-wide association data and large-scale replication identifies additional susceptibility loci for type 2 diabetes. *Nat. Genet.* **2008**, *40*, 638–645. [CrossRef] [PubMed]

24. Scott, L.J.; Mohlke, K.L.; Bonnycastle, L.L.; Willer, C.J.; Li, Y.; Duren, W.L.; Erdos, M.R.; Stringham, H.M.; Chines, P.S.; Jackson, A.U.; et al. A Genome-Wide Association Study of Type 2 Diabetes in Finns Detects Multiple Susceptibility Variants. *Science* **2007**, *316*, 1341–1345. [CrossRef] [PubMed]

25. Sandhu, M.S.; Weedon, M.N.; Fawcett, K.A.; Wasson, J.; Debenham, S.L.; Daly, A.; Lango, H.; Frayling, T.M.; Neumann, R.J.; Sherva, R.; et al. Common variants in *WFS1* confer risk of type 2 diabetes. *Nat. Genet.* **2007**, *39*, 951–953. [CrossRef] [PubMed]

26. Voight, B.F.; Scott, L.J.; Steinthorsdottir, V.; Morris, A.P.; Dina, C.; Welch, R.P.; Zeggini, E.; Huth, C.; Aulchenko, Y.S.; Thorleifsson, G.; et al. Twelve type 2 diabetes susceptibility loci identified through large-scale association analysis. *Nat. Genet.* **2010**, *42*, 579–589. [CrossRef] [PubMed]

27. Saxena, R.; Voight, B.F.; Lyssenko, V.; Burtt, N.P.; de Bakker, P.I.; Chen, H.; Roix, J.J.; Kathiresan, S.; Hirschhorn, J.N.; Daly, M.J.; et al. Genome-wide association analysis identifies loci for type 2 diabetes and triglyceride levels. *Science* **2007**, *316*, 1331–1336. [PubMed]

28. Prokopenko, I.; Langenberg, C.; Florez, J.C.; Saxena, R.; Soranzo, N.; Thorleifsson, G.; Loos, R.J.; Manning, A.K.; Jackson, A.U.; Aulchenko, Y.; et al. Variants in *MTNR1B* influence fasting glucose levels. *Nat. Genet.* **2009**, *41*, 77–81. [CrossRef] [PubMed]

29. Yasuda, K.; Miyake, K.; Horikawa, Y.; Hara, K.; Osawa, H.; Furuta, H.; Hirota, Y.; Mori, H.; Jonsson, A.; Sato, Y.; et al. Variants in *KCNQ1* are associated with susceptibility to type 2 diabetes mellitus. *Nat. Genet.* **2008**, *40*, 1092–1097. [CrossRef] [PubMed]

30. Winckler, W.; Weedon, M.N.; Graham, R.R.; McCarroll, S.A.; Purcell, S.; Almgren, P.; Tuomi, T.; Gaudet, D.; Bostrom, K.B.; Walker, M.; et al. Evaluation of Common Variants in the Six Known Maturity-Onset Diabetes of the Young (MODY) Genes for Association With Type 2 Diabetes. *Diabetes* **2007**, *56*, 685–693. [CrossRef] [PubMed]

31. Wigginton, J.E.; Cutler, D.J.; Abecasis, G.R. A Note on Exact Tests of Hardy-Weinberg Equilibrium. *Am. J. Hum. Genet.* **2005**, *76*, 887–893. [CrossRef] [PubMed]

32. R Development Core Team. R: A language and environment for statistical computing. R foundation for statistical computing: Vienna, Austria, 2011. Available online: http://www.R-project.Org (accessed on 5 March 2014).

33. Purcell, S. Plink (v1.07). Available online: http://pngu.mgh.harvard.edu/purcell/plink/ (accessed on 12 January 2014).

34. Purcell, S.; Neale, B.; Todd-Brown, K.; Thomas, L.; Ferreira, M.A.; Bender, D.; Maller, J.; Sklar, P.; de Bakker, P.I.; Daly, M.J.; et al. PLINK: A Tool Set for Whole-Genome Association and Population-Based Linkage Analyses. *Am. J. Hum. Genet.* **2007**, *81*, 559–575. [CrossRef] [PubMed]

35. Swen, J.J.; Baak-Pablo, R.F.; Guchelaar, H.J.; van der Straaten, T. Alternative methods to a TaqMan assay to detect a tri-allelic single nucleotide polymorphism rs757210 in the *HNF1β* gene. *Clin. Chem. Lab. Med.* **2012**, *50*, 279–284. [CrossRef] [PubMed]

36. Holmkvist, J.; Almgren, P.; Lyssenko, V.; Lindgren, C.M.; Eriksson, K.F.; Isomaa, B.; Tuomi, T.; Nilsson, P.; Groop, L. Common Variants in Maturity-Onset Diabetes of the Young Genes and Future Risk of Type 2 Diabetes. *Diabetes* **2008**, *57*, 1738–1744. [CrossRef] [PubMed]

37. Sladek, R.; Rocheleau, G.; Rung, J.; Dina, C.; Shen, L.; Serre, D.; Boutin, P.; Vincent, D.; Belisle, A.; Hadjadj, S.; et al. A genome-wide association study identifies novel risk loci for type 2 diabetes. *Nature* **2007**, *445*, 881–885. [CrossRef] [PubMed]

38. Waters, K.M.; Stram, D.O.; Hassanein, M.T.; Le Marchand, L.; Wilkens, L.R.; Maskarinec, G.; Monroe, K.R.; Kolonel, L.N.; Altshuler, D.; Henderson, B.E.; et al. Consistent Association of Type 2 Diabetes Risk Variants Found in Europeans in Diverse Racial and Ethnic groups. *PLoS Genet.* **2010**, *6*. [CrossRef] [PubMed]

39. Loizidou, M.A.; Hadjisavvas, A.; Ioannidis, J.P.; Kyriacou, K. Replication of genome-wide discovered breast cancer risk loci in the Cypriot population. *Breast Cancer Res.Treat.* **2011**, *128*, 267–272. [CrossRef] [PubMed]

40. Novembre, J.; Johnson, T.; Bryc, K.; Kutalik, Z.; Boyko, A.R.; Auton, A.; Indap, A.; King, K.S.; Bergmann, S.; Nelson, M.R.; et al. Genes mirror geography within Europe. *Nature* **2008**, *456*, 98–101. [PubMed]

41. Voskarides, K.; Mazieres, S.; Hadjipanagi, D.; Di Cristofaro, J.; Ignatiou, A.; Stefanou, C.; King, R.J.; Underhill, P.A.; Chiaroni, J.; Deltas, C. Y-chromosome phylogeographic analysis of the Greek-Cypriot population reveals elements consistent with Neolithic and Bronze age settlements. *Investig. Genet.* **2016**, *7*, 1. [CrossRef] [PubMed]

42. Omori, S.; Tanaka, Y.; Takahashi, A.; Hirose, H.; Kashiwagi, A.; Kaku, K.; Kawamori, R.; Nakamura, Y.; Maeda, S. Association of *CDKAL1, IGF2BP2, CDKN2A/B, HHEX, SLC30A8,* and *KCNJ11* With Susceptibility to Type 2 Diabetes in a Japanese Population. *Diabetes* **2008**, *57*, 791–795. [CrossRef] [PubMed]

43. Rong, R.; Hanson, R.L.; Ortiz, D.; Wiedrich, C.; Kobes, S.; Knowler, W.C.; Bogardus, C.; Baier, L.J. Association analysis of variation in/near *FTO, CDKAL1, SLC30A8, HHEX, EXT2, IGF2BP2, LOC387761,* and *CDKN2B* with type 2 diabetes and related quantitative traits in Pima Indians. *Diabetes* **2009**, *58*, 478–488. [CrossRef] [PubMed]

44. Kalnina, I.; Zaharenko, L.; Vaivade, I.; Rovite, V.; Nikitina-Zake, L.; Peculis, R.; Fridmanis, D.; Geldnere, K.; Jacobsson, J.A.; Almen, M.S.; et al. Polymorphisms in *FTO* and near *TMEM18* associate with type 2 diabetes and predispose to younger age at diagnosis of diabetes. *Gene* **2013**, *527*, 462–468. [CrossRef] [PubMed]

45. Abbas, S.; Raza, S.T.; Ahmed, F.; Ahmad, A.; Rizvi, S.; Mahdi, F. Association of Genetic polymorphism of *PPARγ-2, ACE, MTHFR, FABP-2* and *FTO* genes in risk prediction of type 2 diabetes mellitus. *J. Biomed. Sci.* **2013**, *20*, 80. [CrossRef] [PubMed]

46. Silbernagel, G.; Renner, W.; Grammer, T.B.; Hugl, S.R.; Bertram, J.; Kleber, M.E.; Hoffmann, M.M.; Winkelmann, B.R.; Marz, W.; Boehm, B.O. Association of *TCF7L2*SNPs with age at onset of type 2 diabetes and proinsulin/insulin ratio but not with glucagon-like peptide 1. *Diabetes Metab. Res. Rev.* **2011**, *27*, 499–505. [CrossRef] [PubMed]

47. Tangjittipokin, W.; Chongjarean, N.; Plengvidhya, N.; Homsanit, M.; Yenchitsomanus, P.T. Transcription factor 7-like 2 (*TCF7L2*) variations associated with earlier age-onset of type 2 diabetes in Thai patients. *J. Genet.* **2012**, *91*, 251–255. [CrossRef] [PubMed]

48. Nemr, R.; Echtay, A.; Dashti, E.A.; Almawi, A.W.; Al-Busaidi, A.S.; Keleshian, S.H.; Irani-Hakime, N.; Almawi, W.Y. Strong Association of Common Variants in the *IGF2BP2* Gene with Type 2 Diabetes in Lebanese Arabs. *Diabetes Res. Clin. Pract.* **2012**, *96*, 225–229. [CrossRef] [PubMed]

49. Nemr, R.; Turki, A.; Echtay, A.; Al-Zaben, G.S.; Daher, H.S.; Irani-Hakime, N.A.; Keleshian, S.H.; Almawi, W.Y. Transcription factor-7-like 2 gene variants are strongly associated with type 2 diabetes in Lebanese subjects. *Diabetes Res. Clin. Pract.* **2012**, *98*, e23–e27. [CrossRef] [PubMed]

50. Ghassibe-Sabbagh, M.; Haber, M.; Salloum, A.K.; Al-Sarraj, Y.; Akle, Y.; Hirbli, K.; Romanos, J.; Mouzaya, F.; Gauguier, D.; Platt, D.E.; et al. T2DM GWAS in the Lebanese population confirms the role of *TCF7L2* and *CDKAL1* in disease susceptibility. *Sci. Rep.* **2014**, *4*, 7351. [CrossRef] [PubMed]

51. Goulielmos, G.N.; Samonis, G.; Apergi, M.; Christofaki, M.; Valachis, A.; Zervou, M.I.; Kofteridis, D.P. *C1q* but not mannose-binding lectin (*Mbl-2*) gene polymorphisms are associated with type 2 diabetes in the genetically homogeneous population of the island of Crete in Greece. *Hum. Immunol.* **2013**, *74*, 878–881. [CrossRef] [PubMed]

52. Papaoikonomou, S.; Tentolouris, N.; Tousoulis, D.; Papadodiannis, D.; Miliou, A.; Papageorgiou, N.; Hatzis, G.; Stefanadis, C. The association of the 174G>C polymorphism of interleukin 6 gene with diabetic nephropathy in patients with type 2 diabetes mellitus. *J. Diabetes Complicat.* **2013**, *27*, 576–579. [CrossRef] [PubMed]

53. Pappa, K.I.; Gazouli, M.; Economou, K.; Daskalakis, G.; Anastasiou, E.; Anagnou, N.P.; Antsaklis, A. Gestational diabetes mellitus shares polymorphisms of genes associated with insulin resistance and type 2 diabetes in the Greek population. *Gynecol. Endocrinol.* **2011**, *27*, 267–272. [CrossRef] [PubMed]

54. Arslan, E.; Acik, L.; Gunaltili, G.; Ayvaz, G.; Altinova, A.E.; Arslan, M. The effect of calpain-10 gene polymorphism on the development of type 2 diabetes mellitus in a Turkish population. *Endokrynol. Pol.* **2014**, *65*, 90–95. [CrossRef] [PubMed]

55. Demirci, H.; Yurtcu, E.; Ergun, M.A.; Yazici, A.C.; Karasu, C.; Yetkin, I. Calpain 10 SNP-44 gene polymorphism affects susceptibility to type 2 diabetes mellitus and diabetic-related conditions. *Genet. Test.* **2008**, *12*, 305–309. [CrossRef] [PubMed]

56. Orkunoglu Suer, F.E.; Mergen, H.; Bolu, E.; Ozata, M. Molecular scanning for mutations in the insulin receptor substrate-1 (*IRS-1*) gene in Turkish with type 2 diabetes mellitus. *Endocr. J.* **2005**, *52*, 593–598. [CrossRef] [PubMed]

57. Gonen, M.S.; Arikoglu, H.; Erkoc Kaya, D.; Ozdemir, H.; Ipekci, S.H.; Arslan, A.; Kayis, S.A.; Gogebakan, B. Effects of single nucleotide polymorphisms in $K_{(ATP)}$ channel genes on type 2 diabetes in a Turkish population. *Arch. Med. Res.* **2012**, *43*, 317–323. [CrossRef] [PubMed]

58. Arikoglu, H.; Ozdemir, H.; Kaya, D.E.; Ipekci, S.H.; Arslan, A.; Kayis, S.A.; Gonen, M.S. The Adiponectin variants contribute to the genetic background of type 2 diabetes in Turkish population. *Gene* **2014**, *534*, 10–16. [CrossRef] [PubMed]

59. Settin, A.; El-Baz, R.; Ismaeel, A.; Tolba, W.; Allah, W.A. Association of *ACE* and *MTHFR* genetic polymorphisms with type 2 diabetes mellitus: Susceptibility and complications. *J. Renin Angiotensin Aldosterone Syst.* **2015**, *16*, 838–843. [CrossRef] [PubMed]

60. Alsaid, A.; El-Missiry, M.; Hatata el, S.; Tarabay, M.; Settin, A. Association of *IL*-4-590 C>T and *IL*-13-1112 C>T Gene Polymorphisms with the Susceptibility to Type 2 Diabetes Mellitus. *Dis. Markers* **2013**, *35*, 243–247. [CrossRef] [PubMed]

61. Amer, M.A.; Ghattas, M.H.; Abo-Elmatty, D.M.; Abou-El-Ela, S.H. Evaluation of glutathione S-transferase P1 genetic variants affecting type-2 diabetes susceptibility and glycemic control. *Arch. Med. Sci.* **2012**, *8*, 631–636. [CrossRef] [PubMed]

62. Ruchat, S.M.; Elks, C.E.; Loos, R.J.; Vohl, M.C.; Weisnagel, S.J.; Rankinen, T.; Bouchard, C.; Perusse, L. Association between insulin secretion, insulin sensitivity and type 2 diabetes susceptibility variants identified in genome-wide association studies. *Acta Diabetol.* **2009**, *46*, 217–226. [CrossRef] [PubMed]

63. Wu, H.H.; Liu, N.J.; Yang, Z.; Tao, X.M.; Du, Y.P.; Wang, X.C.; Lu, B.; Zhang, Z.Y.; Hu, R.M.; Wen, J. *IGF2BP2* and obesity interaction analysis for type 2 diabetes mellitus in Chinese Han population. *Eur. J. Med. Res.* **2014**, *19*, 40. [CrossRef] [PubMed]

64. Yang, B.T.; Dayeh, T.A.; Kirkpatrick, C.L.; Taneera, J.; Kumar, R.; Groop, L.; Wollheim, C.B.; Nitert, M.D.; Ling, C. Insulin promoter DNA methylation correlates negatively with insulin gene expression and positively with $HbA_{(1C)}$ levels in human pancreatic islets. *Diabetologia* **2011**, *54*, 360–367. [CrossRef] [PubMed]

65. Yang, B.T.; Dayeh, T.A.; Volkov, P.A.; Kirkpatrick, C.L.; Malmgren, S.; Jing, X.; Renstrom, E.; Wollheim, C.B.; Nitert, M.D.; Ling, C. Increased DNA methylation and decreased expression of *PDX-1* in pancreatic islets from patients with type 2 diabetes. *Mol. Endocrinol.* **2012**, *26*, 1203–1212. [CrossRef] [PubMed]

66. Volkmar, M.; Dedeurwaerder, S.; Cunha, D.A.; Ndlovu, M.N.; Defrance, M.; Deplus, R.; Calonne, E.; Volkmar, U.; Igoillo-Esteve, M.; Naamane, N.; et al. DNA methylation profiling identifies epigenetic dysregulation in pancreatic islets from type 2 diabetic patients. *EMBO J.* **2012**, *31*, 1405–1426. [CrossRef] [PubMed]

67. Franks, P.W. Gene x environment interactions in type 2 diabetes. *Curr. Diab. Rep.* **2011**, *11*, 552–561. [CrossRef] [PubMed]

68. Dayeh, T.; Volkov, P.; Salo, S.; Hall, E.; Nilsson, E.; Olsson, A.H.; Kirkpatrick, C.L.; Wollheim, C.B.; Eliasson, L.; Ronn, T.; et al. Genome-Wide DNA Methylation Analysis of Human Pancreatic Islets from Type 2 Diabetic and Non-Diabetic Donors Identifies Candidate Genes That Influence Insulin Secretion. *PLoS Genet.* **2014**, *10*, e1004160. [CrossRef] [PubMed]

mRNA Expression and DNA Methylation Analysis of Serotonin Receptor 2A (*HTR2A*) in the Human Schizophrenic Brain

Sern-Yih Cheah [1], Bruce R. Lawford [1,2], Ross McD. Young [3], Charles P. Morris [1] and Joanne Voisey [1,*]

[1] School of Biomedical Sciences, Institute of Health and Biomedical Innovation, Queensland University of Technology, 60 Musk Ave., Kelvin Grove, Queensland 4059, Australia; sern.cheah@hdr.qut.edu.au (S.-Y.C.); lawford.b@gmail.com (B.R.L.); p.morris@qut.edu.au (C.P.M.)

[2] Discipline of Psychiatry, Royal Brisbane and Women's Hospital, Herston, Queensland 4006, Australia

[3] Faculty of Health, Institute of Health and Biomedical Innovation, Queensland University of Technology, 60 Musk Ave., Kelvin Grove, Queensland 4059, Australia; rm.young@qut.edu.au

* Correspondence: j.voisey@qut.edu.au

Academic Editor: Dennis R. Grayson

Abstract: Serotonin receptor 2A (*HTR2A*) is an important signalling factor implicated in cognitive functions and known to be associated with schizophrenia. The biological significance of *HTR2A* in schizophrenia remains unclear as molecular analyses including genetic association, mRNA expression and methylation studies have reported inconsistent results. In this study, we examine *HTR2A* expression and methylation and the interaction with *HTR2A* polymorphisms to identify their biological significance in schizophrenia. Subjects included 25 schizophrenia and 25 control post-mortem brain samples. Genotype and mRNA data was generated by transcriptome sequencing. DNA methylation profiles were generated for CpG sites within promoter-exon I region. Expression, genotype and methylation data were examined for association with schizophrenia. *HTR2A* mRNA levels were reduced by 14% ($p = 0.006$) in schizophrenia compared to controls. Three CpG sites were hypermethylated in schizophrenia (cg5 $p = 0.028$, cg7 $p = 0.021$, cg10 $p = 0.017$) and *HTR2A* polymorphisms rs6314 ($p = 0.008$) and rs6313 ($p = 0.026$) showed genetic association with schizophrenia. Differential DNA methylation was associated with rs6314 and rs6313. There was a strong correlation between *HTR2A* DNA methylation and mRNA expression. The results were nominally significant but did not survive the rigorous Benjamini-Hochberg correction for multiple testing. Differential *HTR2A* expression in schizophrenia in our study may be the result of the combined effect of multiple differentially methylated CpG sites. Epigenetic *HTR2A* regulation may alter brain function, which contributes to the development of schizophrenia.

Keywords: *HTR2A*; schizophrenia association; mRNA expression; DNA methylation; rs6314; rs6313; epigenetics

1. Introduction

The serotonin receptor 2A (*HTR2A*) is an important signalling factor implicated in high-order cognition [1,2]. It is found expressed abundantly in the glutamatergic interneurons and gamma-aminobutyric acid (GABA)-ergic neurons in the prefrontal cortex and hippocampal regions and both neurotransmission systems are known to be involved in the pathogenesis of schizophrenia [3–6]. The serotonin receptor 2A gene (*HTR2A*) was shown to influence prefrontal cognition via binding of receptor agonists including quipazine and 3-trifluoromethylphenylpiperazine [7]. Deficits in prefrontal cognitive function (such as executive roles and working memory) are a core feature of schizophrenia [8]. In a later review, it was suggested that HTR2A antagonists were used as treatments aimed to improve cognitive function, although the efficacy of HTR2A antagonists has yet to be unequivocally established [9]. Overall, this evidence strongly suggests a role for *HTR2A* in the pathogenesis of schizophrenia.

Based on pharmacological and expression studies, results suggest that the downregulated *HTR2A* mRNA expression and reduction of receptor density or activity are associated with schizophrenia [7,10–12]. According to a review by Selvaraj et al., nine frontal cortex studies found 54 patients with decreased HTR2A receptor activity in schizophrenia [13]. However, one cohort of antipsychotic-free patients revealed upregulation of HTR2A receptor density in the prefrontal regions [14,15]. Maple et al. suggested that stress-induced expression of *HTR2A* as an adaptive function and is disrupted in schizophrenia patients. Although the induced expression of *HTR2A* may appear beneficial, along with the majority of the evidence pointing to the increased expression of *HTR2A* in healthy individuals, it is rather paradoxical that antipsychotics block the action of HTR2A [16]. This paradox is further highlighted by the fact that hallucinogens (such as lysergic acid diethylamide) are HTR2A agonists, while atypical antipsychotics are antagonists [16]. While clozapine can be considered an antagonist, it eventually triggers downstream activation of Akt, similar to the effect of serotonin agonists via different mechanisms [17]. Although such inconsistency remains to be clarified, these studies suggest an association between the dysregulation of *HTR2A* mRNA expression and schizophrenia.

The *HTR2A* polymorphism, rs6314 is a non-synonymous DNA variant located in exon 3 that results in a His452Tyr substitution. Studies suggest that rs6314 may have an effect on calcium signalling and mobilisation and altered activation of phospholipases C and D, possibly resulting in reduced receptor activity [18,19], but it is not clear how this impacts on neurotransmission or susceptibility to schizophrenia. Serretti et al. reviewed a number of association studies between rs6314 and schizophrenia and found inconsistent results [20]. Decreased *HTR2A* expression was associated with rs6314 in one study [21] and the polymorphism was also associated with a number of endophenotypes including hippocampal volume and activity [22,23], memory [23,24], and clozapine treatment response in patients with schizophrenia [25].

A synonymous *HTR2A* polymorphism, rs6313 (T102C) is a well-studied variant that was found to be in linkage disequilibrium (LD) with another functional polymorphism (rs6311) known to alter *HTR2A* promoter activity [26]. A number of studies have either found [12,27,28] or failed to find an association [10,29] between rs6313 and schizophrenia. Poorer cognitive performance was found to be associated with the T-allele of rs6313 [30]. However, another study found that poorer visual sustained attention was associated with the C-allele [31]. These different findings may be due to significant ethnic differences between the two studies. Binding activity or receptor density of HTR2A in the brain [32,33] and differential expression in the temporal cortex [12] are associated with the rs6313 polymorphism. Although the biological mechanism of rs6313 and rs6314 remains unclear, the findings suggest that both polymorphisms are good candidates for schizophrenia risk.

DNA methylation is thought to be an important epigenetic mechanism in schizophrenia [34] as environmental influences on *HTR2A* DNA methylation are associated with infant neurobehavioural outcomes [35]. For *HTR2A*, the majority of the DNA methylation activity occurs within the promoter and exon I, but there is little DNA methylation activity in other regions of the gene. *HTR2A* CpG sites were either hypermethylated (near rs6311, at position −1438 of the promoter region) or hypomethylated (near rs6313, at position 102 of exon I) in the prefrontal cortex of patients with schizophrenia, potentially resulting in downregulation of *HTR2A* expression in patients with schizophrenia [10]. Another study also reported similar results in the saliva of patients with schizophrenia [36]. Increased DNA methylation at rs6313 in peripheral leukocytes was reported in major psychosis patients with suicidal tendency [37], suggesting the potential involvement of *HTR2A* promoter hypermethylation in psychosis. Based on the literature, there is strong evidence of schizophrenia-specific DNA methylation changes in *HTR2A* that influence mRNA expression.

While it is clear that DNA methylation alters *HTR2A* mRNA expression [10,38–40] (apart from other known factors such as stress, medication and substance use, nutrition history, comorbidities and other underlying biological factors), *HTR2A* polymorphisms have also been reported to influence mRNA expression, possibly by altering recognition of CpG sites. One study [21] found that *HTR2A* rs6314 was associated with mRNA expression changes, although the effects of promoter DNA methylation were not taken into account. In this study, we investigated the role of *HTR2A* in schizophrenia by examining rs6314 and rs6313 genotype, DNA methylation, and the expression of *HTR2A* mRNA in prefrontal cortex samples of schizophrenia patients.

2. Materials and Methods

2.1. Samples

The brain tissue was provided by the Human Brain and Spinal Fluid Resource Centre, Los Angeles, California (courtesy of James Riehl). The brain tissues were collected from 25 schizophrenia patients and 25 healthy controls. Each sample consisted of a 7-mm coronal section that had been quick-frozen, followed by dissection of the frontal cortex (0.4–1.0 g) from the frozen sections. A summary of sample age, post-mortem interval (PMI) and gender are shown in Table 1, with complete sample information detailed in Supplementary Table S1. PMI represents the time interval from the death of a patient to the quick-freezing of the brain section. All but two of the patients with schizophrenia were using medication at their time of death. Five patients with schizophrenia were identified as suicide victims.

Extraction of RNA and DNA from post-mortem frontal cortex brain tissues was performed at the UCLA Clinical Microarray Core Laboratory (Los Angeles, CA, USA) using the Roche MagNa Pure Compact. There were five samples with missing mRNA data due to quality control failure, data generation failure and insufficient/no RNA content. Two samples had missing DNA methylation data as they were excluded due to protocol constraints (we could only run 48 samples due to the bead chip plate format).

Ethics approval for the project was obtained from the Human Research Ethics Committee of the Queensland University of Technology.

Table 1. Summary of demographic D=details.

	Control (*n* = 23)	Schizophrenia (*n* = 22)	*t*-Test/Chi-Square *p*-Value
Age, mean (s.d.)	70.2 (9.2)	52.5 (22.7)	0.0022
PMI, mean (s.d.)	14.1 (3.2)	24 (10.6)	0.0003
Sex, male (%)	18 (78)	14 (64)	0.2778

s.d., one standard deviation.

2.2. mRNA Data Generation

Sequencing of total RNA for each sample was performed by the Australian Genome Research Facility (AGRF). Quality of extracted RNA was assessed by electrophoresis using Agilent Bioanalyzer RNA 6000 Nano assay (Santa Clara, CA, USA). The percentage of RNA fragments >200 nucleotides was assessed to determine the appropriate input. Samples with <30% of RNA >200 nucleotides were excluded. A total of 46 samples passed the initial quality control (QC). Following sample QC, samples were processed and used to generate *HTR2A* mRNA sequence data with the TruSeq RNA Access Sample Preparation Kit as per the manufacturer's instructions (Illumina, San Diego, CA, USA). Briefly, total mRNA was firstly fragmented, followed by first and subsequently second strand cDNA synthesis. The blunt-ended cDNA were 3'-adenylated and ligated with adapters that hybridise to the flow cell. The adapter-added cDNA were subsequently amplified creating a cDNA library. One sample failed to generate enough library to proceed to capture hybridisation and was excluded from the final analysis. A total of 45 samples were further investigated.

Captured libraries were pooled for sequencing. Each pool of libraries was clustered on the Illumina cBot system (Illumina, San Diego, CA, USA) using HiSeq PE Cluster Kit v4 reagents (Illumina, San Diego, CA, USA) followed by sequencing on the Illumina HiSeq 2500 system with HiSeq SBS Kit v4 reagents with 159 cycles (75 base pair paired end reads). Illumina RTA 1.18.61 (Illumina, San Diego, CA, USA) was used for base calling, quality scoring. Bcl2fastq pipeline 1.8.4 (Illumina, San Diego, CA, USA) was used for de-multiplexing and FASTQ file generation.

2.3. mRNA Data Transformation

The total generated sequence reads in FASTQ files were mapped to the human genome (GRCh38/hg38) using Tophat 2.0.13 [41] to identify sequence reads that align/correspond to the *HTR2A* gene. Mapped RNA sequence reads aligned to *HTR2A* were subsequently counted using the tool feature Counts from the SubRead package 1.4.6p5 [42] to assign raw non-normalised read counts to genes (specifically *HTR2A*). Non-normalised read counts were used for the edgeR package 3.12.0 [43] to perform quality control and normalisation using the TMM (trimmed mean of M values) method [44]. Normalised read counts (expressed as reads per kilobase transcript per million mapped read counts or RPKM) were transformed to $\log_{10}$RPKM using SPSS (Statistical Package for the Social Sciences) 22 (SPSS, Chicargo, IL, USA).

2.4. Genotyping

Total RNA sequence reads were aligned to GRCh38/hg38 using BWA (Burrows-Wheeler Alignment) for Illumina version 1.2.3 (Digital Equipment Corporation, Palo Alto, CA, USA). Genotypes of rs6314 and rs6313 (loci shown in Figure 1) were determined by visualising the aligned sequence reads using IGV (Integrative Genomics Viewer) version 2.3.52 (Broad Instituite, Cambridge, MA, USA). Polymorphism quality was further screened for its span within reads with spans at or near ends of reads excluded due to high sequencing error rates at ends of reads [45].

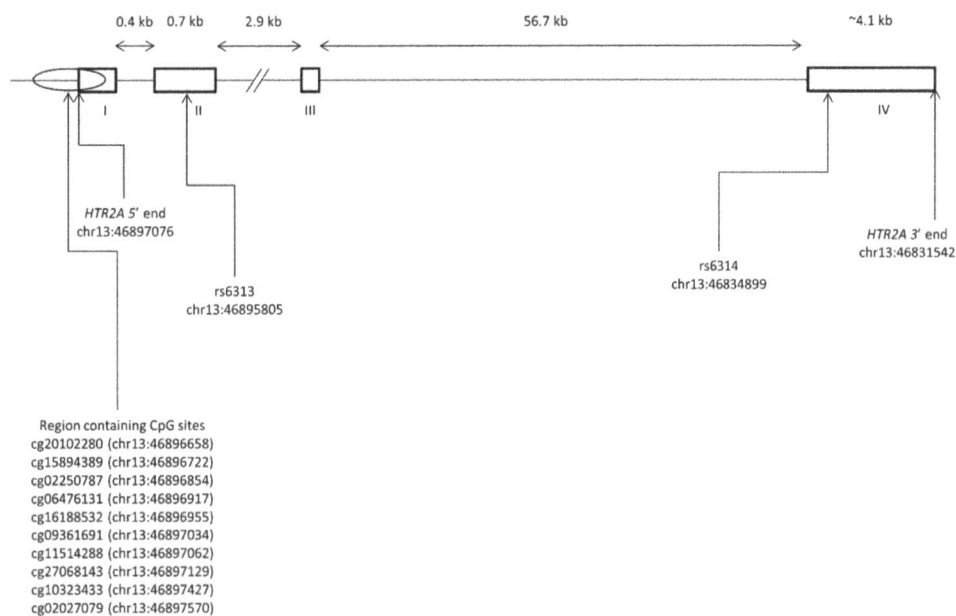

Figure 1. *HTR2A* gene structure showing exons, loci of rs6313 and rs6314, and all 10 selected CpG sites.

2.5. CpG Site Methylation Data Generation and Transformation

Genome-wide DNA methylation was analysed using the Illumina HumanMethylation450 array that assays more than 485,000 CpG sites across the genome exactly as previously described [34,46]. *HTR2A* methylation profiles were generated by selecting CpG sites within regions with active DNA methylation (± 500 bp of transcription start site) including the promoter and exon I of *HTR2A*. These sites were selected for their loci within DNA regions with rich presence of regulatory signals/elements based on the UCSC genome browser. The sites were identified as cg20102280, cg15894389, cg02250787, cg06476131, cg16188532, cg09361691, cg11514288, cg27068143, cg10323433 and cg02027079 as detailed in Table 2 (loci shown in Figure 1). The methylation status for each site was recorded as a β-value that ranged between 0 and 1, where values close to 1 represent high levels of methylation and where values close to 0 represent low levels of methylation. β-values were transformed to logit values or $\log_{10}(\beta/1-\beta)$ values using SPSS 22.

Table 2. Details of selected CpG sites.

CpG Site	Illumina CpG Site Name	Feature	Locus
cg1	cg20102280	Exon I	chr13:46896658
cg2	cg15894389	Exon I	chr13:46896722
cg3	cg02250787	Exon I	chr13:46896854
cg4	cg06476131	Exon I	chr13:46896917
cg5	cg16188532	Exon I	chr13:46896955
cg6	cg09361691	Exon I	chr13:46897034
cg7	cg11514288	Exon I	chr13:46897062
cg8	cg27068143	Promoter	chr13:46897129
cg9	cg10323433	Promoter	chr13:46897427
cg10	cg02027079	Promoter	chr13:46897570

2.6. Data Analysis

Using the $\log_{10}$RPKM values to represent *HTR2A* mRNA expression levels and $\log_{10}(\beta/1-\beta)$ values to represent CpG site DNA methylation levels, analysis of variance (ANOVA) adjusted for age and PMI (placing age and PMI variables as covariates) was performed with SPSS 22 to compare *HTR2A* mRNA expression differences and DNA methylation differences for the CpG sites between schizophrenia and control subjects. ANOVA was also used to examine the effect of rs6314 and rs6313 genotypes on *HTR2A* mRNA expression and DNA methylation of the CpG sites. Linear regression analysis adjusted for age and PMI was performed using SPSS 22 to correlate *HTR2A* expression levels and *HTR2A* methylation levels. *HTR2A* mRNA expression and DNA methylation for the CpG sites were firstly correlated in total samples, followed by correlation in schizophrenia-only and control-only subjects.

3. Results

3.1. Schizophrenia and HTR2A mRNA Expression

The $\log_{10}$RPKM values were compared between schizophrenia and control subjects using ANOVA. There was a 14% reduction in *HTR2A* mRNA expression in patients with schizophrenia compared to controls (*F*-value of 8.564 and *p*-value of 0.006; Figure 2).

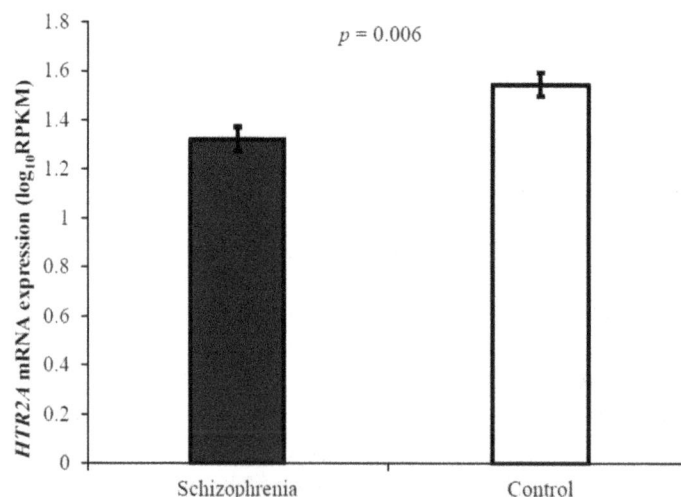

Figure 2. *HTR2A* mRNA expression in patients with schizophrenia compared to controls. ANOVA showing mean expression level and standard error in patients with schizophrenia.

3.2. Schizophrenia and DNA Methylation

Methylation levels (logit β-values) were compared between schizophrenia and control subjects using ANOVA for all CpG sites. Three CpG sites, cg5 (F = 5.194, p = 0.028), cg7 (F = 5.753, p = 0.021) and cg10 (F = 6.155, p = 0.017) were hypermethylated in patients with schizophrenia compared to controls (Figure 3).

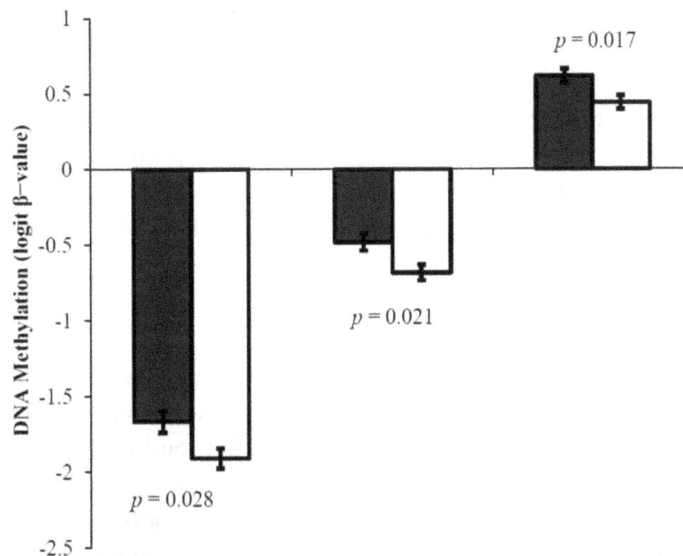

Figure 3. *HTR2A* DNA methylation in patients with schizophrenia compared to controls for three CpG sites, cg5, cg7 and cg10. ANOVA showing mean DNA methylation level and standard error in patients with schizophrenia (solid dark-grey bar) and controls (solid white bar).

3.3. rs6314 and rs6313 Genotypes

Genotype frequencies were generated for schizophrenia and control subjects (Table 3). No T/T genotype was present in our samples for rs6314. For rs6314 we observed a significant difference between the genotype counts of schizophrenia and control subjects (p = 0.008) with an odds ratio (OR) of 5.7 (C/T as high risk genotype) while no association was detected between rs6313 and schizophrenia (p = 0.078). Although rs6313 association was not significant, individuals with the T/T genotype had an OR of 7.0 (p = 0.144). When analysed as a T-recessive trait, rs6313 was found associated with schizophrenia (p = 0.026) with an OR of 8.4 (T as high risk recessive allele). This suggests the T-allele is the schizophrenia risk allele for both rs6313 and rs6314 although results should be interpreted with caution due to the low sample number.

Table 3. Genotype and allele frequencies for rs6314 in schizophrenia and control groups.

	Genotype/Allele	Schizophrenia	Control	Chi-Square p-Value	Odds Ratio
rs6314 Genotype	C/C	10	19		
	C/T	12	4	0.008	5.7
	T/T	0	0		
	Total	22	23		
rs6314 Allele	C	32	42		
	T	12	4	0.019	3.9
	Total	44	46		
rs6313 Genotype	C/C	6	7		
	C/T	9	14	0.078	–
	T/T	6	1		
	Total	21	22		
rs6313 Allele	C	21	28		
	T	21	16	0.201	–
	Total	42	44		

3.4. HTR2A mRNA Expression and Genotypes

ANOVA was performed to test the association between *HTR2A* mRNA expression and rs6314 and rs6313 genotypes. We compared the *HTR2A* mRNA expression between C/C genotypes and C/T genotypes for all samples, i.e., schizophrenia and control samples were pooled (there were no T/T genotypes present in our sample for rs6314). No differential expression was observed for either rs6314 ($F = 0.160$, $p = 0.691$) or rs6313 ($F = 1.690$, $p = 0.198$) genotypes.

3.5. DNA Methylation and Genotypes

For rs6314, ANOVA detected differential methylation between the subjects with C/T genotype and C/C genotype for cg3 (hypermethylated C/T group, $F = 8.645$, $p = 0.005$) and cg8 (hypermethylated C/T group, $F = 5.800$, $p = 0.021$). For rs6313, ANOVA detected differential methylation between C-allele carriers and subjects with the T/T genotype for cg6 (hypomethylated C-allele carriers, $F = 4.184$, $p = 0.048$).

3.6. Correlation between mRNA Expression and DNA Methylation

Using linear regression, we firstly tested the correlation between *HTR2A* mRNA expression and single site DNA methylation for the 10 CpG sites. No significant relationship was detected (Table S2). When all 10 CpG sites were analysed together, the combined effect of ten CpG sites on mRNA expression were not significant ($R^2 = 0.243$, $p = 0.291$) although two CpG sites within the 10-CpG site model, cg2 (B = 0.282, $p = 0.045$) and cg4 (B = -0.328, $p = 0.037$) were significant.

We also tested for a correlation between *HTR2A* mRNA expression and single CpG site methylation for all 10 CpG sites within patients with schizophrenia as well as control subjects (Supplementary Table S2). For patients with schizophrenia, a significant correlation was identified for cg5 (B = 0.525, $R^2 = 0.196$, $p = 0.014$) for single CpG site analysis. When all 10 CpG sites were analysed together in patients with schizophrenia, cg5 (B = 0.615, $p = 0.007$), cg6 (B = -0.695, $p = 0.034$), cg7 (B = 1.066, $p = 0.026$) and cg9 (B = -0.745, $p = 0.049$) were significantly correlated with mRNA expression with an overall $R^2 = 0.537$ and overall significance of $p = 0.008$.

For control subjects, a significant correlation was identified for cg9 (B = 0.666, $R^2 = 0.272$, $p = 0.015$) for single CpG site analysis. When all 10 CpG sites were analysed together in control subjects, no significantly correlation with mRNA expression was detected.

4. Discussion

This paper reports a number of complex associations with schizophrenia for the *HTR2A* gene that are potentially all interacting with each other. We found association between (1) schizophrenia and reduced *HTR2A* mRNA expression; (2) schizophrenia and hypermethylation of *HTR2A* promoter CpG sites (cg5, cg7 and cg10); and (3) schizophrenia and *HTR2A* genotypes for rs6314 and rs6313. Association was also found between DNA methylation and genotypes of both rs6314 and rs6313. There was a strong correlation between DNA methylation and mRNA expression in patients with schizophrenia.

No association was found between rs6314 and mRNA expression. This is consistent with Blasi et al. [21], who only found a trend of reduced *HTR2A* mRNA in prefrontal cortex samples bearing the T-allele of rs6314, but the trend was not significant ($p = 0.06$). Furthermore, they did not examine the effects of DNA methylation. The authors suggested, however, that rs6314 could alter splicing patterns with possible effects on *HTR2A* expression.

The polymorphism rs6314 was associated with differential methylation at CpG sites cg3 and cg8, though it is unclear what mechanism might lead to this considering that rs6314 is located 61 kb from the CpG sites in the promoter-exon I region. This is also consistent with Guhathakurta et al., who showed that rs6314 is unlikely to influence DNA methylation activity in the promoter (such as methylation site at rs6311) due to low LD [47]. Association was also detected between rs6313 and differential methylation of cg6. The rs6313 polymorphism is only 1.2 kb from cg6, so it is possible that it has a proximal effect on the binding of DNA methylases; further studies investigating the biological function of rs6313 on DNA methylation would clarify the role of rs6313 in schizophrenia.

An interesting observation in the mRNA expression–DNA methylation correlation analysis is the significant combined effect of all CpG sites detected when only patients with schizophrenia were analysed. Furthermore, methylation of cg5 and cg9 was correlated with mRNA expression and associated with schizophrenia. No significant combined effect of all CpG sites was detected when control-only subjects or pooled subjects (schizophrenia plus control) were analysed. This suggests that DNA methylation may have an effect on mRNA expression in patients with schizophrenia but not in control subjects. Perhaps there is little to no significant variation of mRNA expression and DNA methylation levels between control subjects for a significant correlation to be detected. However, it is possible that the significant correlation of DNA methylation-mRNA expression detected in schizophrenia may be a net result of DNA methylation alterations, along with other interacting factors (including genes, environmental factors and other underlying biological factors). Indeed, Abdolmaleky et al. found a negative correlation between HTR2A and COMT expression, a positive correlation between HTR2A and RELN expression, and a negative correlation between HTR2A expression and RELN promoter DNA methylation [10]. Overall, our results are consistent with those of Abdolmaleky et al., who reported an inverse relationship between *HTR2A* mRNA expression and promoter region DNA methylation.

Our results showed an association between three hypermethylated CpG sites and schizophrenia. This is consistent with other studies that found hypermethylated DNA within the promoter region of *HTR2A* in schizophrenia brains and saliva [10,36], hypermethylated DNA at rs6313 in peripheral leukocytes in major psychosis patients with suicidal tendency [37] and hypermethylated *HTR2A* promoter CpG islands in the blood of patients with borderline personality disorder [48]. It should be noted that in a study by Abdolmaleky et al. that detected reduced *HTR2A* mRNA expression in brain samples, regions within the promoter were hypermethylated while regions near rs6313 within exon I were hypomethylated. To some extent, this observation is consistent with our findings, where both positive and negative correlations of DNA methylation-RNA expression are present. This suggests that the mechanisms regulating mRNA expression in brain are different in the blood and the saliva. Furthermore, combined CpG-site analysis suggests interaction between multiple CpG-sites resulting in altered *HTR2A* mRNA expression.

A limitation to this study is that, after correcting relevant results for multiple testing, they did not survive their respective thresholds. However, the samples are not completely independent so correction for multiple testing may not be valid.

This study has concentrated on DNA methylation in the promoter and exon I regions of *HTR2A*. Future studies should explore the $-3'$ region of the gene to investigate methylation alteration in schizophrenia. These studies should be conducted in relevant brain tissues as it is the organ of major pathology for schizophrenia. In addition to age and PMI, factors influencing the outcomes of mRNA expression and DNA methylation including stress, medication and substance use, nutrition history, comorbidities, ethnogeographic information and other environmental factors should be included in future studies to further improve the accuracy of the findings.

5. Conclusions

In conclusion, we report that there is an association between schizophrenia and three separate factors: reduced *HTR2A* mRNA expression, hypermethylation of *HTR2A* promoter CpG sites (cg5, cg7 and cg10) and genetic association with *HTR2A* genotypes for rs6314 and rs6313. The association of the T-allele of rs6314 and rs6313 with schizophrenia may be due to an alteration in biological mechanisms other than mRNA expression regulation. The reduced expression of *HTR2A* mRNA in schizophrenia patients may be mainly attributed to DNA methylation of CpG sites within the promoter-exon I region, and possibly influenced by other underlying factors. Therefore, epigenetic *HTR2A* regulation may affect brain function, which contributes to the development of schizophrenia.

Acknowledgments: This study was supported by the Post-graduate Research Award scholarship provided by the School of Biomedical Sciences, Queensland University of Technology. This work was also financially supported by the Queensland State Government; the Nicol Foundation; and the Institute of Health and Biomedical Innovation, QUT. Brain samples were obtained from the Human Brain and Spinal Fluid Resource Centre, California (courtesy of James Riehl). Robert McLeay and Leesa Wockner made contributions to the analysis of the DNA methylation and RNA seq data.

Author Contributions: Sern-Yih Cheah, Joanne Voisey, and Charles P. Morris made a substantial contribution to conception and design, helped with the analysis and interpretation of the data, and drafted and critically reviewed the manuscript. Bruce R. Lawford and Ross McD. Young made a substantial contribution to the conception and design and critically reviewed the manuscript.

References

1. Buhot, M.C. Serotonin receptors in cognitive behaviors. *Curr. Opin. Neurobiol.* **1997**, *7*, 243–254. [CrossRef]

2. Fink, K.B.; Gothert, M. 5-HT receptor regulation of neurotransmitter release. *Pharmacol. Rev.* **2007**, *59*, 360–417. [CrossRef] [PubMed]

3. Jakab, R.L.; Goldman-Rakic, P.S. Segregation of serotonin 5-HT2A and 5-HT3 receptors in inhibitory circuits of the primate cerebral cortex. *J. Comp. Neurol.* **2000**, *417*, 337–348. [CrossRef]

4. Willins, D.L.; Deutch, A.Y.; Roth, B.L. Serotonin 5-HT2A receptors are expressed on pyramidal cells and interneurons in the rat cortex. *Synapse* **1997**, *27*, 79–82. [CrossRef]

5. Kim, Y.K.; Yoon, H.K. Effect of serotonin-related gene polymorphisms on pathogenesis and treatment response in korean schizophrenic patients. *Behav. Genet.* **2011**, *41*, 709–715. [CrossRef] [PubMed]

6. Lang, U.E.; Puls, I.; Muller, D.J.; Strutz-Seebohm, N.; Gallinat, J. Molecular mechanisms of schizophrenia. *Cell. Physiol. Biochem.* **2007**, *20*, 687–702. [CrossRef] [PubMed]

7. Harvey, J.A. Role of the serotonin 5-HT(2A) receptor in learning. *Learn. Mem.* **2003**, *10*, 355–362. [CrossRef] [PubMed]

8. Weickert, T.W.; Goldberg, T.E.; Gold, J.M.; Bigelow, L.B.; Egan, M.F.; Weinberger, D.R. Cognitive impairments in patients with schizophrenia displaying preserved and compromised intellect. *Arch. Gen. Psychiatry* **2000**, *57*, 907–913. [CrossRef] [PubMed]

9. Miyamoto, S.; Duncan, G.E.; Marx, C.E.; Lieberman, J.A. Treatments for schizophrenia: A critical review of pharmacology and mechanisms of action of antipsychotic drugs. *Mol. Psychiatry* **2005**, *10*, 79–104. [CrossRef] [PubMed]

10. Abdolmaleky, H.M.; Yaqubi, S.; Papageorgis, P.; Lambert, A.W.; Ozturk, S.; Sivaraman, V.; Thiagalingam, S. Epigenetic dysregulation of HTR2A in the brain of patients with schizophrenia and bipolar disorder. *Schizophr. Res.* **2011**, *129*, 183–190. [CrossRef] [PubMed]

11. Abi-Dargham, A. Alterations of serotonin transmission in schizophrenia. *Int. Rev. Neurobiol.* **2007**, *78*, 133–164. [PubMed]

12. Polesskaya, O.O.; Sokolov, B.P. Differential expression of the "C" and "T" alleles of the 5-HT2A receptor gene in the temporal cortex of normal individuals and schizophrenics. *J. Neurosci. Res.* **2002**, *67*, 812–822. [CrossRef] [PubMed]

13. Selvaraj, S.; Arnone, D.; Cappai, A.; Howes, O. Alterations in the serotonin system in schizophrenia: A systematic review and meta-analysis of postmortem and molecular imaging studies. *Neurosci. Biobehav. Rev.* **2014**, *45*, 233–245. [CrossRef] [PubMed]

14. Gonzalez-Maeso, J.; Ang, R.L.; Yuen, T.; Chan, P.; Weisstaub, N.V.; Lopez-Gimenez, J.F.; Zhou, M.; Okawa, Y.; Callado, L.F.; Milligan, G.; et al. Identification of a serotonin/glutamate receptor complex implicated in psychosis. *Nature* **2008**, *452*, 93–97. [CrossRef] [PubMed]

15. Muguruza, C.; Moreno, J.L.; Umali, A.; Callado, L.F.; Meana, J.J.; Gonzalez-Maeso, J. Dysregulated 5-HT(2A) receptor binding in postmortem frontal cortex of schizophrenic subjects. *Eur. Neuropsychopharmacol.* **2013**, *23*, 852–864. [CrossRef] [PubMed]

16. Maple, A.M.; Zhao, X.; Elizalde, D.I.; McBride, A.K.; Gallitano, A.L. Htr2a expression responds rapidly to environmental stimuli in an Egr3-dependent manner. *ACS Chem. Neurosci.* **2015**, *6*, 1137–1142. [CrossRef] [PubMed]

17. Schmid, C.L.; Streicher, J.M.; Meltzer, H.Y.; Bohn, L.M. Clozapine acts as an agonist at serotonin 2A receptors to counter MK-801-induced behaviors through a betaarrestin2-independent activation of AKT. *Neuropsychopharmacology* **2014**, *39*, 1902–1913. [CrossRef] [PubMed]

18. Hazelwood, L.A.; Sanders-Bush, E. His452Tyr polymorphism in the human 5-HT2A receptor destabilizes the signaling conformation. *Mol. Pharmacol.* **2004**, *66*, 1293–1300. [PubMed]

19. Ozaki, N.; Manji, H.; Lubierman, V.; Lu, S.J.; Lappalainen, J.; Rosenthal, N.E.; Goldman, D. A naturally occurring amino acid substitution of the human serotonin 5-HT2A receptor influences amplitude and timing of intracellular calcium mobilization. *J. Neurochem.* **1997**, *68*, 2186–2193. [CrossRef] [PubMed]

20. Serretti, A.; Drago, A.; De Ronchi, D. HTR2A gene variants and psychiatric disorders: A review of current literature and selection of snps for future studies. *Curr. Med. Chem.* **2007**, *14*, 2053–2069. [CrossRef] [PubMed]

21. Blasi, G.; De Virgilio, C.; Papazacharias, A.; Taurisano, P.; Gelao, B.; Fazio, L.; Ursini, G.; Sinibaldi, L.; Andriola, I.; Masellis, R.; et al. Converging evidence for the association of functional genetic variation in the serotonin receptor 2A gene with prefrontal function and olanzapine treatment. *JAMA Psychiatry* **2013**, *70*, 921–930. [CrossRef] [PubMed]

22. Filippini, N.; Scassellati, C.; Boccardi, M.; Pievani, M.; Testa, C.; Bocchio-Chiavetto, L.; Frisoni, G.B.; Gennarelli, M. Influence of serotonin receptor 2A His452Tyr polymorphism on brain temporal structures: A volumetric MR study. *Eur. J. Hum. Genet.* **2006**, *14*, 443–449. [CrossRef] [PubMed]

23. Schott, B.H.; Seidenbecher, C.I.; Richter, S.; Wustenberg, T.; Debska-Vielhaber, G.; Schubert, H.; Heinze, H.J.; Richardson-Klavehn, A.; Duzel, E. Genetic variation of the serotonin 2A receptor affects hippocampal novelty processing in humans. *PLoS ONE* **2011**, *6*, e15984. [CrossRef] [PubMed]

24. De Quervain, D.J.; Henke, K.; Aerni, A.; Coluccia, D.; Wollmer, M.A.; Hock, C.; Nitsch, R.M.; Papassotiropoulos, A. A functional genetic variation of the 5-HT2A receptor affects human memory. *Nat. Neurosci.* **2003**, *6*, 1141–1142. [CrossRef] [PubMed]

25. Arranz, M.J.; Collier, D.A.; Munro, J.; Sham, P.; Kirov, G.; Sodhi, M.; Roberts, G.; Price, J.; Kerwin, R.W. Analysis of a structural polymorphism in the 5-HT2A receptor and clinical response to clozapine. *Neurosci. Lett.* **1996**, *217*, 177–178. [CrossRef]

26. Arranz, M.J.; Munro, J.; Owen, M.J.; Spurlock, G.; Sham, P.C.; Zhao, J.; Kirov, G.; Collier, D.A.; Kerwin, R.W. Evidence for association between polymorphisms in the promoter and coding regions of the 5-HT2A receptor gene and response to clozapine. *Mol. Psychiatry* **1998**, *3*, 61–66. [CrossRef] [PubMed]

27. Abdolmaleky, H.M.; Faraone, S.V.; Glatt, S.J.; Tsuang, M.T. Meta-analysis of association between the T102C polymorphism of the 5ht2a receptor gene and schizophrenia. *Schizophr. Res.* **2004**, *67*, 53–62. [CrossRef]

28. Khait, V.D.; Huang, Y.Y.; Zalsman, G.; Oquendo, M.A.; Brent, D.A.; Harkavy-Friedman, J.M.; Mann, J.J. Association of serotonin 5-HT2A receptor binding and the T102C polymorphism in depressed and healthy caucasian subjects. *Neuropsychopharmacology* **2005**, *30*, 166–172. [CrossRef] [PubMed]

29. Yildiz, S.H.; Akilli, A.; Bagcioglu, E.; Ozdemir Erdogan, M.; Coskun, K.S.; Alpaslan, A.H.; Subasi, B.; Arikan Terzi, E.S. Association of schizophrenia with T102C (rs6313) and 1438 a/g (rs6311) polymorphisms of HTR2A gene. *Acta Neuropsychiatr.* **2013**, *25*, 342–348. [CrossRef] [PubMed]

30. Ucok, A.; Alpsan, H.; Cakir, S.; Saruhan-Direskeneli, G. Association of a serotonin receptor 2a gene polymorphism with cognitive functions in patients with schizophrenia. *Am. J. Med. Genet. Part B Neuropsychiatr. Genet.* **2007**, *144B*, 704–707. [CrossRef] [PubMed]

31. Vyas, N.S.; Lee, Y.; Ahn, K.; Ternouth, A.; Stahl, D.R.; Al-Chalabi, A.; Powell, J.F.; Puri, B.K. Association of a serotonin receptor 2A gene polymorphism with visual sustained attention in early-onset schizophrenia patients and their non-psychotic siblings. *Aging Dis.* **2012**, *3*, 291–300. [PubMed]

32. Turecki, G.; Sequeira, A.; Gingras, Y.; Seguin, M.; Lesage, A.; Tousignant, M.; Chawky, N.; Vanier, C.; Lipp, O.; Benkelfat, C.; et al. Suicide and serotonin: Study of variation at seven serotonin receptor genes in suicide completers. *Am. J. Med. Genet. Part B Neuropsychiatr. Genet.* **2003**, *118B*, 36–40. [CrossRef] [PubMed]

33. Parsons, M.J.; D'Souza, U.M.; Arranz, M.J.; Kerwin, R.W.; Makoff, A.J. The -1438a/g polymorphism in the 5-hydroxytryptamine type 2A receptor gene affects promoter activity. *Biol. Psychiatry* **2004**, *56*, 406–410. [CrossRef] [PubMed]

34. Wockner, L.F.; Noble, E.P.; Lawford, B.R.; Young, R.M.; Morris, C.P.; Whitehall, V.L.; Voisey, J. Genome-wide DNA methylation analysis of human brain tissue from schizophrenia patients. *Transl. Psychiatry* **2014**, *4*. [CrossRef] [PubMed]

35. Paquette, A.G.; Lesseur, C.; Armstrong, D.A.; Koestler, D.C.; Appleton, A.A.; Lester, B.M.; Marsit, C.J. Placental HTR2A methylation is associated with infant neurobehavioral outcomes. *Epigenetics* **2013**, *8*, 796–801. [CrossRef] [PubMed]

36. Ghadirivasfi, M.; Nohesara, S.; Ahmadkhaniha, H.R.; Eskandari, M.R.; Mostafavi, S.; Thiagalingam, S.; Abdolmaleky, H.M. Hypomethylation of the serotonin receptor type-2A gene (HTR2A) at T102C polymorphic site in DNA derived from the saliva of patients with schizophrenia and bipolar disorder. *Am. J. Med. Genet. B Neuropsychiatr. Genet.* **2011**, *156B*, 536–545. [CrossRef] [PubMed]

37. De Luca, V.; Viggiano, E.; Dhoot, R.; Kennedy, J.L.; Wong, A.H. Methylation and qtdt analysis of the 5-HT2A receptor 102C allele: Analysis of suicidality in major psychosis. *J. Psychiatr. Res.* **2009**, *43*, 532–537. [CrossRef] [PubMed]

38. Polesskaya, O.O.; Aston, C.; Sokolov, B.P. Allele c-specific methylation of the 5-HT2A receptor gene: Evidence for correlation with its expression and expression of DNA methylase DNMT1. *J. Neurosci. Res.* **2006**, *83*, 362–373. [CrossRef] [PubMed]

39. Falkenberg, V.R.; Gurbaxani, B.M.; Unger, E.R.; Rajeevan, M.S. Functional genomics of serotonin receptor 2a (HTR2A): Interaction of polymorphism, methylation, expression and disease association. *Neuromol. Med.* **2011**, *13*, 66–76. [CrossRef] [PubMed]

40. Paquette, A.G.; Marsit, C.J. The developmental basis of epigenetic regulation of HTR2A and psychiatric outcomes. *J. Cell. Biochem.* **2014**, *115*, 2065–2072. [CrossRef] [PubMed]

41. Trapnell, C.; Roberts, A.; Goff, L.; Pertea, G.; Kim, D.; Kelley, D.R.; Pimentel, H.; Salzberg, S.L.; Rinn, J.L.; Pachter, L. Differential gene and transcript expression analysis of RNA-Seq experiments with tophat and cufflinks. *Nat. Protoc.* **2012**, *7*, 562–578. [CrossRef] [PubMed]

42. Liao, Y.; Smyth, G.K.; Shi, W. The subread aligner: Fast, accurate and scalable read mapping by seed-and-vote. *Nucleic Acids Res.* **2013**, *41*. [CrossRef] [PubMed]

43. McCarthy, D.J.; Chen, Y.; Smyth, G.K. Differential expression analysis of multifactor RNA-Seq experiments with respect to biological variation. *Nucleic Acids Res.* **2012**, *40*, 4288–4297. [CrossRef] [PubMed]

44. Robinson, M.D.; Oshlack, A. A scaling normalization method for differential expression analysis of RNA-Seq data. *Genome Biol.* **2010**, *11*. [CrossRef] [PubMed]

45. Schirmer, M.; Ijaz, U.Z.; D'Amore, R.; Hall, N.; Sloan, W.T.; Quince, C. Insight into biases and sequencing errors for amplicon sequencing with the illumina miseq platform. *Nucleic Acids Res.* **2015**, *43*. [CrossRef] [PubMed]

46. Wockner, L.F.; Morris, C.P.; Noble, E.P.; Lawford, B.R.; Whitehall, V.L.; Young, R.M.; Voisey, J. Brain-specific epigenetic markers of schizophrenia. *Transl. Psychiatry* **2015**, *5*. [CrossRef] [PubMed]

47. Guhathakurta, S.; Singh, A.S.; Sinha, S.; Chatterjee, A.; Ahmed, S.; Ghosh, S.; Usha, R. Analysis of serotonin receptor 2A gene (HTR2A): Association study with autism spectrum disorder in the Indian population and investigation of the gene expression in peripheral blood leukocytes. *Neurochem. Int.* **2009**, *55*, 754–759. [CrossRef] [PubMed]

48. Dammann, G.; Teschler, S.; Haag, T.; Altmuller, F.; Tuczek, F.; Dammann, R.H. Increased DNA methylation of neuropsychiatric genes occurs in borderline personality disorder. *Epigenetics* **2011**, *6*, 1454–1462. [CrossRef] [PubMed]

Permissions

The contributors of this book come from diverse backgrounds, making this book a truly international effort. This book will bring forth new frontiers with its revolutionizing research information and detailed analysis of the nascent developments around the world.

We would like to thank all the contributing authors for lending their expertise to make the book truly unique. They have played a crucial role in the development of this book. Without their invaluable contributions this book wouldn't have been possible. They have made vital efforts to compile up to date information on the varied aspects of this subject to make this book a valuable addition to the collection of many professionals and students.

This book was conceptualized with the vision of imparting up-to-date information and advanced data in this field. To ensure the same, a matchless editorial board was set up. Every individual on the board went through rigorous rounds of assessment to prove their worth. After which they invested a large part of their time researching and compiling the most relevant data for our readers.

The editorial board has been involved in producing this book since its inception. They have spent rigorous hours researching and exploring the diverse topics which have resulted in the successful publishing of this book. They have passed on their knowledge of decades through this book. To expedite this challenging task, the publisher supported the team at every step. A small team of assistant editors was also appointed to further simplify the editing procedure and attain best results for the readers.

Apart from the editorial board, the designing team has also invested a significant amount of their time in understanding the subject and creating the most relevant covers. They scrutinized every image to scout for the most suitable representation of the subject and create an appropriate cover for the book.

The publishing team has been an ardent support to the editorial, designing and production team. Their endless efforts to recruit the best for this project, has resulted in the accomplishment of this book. They are a veteran in the field of academics and their pool of knowledge is as vast as their experience in printing. Their expertise and guidance has proved useful at every step. Their uncompromising quality standards have made this book an exceptional effort. Their encouragement from time to time has been an inspiration for everyone.

The publisher and the editorial board hope that this book will prove to be a valuable piece of knowledge for researchers, students, practitioners and scholars across the globe.

List of Contributors

Navya Laxman
Department of Medical Sciences, Uppsala University, Uppsala 75185, Sweden
Science for Life Laboratory, Department of Medical Sciences, Uppsala University Hospital, Uppsala 75185, Sweden
Science for Life Laboratory, Department of Biochemistry and Biophysics, Stockholm University, Stockholm 17121, Sweden

Andreas Kindmark
Department of Medical Sciences, Uppsala University, Uppsala 75185, Sweden
Science for Life Laboratory, Department of Medical Sciences, Uppsala University Hospital, Uppsala 75185, Sweden

Hans Mallmin and Olle Nilsson
Department of Surgical Sciences, Uppsala University, Uppsala 75185, Sweden

Robert E. Parker, David Knupp, Rim Al Safadi and Shannon D. Manning
Department of Microbiology and Molecular Genetics, Michigan State University, East Lansing, MI 48824, USA

Agnès Rosenau
Infectiologie et Santé Publique ISP, Institut National de la Recherche Agronomique, Université de Tours, Equipe Bactéries et Risque Materno-foetal, UMR1282 Tours, France

Fangfang Zhao
Gansu Key Laboratory of Herbivorous Animal Biotechnology, Faculty of Animal Science and Technology, Gansu Agricultural University, Lanzhou 730070, China

Shaobin Li, Jiqing Wang, Xiu Liu and Yuzhu Luo
Gansu Key Laboratory of Herbivorous Animal Biotechnology, Faculty of Animal Science and Technology, Gansu Agricultural University, Lanzhou 730070, China
International Wool Research Institute, Gansu Agricultural University, Lanzhou 730070, China

Huitong Zhou
Gansu Key Laboratory of Herbivorous Animal Biotechnology, Faculty of Animal Science and Technology, Gansu Agricultural University, Lanzhou 730070, China

International Wool Research Institute, Gansu Agricultural University, Lanzhou 730070, China
Gene-marker Laboratory, Faculty of Agricultural and Life Sciences, Lincoln University, Lincoln 7647, New Zealand

Hua Gong and Jon G. H. Hickford
International Wool Research Institute, Gansu Agricultural University, Lanzhou 730070, China
Gene-marker Laboratory, Faculty of Agricultural and Life Sciences, Lincoln University, Lincoln 7647, New Zealand

Jong-Oh Kim, Jae-Ok Kim, Wi-Sik Kim and Myung-Joo Oh
Department of Aqualife Medicine, College of Fisheries and Ocean Science, Chonnam National University, Yeosu 550-749, Korea

Carrie Mae Long
Immunology and Microbial Pathogenesis Graduate Program, West Virginia University, Morgantown, WV 26505, USA
Centers for Disease Control and Prevention, National Institute for Occupational Safety and Health, Allergy and Clinical Immunology Branch, Morgantown, WV 26505, USA

Ewa Lukomska, Ajay Nayak and Stacey E. Anderson
Centers for Disease Control and Prevention, National Institute for Occupational Safety and Health, Allergy and Clinical Immunology Branch, Morgantown, WV 26505, USA

Nikki B. Marshall
Inovio Pharmaceuticals, Inc., 660 West Germantown Pike, Plymouth Meeting, PA 19462, USA

Loris De Cecco, Marco Giannoccaro, Edoardo Marchesi and Silvana Canevari
Functional Genomics and Bioinformatics, Department of Experimental Oncology and Molecular Medicine, Fondazione IRCCS Istituto Nazionale dei Tumori, Milan 20133, Italy

Paolo Bossi, Federica Favales, Laura D. Locati and Lisa Licitra
Head and Neck Medical Oncology Unit, Fondazione IRCCS Istituto Nazionale dei Tumori, Milan 20133, Italy

Silvana Pilotti
Laboratory of Experimental Molecular Pathology, Department of Diagnostic Pathology and Laboratory, Fondazione IRCCS Istituto Nazionale dei Tumori, Milan 20133, Italy

Michael Griswold
Department of Data Science, School of Population Health, University of Mississippi Medical Center, Jackson, MS 39216, USA

Hao Mei
Department of Data Science, School of Population Health, University of Mississippi Medical Center, Jackson, MS 39216, USA
Shanghai Children's Medical Center, School of Public Health and School of Medicine, Shanghai Jiaotong University, Shanghai 200127, China

Shijian Liu and Fan Jiang
Shanghai Children's Medical Center, School of Public Health and School of Medicine, Shanghai Jiaotong University, Shanghai 200127, China

Lianna Li
Department of Biology, Tougaloo College, Jackson, MI 39216, USA

Thomas Mosley
Department of Neurology, University of Mississippi Medical Center, Jackson, MS 39216, USA

Weihong Sun, Jongphil Kim, Xiuhua Zhao, Tuya Pal, Roohi Ismail-Khan and Monique Sajjad
H. Lee Moffitt Cancer Center and Research Institute, Tampa, FL 33612, USA
Division of Cardiovascular Medicine, University of South Florida, Tampa, FL 33620, USA

Michael Fradley
Division of Cardiovascular Medicine, University of South Florida, Tampa, FL 33620, USA

Olivier Sandre
Laboratory of Organic Polymer Chemistry, LCPO, UMR 5629 CNRS, University of Bordeaux, Bordeaux-INP, Pessac 33600, France

Coralie Genevois and Franck Couillaud
Molecular Imaging and Innovative Therapies in Oncology, IMOTION, EA 7435, University of Bordeaux, 146 rue Léo Saignat, case 127, Bordeaux cedex 33076, France

Eneko Garaio
Department of Electricity and Electronics, University of the Basque Country (UPV/EHU), P.K. 644, Leioa 48940, Spain

Laurent Adumeau and Stéphane Mornet
Institute for Condensed Matter Chemistry of Bordeaux, ICMCB, UPR 9048, CNRS, University of Bordeaux, Pessac F-33600 France

Josephine Galipon and Masaru Tomita
Keio University Institute for Advanced Biosciences, Tsuruoka 997-0017, Japan

Rintaro Ishii and Yutaka Suzuki
Graduate School of Frontier Sciences, The University of Tokyo, Kashiwa 277-8562, Japan

Kumiko Ui-Tei
Graduate School of Frontier Sciences, The University of Tokyo, Kashiwa 277-8562, Japan
Graduate School of Science, The University of Tokyo, Tokyo 113-0032, Japan

Ying Yang and Song-Mei Liu
Center for Gene Diagnosis, Zhongnan Hospital of Wuhan University, Donghu Road 169#, Wuhan 430071, China

Boyang Liu
Department of Geography, Wilkeson Hall, State University of New York at Buffalo, Buffalo, NY 14261, USA

Wei Xia
Department of Clinical Laboratory, Wuhan Children's Hospital (Wuhan Maternal and Child Healthcare Hospital), Tongji Medical College, Huazhong University of Science and Technology, Wuhan 430016, China

Jing Yan
Hubei Meteorological Information and Technology Support Center, Wuhan 430074, China

Huan-Yu Liu
Department of Clinical Medicine, Hubei University of Medicine, Hubei 442000, China

Ling Hu
Department of Neurology, Wuhan Children's Hospital (Wuhan Maternal and Child Healthcare Hospital), Tongji Medical College, Huazhong University of Science and Technology, Wuhan 430016, China

Yang Liu
Research Center for Clinical and Translational Medicine, Beijing 302 Hospital, Beijing 100039, China

Xiaoyao Yin, Naiyang Guan and Zhigang Luo
Science and technology on Parallel and Distributed Processing Laboratory, National University of Defense Technology, Changsha 410073, China

Jing Zhong, Lishan Min, Xing Yao and Licheng Dai
Huzhou Key Laboratory of Molecular Medicine, Huzhou Central Hospital, Huzhou 313000, China

Xiaochen Bo
Beijing Institute of Radiation Medicine, Beijing 100850, China

Hui Bai
Beijing Institute of Radiation Medicine, Beijing 100850, China
No. 451 Hospital of PLA, Xi'an 710054, China

Madeleine Heep, Pia Mach, Philipp Reautschnig, Jacqueline Wettengel and Thorsten Stafforst
Interfaculty Institute of Biochemistry, University of Tübingen, Auf der Morgenstelle 15, 72076 Tübingen, Germany

Christina Votsi and Kyproula Christodoulou
Department of Neurogenetics, The Cyprus Institute of Neurology and Genetics and the Cyprus School of Molecular Medicine, Ayios Dhometios, 2370 Nicosia, Cyprus

Costas Toufexis
Department of Endocrinology and Diabetes, Hippocrateon Private Hospital, Engomi, 2408 Nicosia, Cyprus

Kyriaki Michailidou
Department of Electron Microscopy/Molecular Pathology, The Cyprus Institute of Neurology and Genetics, Ayios Dhometios, 2370 Nicosia, Cyprus

Athos Antoniades, Minas Karaolis and Constantinos S. Pattichis
Department of Computer Science, University of Cyprus, 1678 Nicosia, Cyprus

Nicos Skordis
St. George's University Medical School at the University of Nicosia, Engomi, 2408 Nicosia, Cyprus and Department of Pediatric Endocrinology, Paedi Center for Specialized Pediatrics, Strovolos, 2025 Nicosia, Cyprus

Sern-Yih Cheah, Charles P. Morris and Joanne Voisey
School of Biomedical Sciences, Institute of Health and Biomedical Innovation, Queensland University of Technology, 60 Musk Ave., Kelvin Grove, Queensland 4059, Australia

Bruce R. Lawford
School of Biomedical Sciences, Institute of Health and Biomedical Innovation, Queensland University of Technology, 60 Musk Ave., Kelvin Grove, Queensland 4059, Australia
Discipline of Psychiatry, Royal Brisbane and Women's Hospital, Herston, Queensland 4006, Australia

Ross McD. Young
Discipline of Psychiatry, Royal Brisbane and Women's Hospital, Herston, Queensland 4006, Australia
Faculty of Health, Institute of Health and Biomedical Innovation, Queensland University of Technology, 60 Musk Ave., Kelvin Grove, Queensland 4059, Australia

Index